Studies in Big Data

Volume 156

The series "Studies in Big Data" (SBD) publishes new developments and advances in the various areas of Big Data- quickly and with a high quality. The intent is to cover the theory, research, development, and applications of Big Data, as embedded in the fields of engineering, computer science, physics, economics and life sciences. The books of the series refer to the analysis and understanding of large, complex, and/or distributed data sets generated from recent digital sources coming from sensors or other physical instruments as well as simulations, crowd sourcing, social networks or other internet transactions, such as emails or video click streams and other. The series contains monographs, lecture notes and edited volumes in Big Data spanning the areas of computational intelligence including neural networks, evolutionary computation, soft computing, fuzzy systems, as well as artificial intelligence, data mining, modern statistics and Operations research, as well as self-organizing systems. Of particular value to both the contributors and the readership are the short publication timeframe and the world-wide distribution, which enable both wide and rapid dissemination of research output.

The books of this series are reviewed in a single blind peer review process.

Indexed by SCOPUS, EI Compendex, SCIMAGO, and zbMATH.

All books published in the series are submitted for consideration in Web of Science.

Ildikó Horváth · Borbála Berki · Anna Sudár ·
Ádám Csapó · Péter Baranyi

Cognitive Aspects of Virtual Reality

Springer

Ildikó Horváth
Corvinus Institute for Advanced Studies
Institute of Data Analytics and Information Systems
Corvinus University of Budapest, Hungary
and Hungarian Research Network
HUN-REN
Budapest, Hungary

Borbála Berki
Corvinus Institute for Advanced Studies
Institute of Data Analytics and Information Systems
Corvinus University of Budapest, Hungary
and Hungarian Research Network
HUN-REN
Budapest, Hungary

Anna Sudár
Corvinus Institute for Advanced Studies
Institute of Data Analytics and Information Systems
Corvinus University of Budapest, Hungary
and Hungarian Research Network
HUN-REN
Budapest, Hungary

Ádám Csapó
Corvinus Institute for Advanced Studies
Institute of Data Analytics and Information Systems
Corvinus University of Budapest, Hungary
and Hungarian Research Network
HUN-REN
Budapest, Hungary

Péter Baranyi
Corvinus Institute for Advanced Studies
Institute of Data Analytics and Information Systems
Corvinus University of Budapest, Hungary
and Hungarian Research Network
HUN-REN
Budapest, Hungary

ISSN 2197-6503 ISSN 2197-6511 (electronic)
Studies in Big Data
ISBN 978-3-031-68129-5 ISBN 978-3-031-68130-1 (eBook)
https://doi.org/10.1007/978-3-031-68130-1

This Springer imprint is published by the registered company Springer Nature Switzerland AG
The registered company address is: Gewerbestrasse 11, 6330 Cham, Switzerland

Preface

In an era where technology continually reshapes our perceptions and interactions, Virtual Reality (VR) stands out as a particularly transformative technological framework. This book introduces a novel perspective on the "cognitive aspects of virtual reality" (cVR), drawing from the rich interdisciplinary domains of Cognitive Infocommunications (CogInfoCom), Digital Reality (DR)/Internet of Digital Reality (IoD) and Cognitive Mobility (CogMob). Within these domains, the fusion of cognitive sciences, infocommunications, artificial intelligence (AI) as well as the study of social-technological interactions leads to a novel set of conceptual frameworks within which immersive and interactive capabilities of VR are not only enhanced, but also promise to re-shape human cognitive capabilities in unprecedented ways.

In this book, our exploration begins with developing an understanding of recent advances within CogInfoCom, which help lay down the fundamentals for a deeper integration of VR into human cognitive processes. We then explore the evolution of VR, augmented reality, and digital twins, expanding these into the context of the IoD. This technological integration offers a glimpse into a future where our digital and physical experiences are seamlessly intertwined, enabling new forms of interaction and understanding.

Following this first part, our focus shifts toward CogMob, exploring how VR extends the boundaries of the physical and digital movement. Cognitive Mobility is not just about spatial transitions but also about navigating complex information landscapes and decision-making environments, which in turn is uniquely facilitated by VR. The potential of VR to revolutionize fields such as education, healthcare, and communication is emphasized in particular. Through AI and sensor technologies, VR is evolving from its origins in gaming into a sophisticated tool for complex informational and communicative tasks.

A key contribution of the book is that it reveals the oftentimes beneficial impact that VR can have on human cognitive abilities such as memory, learning, and spatial awareness. Several chapters detail studies carried out in non-immersive VR environments, like desktop VR, and illustrate how digital interactions within an environment offering nothing more than 3D spatial navigation can still empirically improve task performance, enhance sense of immersion and presence, and significantly boost memory retention and learning outcomes.

Another main contribution of the book is that it proposes advanced strategies for navigating and managing cognitive load in VR. Chapters include innovative methods such as the use of Markov Chain Monte Carlo for path generation in VR spaces, and strategic design considerations for the curation and arrangement of content, leading to more personalized workflows that may support faster completion times at a lower cognitive load.

Advancing into the realm of Augmented Communication Channels in VR, the book explores novel forms of meaning representation that VR supports, contrasting these with traditional media. This part of the book focuses on how well-designed VR environments can serve as powerful tools for memory management and learning, using a variety of rich, immersive scenarios to aid the formation of lasting memories and understanding.

The integration potential of VR is particularly exciting. It serves as a platform that not only supports but enhances a variety of other innovative technologies. For instance, AI can be seamlessly incorporated to create more adaptive and responsive environments. The book concludes with a discussion of automated Context Control in VR. The innovative approach to user experience presented in the book focuses on personalizing VR environments through the construction of cognitive profiles and schematic layouts, tailored to individual preferences and learning styles or other personal requirements. By automating the customization of these environments, VR can be transformed to optimize cognitive load and maximize efficiency and effectiveness in various applications, from education to professional workflows.

This book aims to convey not just the current capabilities and applications of VR, but its vast potential as well. It argues for the integration of cognitive sciences with VR technologies, proposing that such a synthesis can lead to more intuitive, immersive, and enriching experiences that extend well beyond traditional uses. As we stand on the brink of this new digital reality, the insights presented here aim to guide researchers, practitioners, and technologists in harnessing the full potential of VR to enrich human capabilities and fundamentally change our interaction with the digital world.

Budapest, Hungary
May 2024

Ildikó Horváth
Borbála Berki
Anna Sudár
Ádám Csapó
Péter Baranyi

Acknowledgements The book is supported by the HUN-REN Research Group for Cognitive Mapping of Decision Support System of the Hungarian Research Network.

Contents

Part I
Related Scientific Fields

In the first part of the book, we outline the foundations of several fields that have emerged as a result of a confluence of technological, social, and legal developments in the past few decades, which in turn influence the development of virtual reality.

Chapter 1 explores the interdisciplinary field of Cognitive Infocommunications (CogInfoCom), which lies at the intersection of cognitive sciences and infocommunication technologies. This chapter presents recent advancements in CogInfoCom, including the concept of Cognitive Entities. Furthermore, the complex relationship between humans and technology is explained in detail.

In Chap. 2, a comprehensive explanation of established concepts such as virtual reality, augmented reality, and digital twins is presented. The chapter then delves into a new era where various technological domains are integrated with human cognitive systems. Here, the concepts of Digital Reality (DR) and the Internet of Digital Reality (IoD) are introduced.

The focus in Chap. 3 expands the horizon towards the evolving landscape of mobility, which now extends beyond physical transportation to encompass digital realms. Through the lens of cognitive frameworks, Cognitive Mobility (CogMob) offers a holistic understanding of mobility systems. Integrating natural and artificial cognitive processes can optimize various domains such as transportation engineering, artificial intelligence, and human-machine communication.

Chapter 4 shifts toward the main theme of this book, as it explores the potential of Virtual Reality (VR) as a tool for communication and information management. Expanding beyond its initial application in gaming,VR has the potential to revolutionize fields such as education and healthcare. This is primarily driven by advancements in AI and sensor technologies. The introduction of the concept of Cognitive Aspects of Virtual Reality (cVR) paves the way for a future where humans, AI, and digital twins collaborate seamlessly within VR environments, ushering in a new era of Digital Reality.

Chapter 1
Cognitive Infocommunications

Abstract Cognitive Infocommunications (CogInfoCom) is an interdisciplinary domain investigating the interplay between cognitive sciences and infocommunication technologies (ICT). That is the scientific field where virtual reality pertains. This chapter discusses recent advances in Cognitive Infocommunications on the technological and human fronts. It presents concepts like Digital Reality and Cognitive Entities, arising from the ongoing technological convergence and the deepening integration of ICT with human cognition. The term cognitive entities refers to the synergetic combination of humans, devices, and the environment. Furthermore, the chapter discusses the current research priorities of CogInfoCom, highlighting the intricate interconnectedness of human-technology interactions and the co-evolution of these dynamics.

Keywords CogInfoCom · Cognitive infocommunications · Cognitive entity · Digital reality · Generation of Cognitive Entities (Gen CE)

1.1 Introduction

Amidst the current era marked by swift technological progress and escalating reliance on digital systems, the domain of cognitive infocommunications (CogInfoCom) [1–4] has arisen as a pivotal area of exploration. CogInfoCom is an interdisciplinary field that explores the relationship between cognitive sciences and information and communication technologies (ICT). The main goal of this field is to understand and improve the interactions between humans and digital systems, both in the immediate and long-term co-evolution. This chapter appraises the foundational concepts within the field and organizes them into an innovative system, spotlighting emergent relationships based on [4]. This initiative is important to establish a common language, prevent division in the field, and make these essential concepts more understandable, especially for those without knowledge in related areas such as AI. This chapter draws upon the employed scientific perspective, mindset, and paradigm from those perspectives referenced within the following foundational articles: [5–8].

I. Horváth et al., *Cognitive Aspects of Virtual Reality*, Studies in Big Data 156,
https://doi.org/10.1007/978-3-031-68130-1_1

In the preceding decade, several transformations and advancements have occurred in digital technologies, infocommunication devices, and the psychological, cognitive, and social context in which users engage with them. The co-evolution of infocommunication technologies and humans appears to have reached a milestone where a unified perspective, concentrating on cognitive entities as a cohesive whole, emerges as a more viable research approach in numerous instances, surpassing existing methodologies that focus on disparate human and technical measures.

The chapter's structure is organized as follows: Sect. 1.2 outlines recent changes in human users arising from the widespread integration of ICT technologies. Subsequently, Sect. 1.3 elaborates on recent technological changes related to CogInfoCom. These changes are intertwined at both the technological and human levels, emphasizing the concept of Cognitive Entities as discussed in Sect. 1.5. This novel focus on the notion of cognitive entities underscores the imperative to reassess other seminal concepts within CogInfoCom. Hence, Sect. 1.6 explains recent focal points of CogInfoCom in light of these developments. The chapter concludes with a summary of key concepts in Sect. 1.7.

1.2 Human Transformations at Physical and Cognitive Level

In the last few decades, notable transformations in human behavior and lifestyles have happened as a consequence of the widespread integration of diverse infocommunication technologies. These transformations have resulted in long-lasting changes, particularly in terms of accessibility to the Internet, which now extends from early childhood to even the most remote locations. When compared with previous generations, this has led to unique patterns of development.

While there is variation in individual usage patterns, ergonomic practices, and personal habits, there has been a general increase in average daily screen time. As of 2021, the average daily screen time across all devices is 6 hours and 58 min, which is an increase of almost 50 min since 2013 [9]. The increasing use of mobile devices for activities such as voice and email communication, social media, multimedia consumption, and gaming has had a noticeable impact on body posture. This effect can be observed in both ergonomic considerations and long-term effects. Similarly, these new ways of using information and communication technology devices have significant implications for the mental and physical well-being of users.

In this section, we provide a summary of somatic, cognitive/psychological, and societal changes that are relevant to the topic.

1.2.1 *Physical Level Transformations*

1.2.1.1 Musculoskeletal Influences

Prolonged downward gaze, such as during text messaging on handheld devices, can lead to pain in the neck and nearby areas [10–12]. Research indicates that using smartphones while standing leads to a more rigid posture compared to sitting [13]. Notably, these symptoms of neck and shoulder pain are not exclusive to adults and also impact teenagers [14]. Beyond phone usage, factors like extended periods of sitting, obesity, and a sedentary lifestyle have been recognized as contributors to the heightened prevalence of musculoskeletal pain [11].

The condition known as De Quervain's syndrome, also called "texting thumb" or "PlayStation thumb," is a common repetitive injury. Studies have shown a higher occurrence of this syndrome among individuals with problematic smartphone use [15].

1.2.1.2 Ocular Influences

Regular and prolonged use of screens can result in various eye problems, including eye strain, dryness, blurred vision, and headaches. These symptoms are frequently linked to conditions like computer vision syndrome and digital eye strain. Furthermore, the blue light emitted from screens has the potential to disturb sleep patterns and may be a contributing factor to long-term eye issues [11, 16].

1.2.1.3 Sleep Influences

Using smartphones before sleep has become common, disrupting sleep patterns. Prior research on screen time [17] has underscored the significance of minimizing screen usage, particularly in the hours preceding bedtime. Exposure to blue light can contribute to sleep disturbances by suppressing the production of melatonin, a hormone that plays a critical role in regulating our sleep-wake cycles. Furthermore, constant connectivity and frequent smartphone notifications can cause anxiety and prevent relaxation, making it difficult to fall asleep. Recent research has shown that limiting exposure to blue light at night can improve the quality of sleep [18]. Furthermore, using smartphones in bed can increase the time it takes to fall asleep, raise average heart rate, decrease heart rate variability, and lead to less total sleep time. These results suggest that sleep quality is compromised, with more frequent periods of wakefulness during the night [19].

1.2.2 Cognitive and Psychological Level Transformations

Continuous technology usage has positive and negative effects on the brain and behavior [20]. Contributing factors encompass information overload, screen time, and media multitasking. In the 21st century, society's reliance on information has made the historical concern of information overload more noticeable [21]. Concurrent media processing, or rapid task-switching, has emerged as a prevalent characteristic of modern technology [22]. In the following sections, we will explore how these changes affect human cognition.

1.2.2.1 Cognitive Development

Recent investigations into the influence of technology on children's cognitive development present intriguing yet conflicting findings. In general, the escalating screen time among children has raised concerns. Even toddlers below the age of 2 now spend more than an hour daily interacting with screens, contributing to a decline in reading activities. This trend correlates with adverse outcomes, including compromised language development, executive functioning, and lowered connectivity among brain regions crucial for word recognition, language, and cognitive control. Moreover, prolonged screen time predicts behavioral issues and a less developed theory of mind [20, 23]. Theory of mind is the ability to understand and attribute mental states, such as beliefs, intentions, and emotions, to oneself and others. It also involves recognizing that others may have perspectives and knowledge different from one's own.

On the one hand, studies have found that excessive screen time can have negative impacts on children's attention and executive control abilities. However, research has also identified various factors that can help reduce or even reverse these negative effects. One such factor is high-quality television programming that is free from commercial interruptions, which has been found to have positive associations with attention and executive control measures. On the other hand, commercial breaks can disrupt children's engagement, highlighting the importance of providing uninterrupted content [24, 25]. Educational cartoons and programs have also demonstrated connections to favorable learning outcomes and improved executive function [26]. Additionally, the interactive nature of touchscreen devices promotes active engagement and control over information flow, thereby enhancing children's ability to learn from video content [27].

These findings underscore the paramount importance of content quality and interactivity in shaping the impact of technology on children's attention, learning, and cognitive development [22].

Furthermore, technology can serve as a cognitive workout for aging brains. Engaging in mentally stimulating activities such as online searches has demonstrated potential in delaying cognitive decline [28]. Consequently, serious gaming emerges as a pivotal tool in alleviating the effects of aging, encompassing both physical and cognitive aspects [29–31].

1.2.2.2 Attention

There is a widely discussed issue of reduced attention span, which was first brought to light by Carr [32] based on his own observations. Some users report experiencing declines in both the duration and intensity of their attention, but the exact link with technology remains uncertain. However, experts generally agree that there is a causal relationship. The constant influx of information can disrupt sustained concentration, leading to multitasking across various sources of information [33].

Repetitive attentional shifts and multitasking might directly impede human executive functions [34] and could be associated with shifts in attentional focus from local to global aspects [23]. Additionally, the constant use of technological devices limiting offline interactions could influence attention span. This lack of offline engagement may impede the brain's default mode, potentially leading to attentional issues [35].

Furthermore, individuals with Internet Use Disorders exhibit structural changes in brain regions linked to attentional control when compared to healthy controls [36]. They also have a threefold increased likelihood of Attention Deficit Hyperactivity Disorder (ADHD) [37]. Excessive Internet usage is associated with reduced gray matter in the anterior cingulate cortex and other prefrontal regions crucial for sustained concentration and distraction avoidance [38] and media multitasking [39].

Conversely, enhanced visual and spatial attention has been observed in users who play action video games for more than four days a week for six months [40]. Similarly, surgeons who engaged in video games made 37% fewer surgical errors, with a 27% faster response time compared to non-gaming counterparts in one study [41].

1.2.2.3 Memory

An increasing concern revolves around the excessive dependence on online information as it is often used as perpetually accessible external storage for memory [37, 42]. Also, specific brain activation patterns, that are crucial for the stable retention of information, appear to be absent [43, 44]. This leads to questions about how these phenomena could affect memory and cognitive abilities. For example, the so-called "Google effect" was observed. This effect refers to the tendency for individuals to rely on external sources for factual information, rather than memorizing it [43]. This implies a shift that individuals increasingly rely on the availability and accessibility of online information rather than their own memory.

1.2.2.4 Digital and Mutlitasking Skills

A study evaluated the verbal and visuoperceptual cognitive performance of 191 school-aged children. They were classified as either high-digital users or low-digital users, using standardized tests and a self-report questionnaire [45]. The findings indicate that increased digital exposure has a positive impact on cognitive development. High-digital users surpassed their low-digital counterparts in tasks related to naming, semantic understanding, visual memory, and logical reasoning. These results are consistent with prior research, emphasizing the substantial contribution of technology in augmenting cognitive skills and fostering smart learning among children.

Multitasking is one of the most important digital skills in today's fast-paced world. Yet, despite its current value, multitasking can have a negative impact on human cognition. As we age, our ability to multitask declines linearly. However, studies have found that playing certain computer games can improve multitasking skills [20, 46, 47].

1.2.2.5 Emotional and Social Intelligence

Elevated screen time is frequently linked to diminished in-person communication [20]. A study conducted on preteens aimed to investigate the impact of restricting screen-based media to provide more possibilities for face-to-face interactions. The aim was to find out if such a restriction could improve their ability to recognize nonverbal emotional and social cues. The research involved 51 schoolchildren who spent five days in a technology-free nature camp, while 54 matched controls continued with their regular screen time. Assessments conducted before and after attending the nature camp showed that participants significantly improved in recognizing nonverbal emotional and social cues. The findings indicate that taking a break from screen-based media and digital communication tools can positively influence both emotional and social intelligence [48].

Another comprehensive review [49] has highlighted the complex role of digital games in the cognitive and socio-emotional development of children. The review has identified both positive and negative outcomes, including pro-social behaviors, anti-social behaviors, and isolation. The review emphasizes the potential impact of technology addiction on social interactions and highlights the importance of exploring mental health apps as possible interventions [20].

1.2.3 Social Level of Transformations

It's important to take into account the social environment that shapes people's daily lives, in addition to the technological and cognitive changes mentioned earlier. Many of the individual changes also represent a societal shift, which has been made possible by the widespread availability of smartphones and internet access, enabling people to connect limitlessly. The ability to communicate instantly creates a feeling of being connected, leading to the formation of digital communities through new forms of communication and interaction.

The full impact of the combination of social changes and technological advancements is currently not fully understood. As our social interactions become more complex due to the integration of cognitive and technological innovations, there is a lack of sufficient data to identify cause-and-effect relationships. This lack of data leaves most researchers and practitioners to speculate about the outcomes of these shifts. Nevertheless, drawing inspiration from Heidegger, philosophical inquiries have probed the neutrality of technology. However, some philosophers, such as Byung-Chul Han from South Korea, have looked into the impact of technology on our lives. He argues that the current socio-technical landscape has caused burnout, narcissism, the loss of traditional rituals, and a sense of alienation. He believes that we are creating a negative utopia where we are constantly under surveillance, not by others but by ourselves [50].

1.3 Digital Transformations

In addition to the human-centric changes detailed in the preceding section, there have been noteworthy improvements in digital technologies in recent years. The cognitive capabilities of information and communication systems [51], also known as "smart" systems, have been enhanced due to the digital transformations of the last decade. These systems have become more sophisticated and now offer advanced features.

Due to the fast pace of technological advancements, adopting new technology has become increasingly important. Initially, people may have high expectations for these new technologies, which may be unrealistic at first, but they tend to stabilize to a more reasonable level over time. Criticisms are crucial in shaping future developments, addressing shortcomings, and influencing the trajectory of improvement. Evaluations provide valuable feedback that helps refine and optimize emerging technologies towards their full potential.

1.3.1 Advancements in Robotics

There have been significant advancements in the field of robotics recently, which have expanded the capabilities of robotic systems. One area of significant progress is autonomous navigation in complex and dynamic environments, largely due to the integration of advanced sensor technologies and artificial intelligence algorithms. This contributes not only to the navigation of mobile robots but also navigation of autonomous vehicles and unmanned aerial vehicles [52]. Guiding ground vehicles through unstructured surroundings is now possible thanks to learning-based approaches that use perception to enable contextually-aware navigation [53].

Human-robot collaboration is a significant challenge in the field of robotics. The collaboration between humans and robots can significantly improve performance and flexibility, provided all safety concerns are addressed for human operators. Collaborative robots, also known as cobots, have been introduced to ease the interaction between humans and robots within shared spaces. As a result, new possibilities in manufacturing, logistics, and healthcare have been unlocked [54].

Medical robotics has made significant progress in various subfields, including robotic laparoscopy, minimally invasive surgery, assistive wearable robots, therapeutic rehabilitation robots, capsule robots, and soft robotics. These developments have improved precision and patient outcomes [55, 56]. Through the integration of perception, navigation, and insights from cognitive science [57], robotics persistently advances, continually pushing the frontiers of what robots can accomplish across diverse domains.

1.3.2 Brain-Computer Interfaces

Brain-computer interfaces (BCIs) allow for direct communication between the brain and external devices. This technology receives, analyzes, and transfers signals generated by the brain into output commands that can accomplish specific tasks. BCIs are especially beneficial for individuals with disabilities, and have a wide range of applications across various fields. By receiving and interpreting brain signals, BCIs can bypass impaired neuromuscular pathways, making it possible for paralyzed patients to perform tasks and interact with their environment [58, 59].

Rapid advancements in neurotechnology and artificial intelligence (AI) have propelled brain signals utilized in brain-computer communication beyond sensory levels, now encompassing higher-level cognitive functions such as goal-directed intentions [60, 61]. The progress in flexible, stretchable, and soft electronics has further enhanced the compatibility of BCIs with the mechanical properties of the brain. Moreover, this could have an impact on neuroscience, as it can be used to record or modulate large-scale neural activities, which allows us to understand more complex and dynamic neural activities, and map the functional connections of the brain [62].

Based on The I3 evolutionary model for BCIs, the development, and evolution of this technology can be divided into three phases: consisting of interface, interaction, and intelligence stages. The first stage is the interface stage, which stands for the interface that allows direct communication for disabled patients to a computer. The next stage is marked by interaction, where a close-loop BCI system promotes the rehabilitation of human functions besides effective device control. The third one is the stage of intelligence, as AI technology has the potential to augment human intelligence through BCIs: as a non-medical application, it can improve human perceptive abilities, information processing, or decision-making abilities [60].

Brain-computer interface technology is a great development for facilitating communication and co-adaptation between the brain and computer, playing a pivotal role in human-computer co-evolution. Moreover, the integration of AI technology has the potential to augment human intelligence through BCIs, marking a significant stride towards the convergence of biological and artificial intelligence.

1.3.3 Spatial Technologies: Immersive and Non-immersive Virtual Realities

In the last decade, significant advancements have been made in Virtual Reality (VR), Augmented Reality (AR), and Extended Reality (XR), leading to a transformative era that has reshaped our perceptions and interactions with digital content. Advancements in VR and AR devices have been substantial, encompassing enhancements in display resolution, tracking accuracy, ergonomics, and the integration of wearable sensors [63–65]. These technological improvements have found diverse applications across industries, including healthcare [66, 67], education [68–71], architecture [72], tourism [73], and entertainment. In industrial settings, the coexistence of digital twins alongside physical entities has become a reality, with virtual models interconnected to physical counterparts through sensors, facilitating the generation of real-time data [74–76]. Digital twins are instrumental in implementing virtual safety training as well [77].

Collaboration and the user-friendliness of content creation tools are current focal points in this evolving field [78]. Efforts are being directed towards enabling users to effortlessly create or edit virtual environments [79–81], and interact with each other through different avatars [82]. Additionally, the personalization of virtual spaces has emerged as a notable area of interest [83, 84], highlighting the growing importance of tailoring virtual experiences to individual preferences and needs.

1.3.4 Personalized Recommendation Systems

Personalized recommendation systems have become ubiquitous across diverse domains, striving to enrich user experiences by providing relevant and tailored suggestions. These systems leverage user preferences and behaviors, analyzing factors such as past choices, browsing history, and feedback to generate personalized recommendations that align with individual interests [85]. As the demand for heightened personalization intensifies, certain systems exceed traditional approaches to hyperpersonalization, characterized by real-time matching to deliver customized experiences [86]. It is important to understand that hyperpersonalization, although useful, can also cause negative reactions [87]. These systems have a significant impact on the accessibility of information, effectively acting as digital nudges. Personalized recommendations shape our decision-making processes by influencing our choices based on presented information [88]. While often viewed as beneficial, several authors caution that optimizing a cost function in this manner may lead to unexpected effects, such as modifying data to induce uniformization of human behaviors rather than precisely understanding preferences [89]. Dörfler [90] underscores that, while pattern recognition and its application in personalization can be highly effective from a business perspective, it may not align as seamlessly with broader objectives. There is a notable discrepancy where, instead of offering individuals what they desire, these systems tend to dictate what users *should* want. This observation challenges the narrative presented by vendors and prompts a critical examination of the societal implications of personalized recommendation systems.

1.3.5 Chatbots: Revolutionizing Conversational Experiences

Interest in chatbots has experienced rapid growth since 2016, driven by significant advancements in Natural Language Processing (NLP) that enable these digital entities to provide more interactive and engaging experiences [91]. Operating as computer programs that simulate human-like conversations and understand human languages, chatbots, and virtual assistants have found applications in diverse fields, including marketing, education, healthcare, cultural heritage, and entertainment. Chatbots can provide intelligent features across categories like contextualizing data, personalizing information, and abstracting content for decision-making by combining NLP and cognitive computing [92]. The effectiveness and acceptability of chatbots depend on the context. Factors like empathy are essential in healthcare settings, while pleasure is crucial in entertainment applications [93]. In essence, chatbots equipped with NLP capabilities significantly enhance interactions and provide tailored experiences across various domains.

1.3.6 Advancements of Large Language Models

The advancement of AI research has been driven by the development of large language models (LLMs) based on the Transformer architecture [94]. Trained on extensive web-text data, LLMs demonstrate exceptional proficiency in various natural language processing tasks. GPT-4, as a representative example, signifies progress towards artificial general intelligence (AGI), showcasing the ability to comprehend and connect diverse topics while performing a wide range of tasks. However, it is important to recognize that the current abilities of technology are still not as advanced as those of humans, particularly in tasks requiring reasoning across multiple domains [95].

A novel approach to studying LLMs has recently emerged, diverging from conventional machine learning techniques and drawing inspiration from traditional psychology [95]. An illustrative example of this approach involves the examination of Theory of Mind (ToM) in LLMs [96–98], which pertains to the capacity to understand and attribute mental states to others. Current findings suggest that existing LLMs often rely on shallow heuristics rather than robust ToM abilities [97]. A parallel investigation applies a similar approach with the introduction of the "Baby Intuitions Benchmark (BIB)" tasks, probing commonsense psychology in LLMs. Results from these tasks indicate that current neural-network models fall short in capturing infants' knowledge in various aspects [99]. These studies underscore the evolving nature of research on LLMs, encompassing both technological advancements and a deeper understanding of their cognitive underpinnings.

1.4 Digital Reality: The Convergence of Technologies

The advancements in various technologies have led to a crucial dimension of their interconnection and convergence. Information technologies that were once separated are now closely linked. Technologies such as Virtual Reality, Augmented Reality, Mixed Reality, Digital Twins, Artificial Intelligence, 5G networks, and the ubiquitous 2D Web are all experiencing widespread adoption. Their integration has the potential to create an entirely new reality. This convergence of technologies blurs the lines between the physical and digital realms, as well as between digital representations and simulations, giving rise to the concept of Digital Reality [7, 100]. Digital Reality represents a "*high-level integration of virtual reality (including augmented reality, virtual and digital simulations, and twins), artificial intelligence, and 2D digital environments which creates a highly contextual reality for humans in which previously disparate realms of human experience are brought together*" [7].

The integration of various technologies has become seamless from the user's perspective, resulting in a mutually beneficial relationship that leads to a united digital experience. The once-distinct boundaries between technologies are gradually disappearing, making it easier to connect and align them for more intuitive usage. The introduction of Language and Learning Models (LLMs) amplifies this experience, with language acting as a universal interface, resembling a meta-dimension (metaD). This meta-dimension facilitates natural communication for human users, adding an extra layer of cohesion to the immersive digital reality.

1.5 Cognitive Entity

With the advancements in both human and technological domains, it is unsurprising that humans and technology have become increasingly intertwined in recent times [3, 101]. As both humans and technology continue to progress, their collaboration holds the potential to result in a new and unique way of thinking, characterized by a qualitatively distinct experience. The concept of cognitive entities, defined as a "*synergetic combination of humans, devices, infrastructure, and environment that is identifiable from the perspective of some (high-level) cognitive capability*" [3] was introduced to encapsulate the intertwined ensembles of human entities and information and communication technology (ICT) capabilities that have coevolved [3, 102]. The previously presented changes both in humans (Sect. 1.2) and digital technologies (Sect. 1.3), alongside their intertwined nature (Sect. 1.4) enhance the role of cognitive entities (Fig. 1.1).

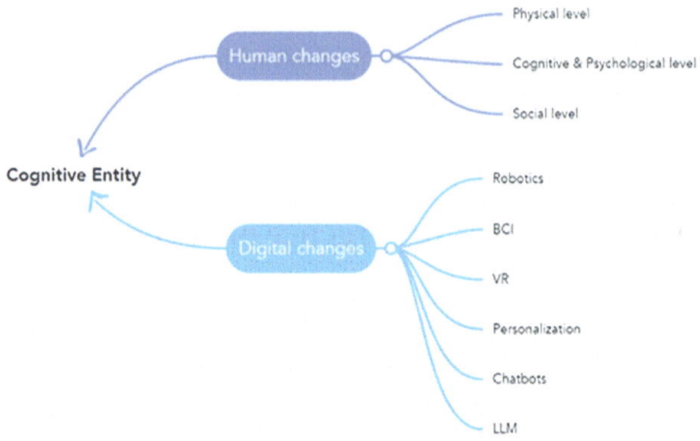

Fig. 1.1 The recent changes in both the human and technological sides have contributed to an enhanced focus on the concept of cognitive entities

While humans inherently qualify as cognitive entities, there exists a theoretical perspective, as posited by numerous philosophers, that machines lack cognitive capabilities in the absence of human involvement. John Searle, for instance, introduces the dichotomy between "observer-independent" and "observer-relative" intelligence. He explains that instances like one chess grandmaster besting another signify intelligence in an observer-independent context, whereas scenarios like Deep Blue surpassing Kasparov exemplify intelligence in an observer-relative sense. In this latter case, the human user explains the interpretation of the computer's outputs as chess moves, even though the computer itself remains unaware of the nature of its engagement in chess play [103, 104]. Building upon the foundation established by Hubert Dreyfus and John Searle, Dörfler [90] explores the domains in which artificial intelligence (AI) can imitate or exceed human performance, and distinguishes the unique domains that are inherently linked to human faculties.

The collaboration between humans and infocommunication technology can enhance efficiency significantly. Consider mathematicians working in tandem with programs, fostering a synergistic brainstorming process. In such situations, the absence of a counterexample from the machine can act as a suggestion to explore the idea further. Conversely, if the machine quickly identifies a valid counterexample, it can indicate that the idea might not be worth pursuing. This collaborative approach maximizes the strengths of both parties by combining human intuition and machine processing power.

The conceptualization of "generation CE", signifying the cohort characterized as the generation of cognitive entities, was introduced in the year 2015 and initially expounded upon in [3], with subsequent elaboration provided in [102]. Since its inception, the relationship between humans and ICT has progressively tightened, culminating in an inseparable relationship. New abilities in the digital world have emerged, including interactive communication facilitated by generative artificial intelligence. Moreover, individuals who have demonstrated higher digital proficiency have shown superior performance compared to those with limited digital understanding in cognitive areas such as naming, semantic and visual memory, and logical reasoning [45].

Generation CE, inclusive of individuals born after the year 2010, has grown up within an environment intricately interwoven with infocommunication technology since their formative years, fostering a co-evolution with technology. In contrast to previous generations who were forced to adapt to emerging technological paradigms, members of Generation CE exhibit an inherent and deeper affinity with ICT due to early and prolonged exposure. The everyday experiences of this generation are seamlessly integrated with ICT, providing them with extended cognitive abilities and easy access to information through handheld devices. In addition, their social environment is partly digital, where online communities hold a comparable significance to their real-life counterparts in shaping their social structure.

1.5.1 Levels of Interconnectedness

The concept of CogInfoCom emphasizes the merging of humans and technology, resulting in complex forms of entanglement. This convergence requires a similar integration between technology and cognitive sciences. This merging has been analyzed by various authors and can be observed at three different interaction levels [101, 102, 105, 106].

The first level of interconnectedness is characterized by low-level direct relationships, including both invasive and non-invasive interfaces such as brain-computer interfaces. Although this level enables direct sensing and control, it is often hindered by the need for implanted or wearable sensors, which can pose operational challenges.

Personal informatics devices allow communication using various sensory modalities, entangling human and non-human elements on a second level. Determining the "communication language" becomes pivotal, with considerations encompassing information semantics, modalities employed, application context, and users' cognitive capacities. The challenge is to convey a broad spectrum of semantic concepts within the confines of human sensory modalities.

At the collective level of multi-user interactions, applications leverage collective behaviors to enhance individual user interactions or analyze communal events. Such applications engage in data mining and analysis across diverse data sources, including social communication platforms.

The advantages of close collaboration and partnership are obvious and provide a more efficient way to handle related tasks. However, in certain situations, there may be potential drawbacks to such close associations. For instance, turn-by-turn navigational aids commonly featured in GPS devices can adversely impact human spatial abilities. This negative impact can be reduced by promoting consciousness in decision-making about the use of navigational aids. Furthermore, optimizing the interaction design of these devices to facilitate active information processing and incorporating landmarks as reference points in GPS aids can enhance environmental awareness [107]. Hence, it is crucial to achieve a balanced approach and effectively utilize the benefits of connectivity.

1.5.2 Joint Assessment

Taking a broader perspective, the collaboration between humans and ICT enables joint measurement of the human-ICT combo. In the ongoing co-evolution of humans and ICT, novel methodological considerations have arisen, given the integration of these entities into our daily existence. Traditionally, assessments of human cognitive abilities and artificial cognitive capabilities were different: employing diverse methodologies. Although instances of convergence have been observed, as presented in Sect. 1.3.6, the study of Stojnic and colleagues [99] compared infants and neural network models in predicting people's actions, on the Baby Intuitions Benchmark (BIB).

To accurately measure the cognitive abilities of human-ICT collaborations, it is necessary to evaluate them as a whole rather than separately. The previously discussed Google effect [43] underscores the necessity for this shift in perspective, wherein the ubiquity of internet access has led individuals to recall the sources where information can be accessed rather than the information itself. Measuring cognitive entities, in addition to natural cognitive capabilities, allows for measuring general cognitive proficiencies in everyday settings, due to the constant proximity of individuals to smart devices.

1.6 Current Concepts Under CogInfoCom

In light of the significant changes detailed earlier, reevaluating the CogInfoCom concept and its subfields was imperative, with a heightened focus on the role of cognitive entities [4]. Originally Cognitive infocommunications (CogInfoCom) was defined as the exploration of the "*link between the research areas of infocommunications and cognitive sciences, as well as the various engineering applications which have emerged as a synergic combination of these sciences*" [1, 3, 108, 109]. It directs attention towards the seamless integration of these disciplines and their practical implementations. CogInfoCom's primary objective revolves around investigating the co-evolution of cognitive processes with infocommunication devices. This exploration aims to extend human cognitive capabilities beyond geographical constraints, fostering collaboration between natural and artificial cognitive systems. The overarching goal is to enhance the effectiveness of engineering applications through cognitive collaboration.

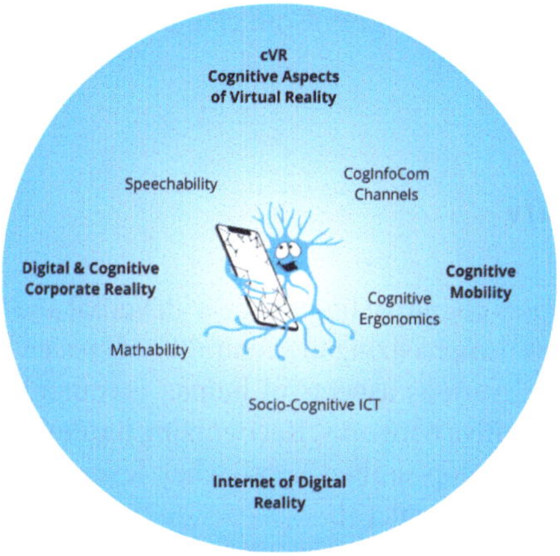

Fig. 1.2 This graphical illustration represents the current emphasis within CogInfoCom. The outer section of the circle shows emerging and innovative directions closely linked to established concepts in CogInfoCom. Please note that this illustration is not comprehensive; other important aspects within the domain of CogInfoCom require exploration beyond those depicted in this illustration. These are further explored in the following chapters of this part

It is important to emphasize that the ideas presented are closely linked and dependent on each other. They exist together and have their own unique significance and impact, rather than existing alone. These concepts have a symbiotic relationship with each other. The visual representation of the current focus areas of CogInfoCom is in Fig. 1.2. The inner circle highlights core concepts within CogInfoCom, while the outer circle encompasses emerging directions, such as the cognitive aspects of virtual reality, which serves as the focal point of this book. This progression is attributed to a significant technological breakthrough, facilitating rapid integration into applied environments. Notably, these emerging directions offer a more tangible manifestation and align seamlessly with ongoing research domains, responding promptly to business and industrial demands.

1.6.1 CogInfoCom Channels

The CogInfoCom channels [110], focuses on the intricate communication challenges encountered by cognitive entities when transferring new information. This conceptual framework integrates both structural and semantic elements to define sets of sensory messages associated with high-level concepts, employing icon-like and message-like design elements across diverse modalities from tactile to auditory modality [111–116]. A toolset based on concept algebra and the spiral discovery method [117] is also involved in mapping semantic meaning. As this tuning model is proposed for the designer of CogInfoCom channels a practical trade-off between complexity and interpretability.

1.6.2 Speechability

"Speechability" integrates cognitive linguistics with verbal and non-verbal signals in human-ICT interactions. Emphasizing the centrality of speech in social interactions, it advocates considering various aspects of human communication to enhance the social capabilities of cognitive networks. Recognizing the embodied nature of speech, Speechability aims to develop unified approaches leveraging the convergence of human and machine embodiments [3].

Speech, being a multifaceted modality, encompasses verbal elements along with prosody, gestures, facial expressions, and more. In human communication, speech facilitates information sharing, establishes emotional foundations, and fosters trust. The term *speechability* pertains to the perspective of artificial devices proficient in speech generation (as discussed in Sect. 1.3.6 on Large Language Models), although it is distinct from the nuanced capabilities inherent in human speech [118].

Beyond speech generation and recognition [119–122], it extends to applications like language learning [123], speech rehabilitation and early diagnosis [124, 125]. Previous research encompassed different aspects of speechability, such as minimalist grammars [126], multimodal cues [127, 128], conversational fillers [129], phonological and spectral analysis, and also corpus-based techniques [130, 131].

1.6.3 Cognitive Ergonomics

Cognitive ergonomics, closely aligned with Human-Computer Interaction (HCI), explores the challenges posed by multi-sensory usability in the context of increasing augmentative reality mediation. Research groups investigated the adaptability of ergonomic approaches, revealing synergies with CogInfoCom [132–137].

The shift from HCI to cognitive infocommunications is an important step, as it is based on the understanding of human behavior, cognition, alongside its limits and needs in interface design [138]. Studies have also been carried out to examine the linkage and interaction among entities in the field of infocommunication [139, 140]. Various researchers have contributed to different aspects of user interface design [141–144]. For example, traditional Japanese gestures and joystick guidance methods have been implemented to improve the interaction between pilots and aircraft [141] Frameworks for evaluating usability in collaborative 3D virtual environments have been proposed [133], and field studies were conducted on collaborative design software [134].

1.6.4 Socio-Cognitive ICT

Professor Hassan Charaf and his research group at the Budapest University of Technology and Economics have introduced the term "socio-cognitive ICT" to describe applications that require both social and cognitive capabilities. This term is increasingly used to characterize such applications [145–147]. The concept of socio-cognitive ICT signifies the intersection of social and cognitive dimensions within computer networks. It focuses on the cognitive properties emerging from user interactions and explores the potential enhancement of social capabilities through the analysis and management of information flow. This framework enables the development of applications aimed at understanding crowd-generated phenomena and optimizing information dissemination in critical situations, bridging the gap between the social and cognitive aspects of ICT [148, 149].

In the context of Socio-Cognitive ICT, applications are developed to augment the social capabilities of cognitive networks. This involves the analysis of user activities, including content and metadata-based analysis, to gain insights into phenomena such as social-political tendencies, the spread of ideas, and epidemics [150]. Additionally, the organizational management of cognitive networks is employed to optimize information flow in critical scenarios, such as disaster recovery and workflow management.

The term "socio-cognitive ICT" reflects the integration of social and cognitive elements within these applications. It signifies a holistic approach to network management and user experience, acknowledging the entanglement between human interactions and information and communication technologies. The concept is particularly relevant in the evolving landscape of technology, where advancements in communication, information technology, media, Internet of Things (IoT), Augmented Reality (AR), Virtual Reality (VR), and artificial intelligence play crucial roles in shaping smarter environments, including Smart Cities and Smart Factories [151].

In virtual reality, both social and cognitive aspects of user interaction have a significant role, thus the holistic approach of socio-cognitive ICT and cVR is closely related.

1.6.5 Mathability

Mathability forms a domain within CogInfoCom that systematically investigates the cognitive capacities relevant to mathematics, encompassing a spectrum from elementary arithmetic operations to advanced symbolic reasoning, with the objective not solely of tracing these capabilities but also of quantifying them within natural and artificial cognitive systems [8].

The need to evaluate the performance of artificial systems is increasing. This evaluation should not only be based on their ability to process data and store memory but also on more advanced metrics that reveal the complexity achieved in the systems' action-reaction processes. Additionally, it addresses the extension of established measurement protocols from human subjects to software tools, acknowledging the prevalent integration of these tools into daily routines, and emphasizes the significance of evaluating and rating mathematical abilities in software products [152]. Mathability assumes great significance in both human and artificial research, providing a comprehensive framework for understanding, quantifying, and potentially enhancing mathematical proficiency [153–158].

1.6.6 Cognitive Mobility—CogMob

Cognitive Mobility (CogMob) is a multidisciplinary field that integrates mobility, transportation, vehicle engineering, social sciences, artificial intelligence, and cognitive infocommunications. Its overarching goal is to comprehensively understand and optimize mobility as a fusion of artificial and human cognitive systems, presenting a holistic perspective on the broader scale of mobility. With a focus on engineering applications within the mobility domain, CogMob explores interconnected research areas, aiming to reveal the emergent cognitive capabilities of the unified CogMob system. Two critical dimensions within CogMob include the ranking of mobility and the constituent elements of mobility, contributing to a comprehensive understanding and modeling of mobility dynamics [159, 160].

1.6.7 Digital and Cognitive Corporate Reality

Digital and Cognitive Corporate Reality (DCR) is a scientific discipline that integrates Corporate Management, Business Science, Internet of Digital and Cognitive Reality, and Cognitive Infocommunications to achieve a higher conceptual level through a holistic perspective [6].

DCR encompasses the co-evolved natural and artificial cognitive capabilities across individual, social, and network levels, exploring interactions within digital corporate ecosystems and various approaches in the Internet of Digital and Cognitive Reality. The research in DCR aims to develop theoretical frameworks and practical solutions, addressing aspects such as digital corporate ecosystems, Internet of Digital and Cognitive Reality, and hybrid human, organizational, and artificial cognitive capabilities [161–165].

1.6.8 cVR—Cognitive Aspects of Virtual Reality

Among the current focal concepts of cognitive infocommunications, the cognitive aspects of virtual reality also appear, as the spatial technological advancements emphasized its role (presented in Sect. 1.3.3). Interacting with digital content became more intuitive with the development in display resolution, tracking accuracy, and ergonomic aspects. Furthermore, these developments allow a more intuitive interaction, the synergic combination with humans as cognitive entities.

Cognitive Aspects of Virtual Reality (cVR) is defined [5] is a field that explores the evolution from 2D graphical user interfaces to integrated 3D spaces, incorporating VR/AR/MR systems, human spatial cognition, Web 2.0/3.0, and AI. It focuses on how this transition enhances and benefits human psychological, cognitive, and social capabilities. Furthermore, emphasizes the understanding of geometric, temporal, and semantic relationships[166–174].

Additionally, cVR explores the implications of these changes on human and AI capabilities across sectors like education [70, 175–186], commerce [187, 188], healthcare [67, 189–193], and industrial production [194–196].

1.7 Conclusions

This chapter summarized the recent changes and advancements of the past decade, in both human capabilities and digital/infocommunication technologies. Consequently, there has been a shift not only in human psychology, cognition, and the social context but also in the emergence of the novel concept of Cognitive Entities, blurring the once clear distinction between human and digital capabilities. These changes and intertwining resulted in new focal points and concepts in the field of cognitive infocommunications. The convergence of technologies has given rise to concepts such as Digital Reality, Cognitive Mobility, and Cognitive Aspects of Virtual Reality.

References

1. Baranyi, P., Csapó, Á.: Definition and synergies of cognitive infocommunications. Acta Polytech. Hung. **9**(1), 67–83 (2012)
2. Csapo, A., Baranyi, P.: An application-oriented review of coginfocom: the state-of-the-art and future perspectives. In: 2012 IEEE 10th International Symposium on Applied Machine Intelligence and Informatics (SAMI). IEEE (2012)
3. Baranyi, P., Csapo, A., Sallai, G.: Cognitive infocommunications (coginfocom). Springer (2015)
4. Baranyi, P., Csapó, Á.B., Berki, B.: Concepts of cognitive infocommunications. Infocommunications J. (2024). Accepted, in press
5. Horváth, I., Csapó, Á.B., Berki, B., Sudár, A., Baranyi, P.: Definition, background and research perspectives behind 'Cognitive Aspects of Virtual Reality' (cVR). Infocommunications J. **15**, no. Special Issue, 9–14 (2023)

6. Kő, A., Szabó, I., Csapó, Á.B., Kovács, T., Lőrincz, L., Baranyi, P.: Digital & cognitive corporate reality. Infocommunications J. **15**, no. Special Issue, 2–10 (2023)
7. Baranyi, P., Csapó, Á., Budai, T., Wersényi, G.: Introducing the concept of internet of digital reality - part I. Acta Polytech. Hung. **18**(7), 225–240 (2021)
8. Baranyi, P., Gilányi, A.: Mathability: emulating and enhancing human mathematical capabilities. In: 2013 IEEE 4th International Conference on Cognitive Infocommunications (CogInfoCom), pp. 555–558. IEEE (2013)
9. Zauderer, S.: Cross river therapy - screen time statistics (2023). https://www.crossrivertherapy.com/research/screen-time-statistics. Accessed 12 June 2023
10. Khattak, S., Gul, M., Ali Kakar, H., Ullah, G., Ur Rahman, M.: The cost of long-term use of smart phones in the form of text neck syndrome; a systematic review. Rehman J. Health Sci. **1**(1), 3–5 (2019)
11. Wade, A.: The impact of digital media use on the musculoskeletal and visual systems. In: Technology and Adolescent Health, pp. 365–380. Elsevier (2020)
12. Neupane, S., Ali, U.T.I., Mathew, A.: Text neck syndrome - systematic review. Imper. J. Interdiscip. Res. **3**(7), 141–148 (2017)
13. D'Anna, C., Schmid, M., Conforto, S.: Linking head and neck posture with muscular activity and perceived discomfort during prolonged smartphone texting. Int. J. Ind. Ergon. **83**, 103134 (2021)
14. Wirth, B., Potthoff, T., Rosser, S., Humphreys, B.K., de Bruin, E.D.: Physical risk factors for adolescent neck and mid back pain: a systematic review. Chiropr. & Man. Therap. **26**(1) (2018)
15. Benites-Zapata, V.A., Jiménez-Torres, V.E., Ayala-Roldán, M.P.: Problematic smartphone use is associated with de quervain's tenosynovitis symptomatology among young adults. Musculoskeletal Sci. Pract. **53**, 102356 (2021)
16. Kaur, K., Gurnani, B., Nayak, S., Deori, N., Kaur, S., Jethani, J., Singh, D., Agarkar, S., Hussaindeen, J.R., Sukhija, J., Mishra, D.: Digital eye strain - a comprehensive review. Ophthalmol. Therapy **11**(5), 1655–1680 (Jul2022)
17. Hale, L., Guan, S.: Screen time and sleep among school-aged children and adolescents: A systematic literature review. Sleep Med. Rev. **21**, 50–58 (2015)
18. Randjelović, P., Stojanović, N., Ilić, I., Vučković, D.: The effect of reducing blue light from smartphone screen on subjective quality of sleep among students. Chronobiol. Int. **40**(3), 335–342 (2023)
19. Kheirinejad, S., Visuri, A., Ferreira, D., Hosio, S.: "Leave your smartphone out of bedâŁž: quantitative analysis of smartphone use effect on sleep quality. Pers. Ubiquitous Comput. **27**(2), 447–466 (2022)
20. Small, G.W., Lee, J., Kaufman, A., Jalil, J., Siddarth, P., Gaddipati, H., Moody, T.D., Bookheimer, S.Y.: Brain health consequences of digital technology use. Dialogues Clin. Neurosci. **22**(2), 179–187 (2020)
21. Bawden, D., Robinson, L.: Information Overload: An Introduction (2020)
22. Vedechkina, M., Borgonovi, F.: A review of evidence on the role of digital technology in shaping attention and cognitive control in children. Front. Psychol. **12** (2021)
23. Konok, V., Liszkai-Peres, K., Bunford, N., Ferdinandy, B., Jurányi, Z., Ujfalussy, D.J., Réti, Z., Pogány, Á., Kampis, G., Miklósi, Á.: Mobile use induces local attentional precedence and is associated with limited socio-cognitive skills in preschoolers. Comput. Hum. Behav. **120**, 106758 (2021)
24. Hudon, T.M., Fennell, C.T., Hoftyzer, M.: Quality not quantity of television viewing is associated with bilingual toddlers' vocabulary scores. Infant Behav. Develop. **36**(2), 245–254 (Apr2013)
25. Nathanson, A.I., Aladé, F., Sharp, M.L., Rasmussen, E.E., Christy, K.: The relation between television exposure and executive function among preschoolers. Develop. Psychol. **50**(5), 1497–1506 (2014)
26. Fisch, S.M.: Children's Learning From Educational Television. Routledge (2014)

27. Kirkorian, H.L., Choi, K., Pempek, T.A.: Toddlers' word learning from contingent and noncontingent video on touch screens. Child Develop. **87**(2), 405–413 (2016)
28. Small, G.W., Moody, T.D., Siddarth, P., Bookheimer, S.Y.: Your brain on google: Patterns of cerebral activation during internet searching. Amer. J. Geriatr. Psychiatry **17**(2), 116–126 (2009)
29. Janhunen, M., Katajapuu, N., Paloneva, J., Pamilo, K., Oksanen, A., Keemu, H., Karvonen, M., Luimula, M., Korpelainen, R., Jämsä, T., et al.: Effects of a home-based, exergaming intervention on physical function and pain after total knee replacement in older adults: a randomised controlled trial. BMJ Open Sport & Exerc. Med. **9**(1), e001416 (2023)
30. Nguyen, T.T.H., Ishmatova, D., Tapanainen, T., Liukkonen, T.N., Katajapuu, N., Makila, T., Luimula, M.: Impact of serious games on health and well-being of elderly: a systematic review. In: Proceedings of the Annual Hawaii International Conference on System Sciences. HICSS. Hawaii International Conference on System Sciences (2017)
31. Li, J., Xu, X., Pham, T.P., Theng, Y.-L., Katajapuu, N., Luimula, M.: Exergames designed for older adults: a pilot evaluation on psychosocial well-being. Games Health J. **6**(6), 371–378 (2017)
32. Carr, N.: The Shallows: What the Internet Is Doing to Our Brains. WW Norton & Company (2011)
33. Firth, J., Torous, J., Stubbs, B., Firth, J.A., Steiner, G.Z., Smith, L., Alvarez-Jimenez, M., Gleeson, J., Vancampfort, D., Armitage, C.J., Sarris, J.: The "online brainâŁž: how the internet may be changing our cognition. World Psychiatry **18**(2), 119–129 (2019)
34. Nikkelen, S.W.C., Valkenburg, P.M., Huizinga, M., Bushman, B.J.: Media use and ADHD-related behaviors in children and adolescents: a meta-analysis. Develop. Psychol. **50**(9), 2228–2241 (2014)
35. Greicius, M.D., Krasnow, B., Reiss, A.L., Menon, V.: Functional connectivity in the resting brain: a network analysis of the default mode hypothesis. Proc. Natl. Acad. Sci. **100**(1), 253–258 (2002)
36. Park, B., Han, D.H., Roh, S.: Neurobiological findings related to internet use disorders. Psychiat. Clin. Neurosci. **71**(7), 467–478 (2016)
37. Firth, J.A., Torous, J., Firth, J.: Exploring the impact of internet use on memory and attention processes. Int. J. Environ. Res. Publ. Health **17**(24), 9481 (2020)
38. Kühn, S., Gallinat, J.: Brains online: structural and functional correlates of habitual internet use. Add. Biol. **20**(2), 415–422 (2014)
39. Loh, K.K., Kanai, R.: Higher media multi-tasking activity is associated with smaller gray-matter density in the anterior cingulate cortex. PLoS ONE **9**(9), e106698 (2014)
40. Green, C.S., Bavelier, D.: Action video game modifies visual selective attention. Nature **423**(6939), 534–537 (2003)
41. Rosser, J.C., Lynch, P.J., Cuddihy, L., Gentile, D.A., Klonsky, J., Merrell, R.: The impact of video games on training surgeons in the 21st century. Arch. Surg. **142**(2), 181–186 (2007)
42. Ward, A.F.: Supernormal: how the internet is changing our memories and our minds. Psychol. Inquiry **24**(4), 341–348 (Oct2013)
43. Sparrow, B., Liu, J., Wegner, D.M.: Google effects on memory: cognitive consequences of having information at our fingertips. Science **333**(6043), 776–778 (2011)
44. Dong, G., Potenza, M.N.: Behavioural and brain responses related to internet search and memory. Eur. J. Neurosci. **42**(8), 2546–2554 (2015)
45. Giacomo, D.D., Ranieri, J., Lacasa, P.: Digital learning as enhanced learning processing? cognitive evidence for new insight of smart learning. Front. Psychol. **8** (2017)
46. Abbott, A.: Gaming improves multitasking skills. Nature **501**(7465), 18 (2013)
47. van der Schuur, W.A., Baumgartner, S.E., Sumter, S.R., Valkenburg, P.M.: The consequences of media multitasking for youth: a review. Comput. Hum. Behav. **53**, 204–215 (2015)
48. Uhls, Y.T., Michikyan, M., Morris, J., Garcia, D., Small, G.W., Zgourou, E., Greenfield, P.M.: Five days at outdoor education camp without screens improves preteen skills with nonverbal emotion cues. Comput. Hum. Behav. **39**, 387–392 (Oct2014)

49. Bochicchio, V., Maldonato, N.M., Valerio, P., Vitelli, R., dell'Orco, S., Scandurra, C.: A review on the effects of digital play on children's cognitive and socio-emotional development. In: 2018 9th IEEE International Conference on Cognitive Infocommunications (CogInfoCom). IEEE (2018)
50. Han, B.-C.: The Burnout Society. In: Butler, E. (ed.) Stanford Briefs, an imprint of Stanford University Press. California, Stanford (2015)
51. Baranyi, P., Csapo, A., Sallai, G.: Cognitive Capabilities in the Future Internet. Springer International Publishing (2015), pp. 173–185
52. Golroudbari, A.A., Sabour, M.H.: Recent advancements in deep learning applications and methods for autonomous navigation: a comprehensive review (2023)
53. Guastella, D.C., Muscato, G.: Learning-based methods of perception and navigation for ground vehicles in unstructured environments: a review. Sensors **21**(1), 73 (2020)
54. Matheson, E., Minto, R., Zampieri, E.G.G., Faccio, M., Rosati, G.: Human-robot collaboration in manufacturing applications: a review. Robotics **8**(4), 100 (2019)
55. Dupont, P.E., Nelson, B.J., Goldfarb, M., Hannaford, B., Menciassi, A., O'Malley, M.K., Simaan, N., Valdastri, P., Yang, G.-Z.: A decade retrospective of medical robotics research from 2010 to 2020. Sci. Robot. **6**(60) (2021)
56. Zhang, Y., Lu, M.: A review of recent advancements in soft and flexible robots for medical applications. Int. J. Med. Robot. Comput. Assist. Surg. **16**(3) (2020)
57. Ho, M.K., Griffiths, T.L.: Cognitive science as a source of forward and inverse models of human decisions for robotics and control. Ann. Rev. Control Robot. Auton. Syst. **5**(1), 33–53 (2022)
58. Mudgal, S.K., Sharma, S.K., Chaturvedi, J., Sharma, A.: Brain computer interface advancement in neurosciences: applications and issues. Interdiscip. Neurosurg. **20**, 100694 (2020)
59. Benabid, A.L., Costecalde, T., Eliseyev, A., Charvet, G., Verney, A., Karakas, S., Foerster, M., Lambert, A., Morinière, B., Abroug, N., Schaeffer, M.-C., Moly, A., Sauter-Starace, F., Ratel, D., Moro, C., Torres-Martinez, N., Langar, L., Oddoux, M., Polosan, M., Pezzani, S., Auboiroux, V., Aksenova, T., Mestais, C., Chabardes, S.: An exoskeleton controlled by an epidural wireless brain-machine interface in a tetraplegic patient: a proof-of-concept demonstration. Lancet Neurol. **18**(12), 1112–1122 (2019)
60. Gao, X., Wang, Y., Chen, X., Gao, S.: Interface, interaction, and intelligence in generalized brain-computer interfaces. Trends Cogn. Sci. **25**(8), 671–684 (2021)
61. Zander, T.O., Kothe, C.: Towards passive brain-computer interfaces: applying brain-computer interface technology to human-machine systems in general. J. Neural Eng. **8**(2), 025005 (2011)
62. Tang, X., Shen, H., Zhao, S., Li, N., Liu, J.: Flexible brain-computer interfaces. Nat. Electr. **6**(2), 109–118 (2023)
63. Kim, H., Kwon, Y.-T., Lim, H.-R., Kim, J.-H., Kim, Y.-S., Yeo, W.-H.: Recent advances in wearable sensors and integrated functional devices for virtual and augmented reality applications. Adv. Funct. Mater. **31**(39), 2005692 (Oct2020)
64. Zhan, T., Yin, K., Xiong, J., He, Z., Wu, S.-T.: Augmented reality and virtual reality displays: perspectives and challenges. iScience **23**(8), 101397 (2020)
65. Szabó, P., Baranyi, P., Sik-Lanyi, C.: Dos-windows-virtual desktop. In: 2023 IEEE 2nd International Conference on Cognitive Aspects of Virtual Reality (CVR). IEEE (2023)
66. Carroll, J., Hopper, Farrelly, L.A.M., Lombard-Vance, R., Bamidis, P.D., Konstantinidis, E.I.: A scoping review of augmented/virtual reality health and wellbeing interventions for older adults: Redefining immersive virtual reality. Front. Virtual Real. **2** (2021)
67. Emmelkamp, P.M.G., Meyerbröker, K.: Virtual reality therapy in mental health. Ann. Rev. Clin. Psychol. **17**(1), 495–519 (2021)
68. Horváth, I.: Evolution of teaching roles and tasks in VR/AR-based education. In: 2018 9th IEEE International Conference on Cognitive Infocommunications (CogInfoCom), pp. 355–360. IEEE (2018)
69. Makransky, G., Petersen, G.B.: Investigating the process of learning with desktop virtual reality: a structural equation modeling approach. Comput. & Educ. **134**, 15–30 (2019). http://www.sciencedirect.com/science/article/pii/S0360131519300259

70. Horváth, I.: Experiences and guidelines for developing 3D VR curricula. In: Accentuated Innovations in Cognitive Info-Communication, pp. 99–123. Springer International Publishing (2022)
71. Horváth, I., Vér, C., Balogh, B.: Empowering circular economy goals through cognitive info-communications in dual training programs. In: 2023 14th IEEE International Conference on Cognitive Infocommunications (CogInfoCom), pp. 000 051–000 056 (2023)
72. Schiavi, B., Havard, V., Beddiar, K., Baudry, D.: BIM data flow architecture with AR/VR technologies: use cases in architecture, engineering and construction. Autom. Construct. **134**, 104054 (2022)
73. Verma, S., Warrier, L., Bolia, B., Mehta, S.: Past, present, and future of virtual tourism-a literature review. Int. J. Inf. Manag. Data Insights **2**(2), 100085 (2022)
74. Pirker, J., Loria, E., Safikhani, S., Kunz, A., Rosmann, S.: Immersive virtual reality for virtual and digital twins: a literature review to identify state of the art and perspectives. In: 2022 IEEE Conference on Virtual Reality and 3D User Interfaces Abstracts and Workshops (VRW). IEEE (2022)
75. Vámos, Á.N., Fülöp, I.M., Reskó, B., Baranyi, P.Z.: Collaboration in virtual reality of intelligent agents. Acta Electrotechnica et Informatica **10**(2), 21–27 (2010). https://eprints.sztaki.hu/6255/
76. Baranyi, P.Z., Solvang, B., Hashimoto, H., Korondi, P.: 3D internet for cognitive info-communication. In: CINTI 2009. 10th International Symposium of Hungarian Researchers on Computational Intelligence and Informatics, Budapest, 2009, pp. 229–243. Budapesti Műszaki Fáiskola, Budapest, Magyar Fuzzy Társaság (2009). https://eprints.sztaki.hu/5928/
77. Kaarlela, T., Pieska, S., Pitkaaho, T.: Digital twin and virtual reality for safety training. In: 2020 11th IEEE International Conference on Cognitive Infocommunications (CogInfoCom). IEEE (2020)
78. Galambos, P., Weidig, C., Baranyi, P., Aurich, J.C., Hamann, B., Kreylos, O.: VirCA NET: a case study for collaboration in shared virtual space. In: 2012 IEEE 3rd International Conference on Cognitive Infocommunications (CogInfoCom), pp. 273–277. IEEE (2012)
79. Chamusca, I.L., Ferreira, C.V., Murari, T.B., Apolinario, A.L., Winkler, I.: Towards sustainable virtual reality: Gathering design guidelines for intuitive authoring tools. Sustainability **15**(4), 2924 (2023)
80. Setti, T., Csapo, A.B.: A canonical set of operations for editing dashboard layouts in virtual reality. Front. Comput. Sci. **3**, 64 (2021)
81. Setti, T., Csapó, Á.B.: Cognitive aspects of 2D content integration and management in 3D virtual reality spaces. Infocommunications J. **15**, no. Special Issue, 37–45 (2023)
82. Persa, G., Csapó, Á.B.: Aggregated avatars for qualitative feedback on interpersonal interaction patterns in 3D environments. In: 2023 IEEE 2nd International Conference on Cognitive Aspects of Virtual Reality (CVR). IEEE (2023)
83. Horváth, I., Csapó, Á.B.: Motivations and tools relevant to personalized workspaces in VR environments. Electronics **12**(9), 2059 (2023). https://www.mdpi.com/2079-9292/12/9/2059
84. Horváth, I., Csapó, Á.B.: Structured template language and generative AI driven content management for personalized workspace reconfiguration. In: 2023 IEEE 2nd International Conference on Cognitive Aspects of Virtual Reality (CVR). IEEE (2023)
85. Shah, K., Salunke, A., Dongare, S., Antala, K.: Recommender systems: an overview of different approaches to recommendations. In: 2017 International Conference on Innovations in Information, Embedded and Communication Systems (ICIIECS). IEEE (2017)
86. Valdez Mendia, J.M., Flores-Cuautle, J.J.A.: Toward customer hyper-personalization experience – a data-driven approach. Cogent Business & Manag. **9**(1) (2022)
87. Hanson, J., Wei, M., Veys, S., Kugler, M., Strahilevitz, L., Ur, B.: Taking data out of context to hyper-personalize ads: crowdworkers' privacy perceptions and decisions to disclose private information. In: Proceedings of the 2020 CHI Conference on Human Factors in Computing Systems, CHI'20. ACM (2020)
88. Jesse, M., Jannach, D.: Digital nudging with recommender systems: Survey and future directions. Comput. Hum. Behav. Rep. **3**, 100052 (2021)

89. Russell, S.: Human Compatible. Penguin, New York (2019)
90. Dörfler, V.: What Every CEO Should Know About AI. Cambridge University Press (2022)
91. Adamopoulou, E., Moussiades, L.: An overview of chatbot technology. In: IFIP Advances in Information and Communication Technology. Springer International Publishing, pp. 373–383 (2020)
92. Rapp, A., Curti, L., Boldi, A.: The human side of human-chatbot interaction: A systematic literature review of ten years of research on text-based chatbots. Int. J. Human-Comput. Stud. **151**, 102630 (2021)
93. Sheth, A., Yip, H.Y., Iyengar, A., Tepper, P.: Cognitive services and intelligent chatbots: Current perspectives and special issue introduction. IEEE Internet Comput. **23**(2), 6–12 (2019)
94. Vaswani, A., Shazeer, N., Parmar, N., Uszkoreit, J., Jones, L., Gomez, A.N., Kaiser, Ł., Polosukhin, I.: Attention is all you need. In: Guyon, I., Luxburg, Bengio, S., Wallach, H., Fergus, R., Vishwanathan, S., Garnett, R. (eds.) Advances in Neural Information Processing Systems, vol. 30. Curran Associates, Inc. (2017)
95. Bubeck, S., Chandrasekaran, V., Eldan, R., Gehrke, J., Horvitz, E., Kamar, E., Lee, P., Lee, Y.T., Li, Y., Lundberg, S., Nori, H., Palangi, H., Ribeiro, M.T., Zhang, Y.: Sparks of artificial general intelligence: early experiments with GPT-4 (2023)
96. Ullman, T.: Large language models fail on trivial alterations to theory-of-mind tasks (2023)
97. Shapira, N., Levy, M., Alavi, S.H., Zhou, X., Choi, Y., Goldberg, Y., Sap, M., Shwartz, V.: Clever hans or neural theory of mind? Stress testing social reasoning in large language models (2023)
98. Kosinski, M.: Theory of mind may have spontaneously emerged in large language models (2023)
99. Stojnić, G., Gandhi, K., Yasuda, S., Lake, B.M., Dillon, M.R.: Commonsense psychology in human infants and machines. Cognition **235**, 105406 (2023)
100. Wersényi, G., Csapó, Á., Budai, T., Baranyi, P.: Internet of digital reality: Infrastructuralback-ground - part ii. Acta Polytech. Hung. **18**(8), 91–104 (2021)
101. Pang, A.S.-K.: The Distraction Addiction, 1st edn. Little, Brown and Company, New York (2013)
102. Baranyi, P., Csapo, A.B.: Revisiting the concept of generation CE - generation of cognitive entities. In: 2015 6th IEEE International Conference on Cognitive Infocommunications (CogInfoCom), pp. 583–586. IEEE (2015)
103. Searle, J.R.: I married a computer. The New York Review of Books, vol. 8 (1999)
104. N. A. of Sciences: In the Light of Evolution: Volume VII: The Human Mental Machinery. In: Cela-Conde, C.J., Lombardo, R.G., Avise, J.C., Ayala, F.J. (eds.) The National Academies Press, Washington, DC (2014). https://nap.nationalacademies.org/catalog/18573/in-the-light-of-evolution-volume-vii-the-human-mental
105. Romportl, J., Zackova, E., Kelemen, J. (eds.): Beyond artificial intelligence. Topics in Intelligent Engineering and Informatics, no. 9. Springer International Publishing, Cham (2015)
106. Hodder, I.: The entanglements of humans and things: a long-term view. New Lit. Hist. **45**(1), 19–36 (2014)
107. Berki, B.: Overview of the relationship between human spatial abilities and GPS usage. In: 2022 IEEE 1st International Conference on Cognitive Mobility (CogMob), pp. 21–24. IEEE (2022)
108. Csapo, A., Baranyi, P.: A unified terminology for coginfocom applications. In: 2011 2nd International Conference on Cognitive Infocommunications (CogInfoCom), pp. 1–6 (2011)
109. Baranyi, P., Persa, G., Csapo, A.: Definition of cognitive infocommunications and an architectural implementation of cognitive infocommunications systems. Int. J. Electr. Commun, Eng (2011)
110. Csapó, Á.B., Baranyi, P.Z.: A unified terminology for the structure and semantics of CogInfoCom channels. Acta Polytech. Hung. **9**(1), 85–105 (2012)
111. Galambos, P.: Vibrotactile feedback for haptics and telemanipulation: survey, concept and experiment. Acta Polytech. Hung. **9**(1), 41–65 (2012)

112. Hildebrandt, T., Hermann, T., Rinderle-Ma, S.: A sonification system for process monitoring as secondary task. In: 2014 5th IEEE Conference on Cognitive Infocommunications (CogInfoCom). IEEE (2014)
113. Yang, J., Hermann, T.: A zen garden interface for the interactive control of sonic ambiences in smart environment. In: 2015 6th IEEE International Conference on Cognitive Infocommunications (CogInfoCom). IEEE (2015)
114. Hammerschmidt, J., Hermann, T., Walender, A., Kromker, N.: InfoPlant: multimodal augmentation of plants for enhanced human-computer interaction. In: 2015 6th IEEE International Conference on Cognitive Infocommunications (CogInfoCom). IEEE (2015)
115. Endrstova, B., Macik, M., Treml, L.: Reprobooktor: a concept of audiobook player for visually impaired older adults. In: 2018 9th IEEE International Conference on Cognitive Infocommunications (CogInfoCom). IEEE (2018)
116. Macik, M., Gintner, V., Palivcova, D., Maly, I.: Tactile symbols for visually impaired older adults. In: 2018 9th IEEE International Conference on Cognitive Infocommunications (CogInfoCom). IEEE (2018)
117. Csapo, A., Baranyi, P.: The spiral discovery method: An interpretable tuning model for CogInfoCom channels. J. Adv. Comput. Intell. **16**(2), 358–367 (2012)
118. Baranyi, P., Csapo, A., Varlaki, P.: An overview of research trends in CogInfoCom. In: IEEE 18th International Conference on Intelligent Engineering Systems INES 2014. IEEE (2014)
119. Wirsching, G., Lorenz, R.: Towards meaning-oriented language modeling. In: 2013 IEEE 4th International Conference on Cognitive Infocommunications (CogInfoCom). IEEE (2013)
120. Kroger, B.J., Bekolay, T., Eliasmith, C.: Modeling speech production using the neural engineering framework. In: 2014 5th IEEE Conference on Cognitive Infocommunications (CogInfoCom). IEEE (2014)
121. Ondas, S., Juhar, J., Kiktova, E., Zimmermann, J.: Anticipation in speech-based human-machine interfaces. In: 2018 9th IEEE International Conference on Cognitive Infocommunications (CogInfoCom). IEEE (2018)
122. Gnjatovic, M., Macek, N.: An entropy minimization approach to dialogue segmentation. In: 2020 11th IEEE International Conference on Cognitive Infocommunications (CogInfoCom). IEEE (2020)
123. Leung, W.-K., Yuen, K.-W., Wong, K.-H., Meng, H.: Development of text-to-audiovisual speech synthesis to support interactive language learning on a mobile device. In: 2013 IEEE 4th International Conference on Cognitive Infocommunications (CogInfoCom). IEEE (2013)
124. Sztahó, D., Tulics, M.G., Vicsi, K., Valálik, I.: Automatic estimation of severity of parkinson's disease based on speech rhythm related features. In: 2017 8th IEEE International Conference on Cognitive Infocommunications (CogInfoCom), pp. 000 011–000 016 (2017)
125. Sztaho, D., Orosz, A., Valalik, I.: Articulation correctness measurement of parkinson's disease using low resource-intensitve segmentation methods. In: 2020 11th IEEE International Conference on Cognitive Infocommunications (CogInfoCom). IEEE (2020)
126. Römer, R., Beim Graben, P., Huber-Liebl, M., Wolff, M.: Unifying physical interaction, linguistic communication, and language acquisition of cognitive agents by minimalist grammars. Front. Comput. Sci. **4** (2022)
127. McLaren, L., Koutsombogera, M., Vogel, C.: Gaze, dominance and dialogue role in the MULTISIMO corpus. In: 2020 11th IEEE International Conference on Cognitive Infocommunications (CogInfoCom). IEEE (2020)
128. Vogel, C., Koutsombogera, M., Reverdy, J.: Aspects of dynamics in dialogue collaboration. Electronics **12**(10), 2210 (2023)
129. Navarretta, C.: Speech pauses and pronominal anaphors. Front. Comput. Sci. **3** (2021)
130. Esposito, A., Raimo, G., Maldonato, M., Vogel, C., Conson, M., Cordasco, G.: Behavioral sentiment analysis of depressive states. In: 2020 11th IEEE International Conference on Cognitive Infocommunications (CogInfoCom). IEEE (2020)
131. Laki, L.J., Yang, Z.G.: Sentiment analysis with neural models for hungarian. Acta Polytech. Hung. **20**(5), 109–128 (2023)

132. Koles, M., Hercegfi, K., Logo, E., Tovolgyi, S., Szabo, B., Hamornik, B.P., Komlodi, A., Baranyi, P.Z., Galambos, P., Persa, G.: Collaboration experience in immersive vr environment in the frame of the neurocogspace project. In: 2014 5th IEEE Conference on Cognitive Infocommunications (CogInfoCom). IEEE (2014)
133. Geszten, D., Komlódi, A., Hercegfi, K., Hámornik, B., Young, A., Köles, M., Lutters, W.G.: A content-analysis approach for exploring usability problems in a collaborative virtual environment. Acta Polytech. Hung. **15**(5), 67–88 (2018)
134. Geszten, D., Hámornik, B.P., Hercegfi, K.: Finding treasure in the wild: a field study of a collaborative design software. In: Accentuated Innovations in Cognitive Info-Communication, pp. 185–202. Springer International Publishing (2022)
135. Wortelen, B., Feuerstack, S., Harre, M.-C., Senturk, S., You, F., Wang, J.: Engineering optimized control panel designs for electric driving. In: 2019 10th IEEE International Conference on Cognitive Infocommunications (CogInfoCom). IEEE (2019)
136. Szucs, V., Sik-Lanyi, C., Guzsvinecz, T.: Presenting the user's focus in needs & development (UFIND) method and its comparison to other design methods. In: 2020 11th IEEE International Conference on Cognitive Infocommunications (CogInfoCom). IEEE (2020)
137. Vittayaphorn, N., Lohaburananont, G., Bhumtakwong, J., Udompongsanont, K., Uchida, M., Asavarojkul, N., Rodphol, P., Sajjapong, P., Boonribsong, R., Chanthateyanonth, S., Julerttrakul, T., Pyae, A.: Design and development of a user-centered mobile application for intermodal public transit in Bangkok: a design thinking approach. Infocommunications J. **15**, no. Special Issue, 41–52 (2023)
138. Torok, A.: From human-computer interaction to cognitive infocommunications: a cognitive science perspective. In: 2016 7th IEEE International Conference on Cognitive Infocommunications (CogInfoCom), pp. 433–438. IEEE (2016)
139. Putnoki, A.M., Mattyasovszky-Philipp, D., Molnár, B.: Cognitive infocommunication and carbon-carbon interaction. In: 2022 13th IEEE International Conference on Cognitive Infocommunications (CogInfoCom). IEEE (2022)
140. Molnár, B., Mattyasovszky-Philipp, D.: Cognitive resonance and the architecture issues of cognitive information systems. In: Accentuated Innovations in Cognitive Info-Communication, pp. 29–56. Springer International Publishing (2022)
141. Ito, A., Nakada, K.: UI design based on traditional japanese gesture. In: 2019 10th IEEE International Conference on Cognitive Infocommunications (CogInfoCom). IEEE (2019)
142. Muramatsu, Y., Niitsuma, M., Thomessen, T.: Perception of tactile sensation using vibrotactile glove interface. In: 2012 IEEE 3rd International Conference on Cognitive Infocommunications (CogInfoCom). IEEE (2012)
143. Zikmund, P., Macik, M., Dubnicky, L., Horpatzska, M.: Comparison of joystick guidance methods. In: 2019 10th IEEE International Conference on Cognitive Infocommunications (CogInfoCom). IEEE (2019)
144. Mikovec, Z., Slavik, P., Zara, J.: Cultural heritage, user interfaces and serious games at CTU prague. In: 2009 15th International Conference on Virtual Systems and Multimedia. IEEE (2009)
145. Szegletes, L., Koles, M., Forstner, B.: The design of a biofeedback framework for dynamic difficulty adjustment in games. In: 2014 5th IEEE Conference on Cognitive Infocommunications (CogInfoCom). IEEE (2014)
146. Fekete, K., Csorba, K.: Power extraction of human behavior in mobile environment. In: 2014 5th IEEE Conference on Cognitive Infocommunications (CogInfoCom). IEEE (2014)
147. Kovesdan, G., Asztalos, M., Lengyel, L.: Fast android application development with component modeling. In: 2014 5th IEEE Conference on Cognitive Infocommunications (CogInfoCom). IEEE (2014)
148. Blázovics, L., Forstner, B., Charaf, H., Fitzek, F.H.P.: On the benefits of cognitive infocommunication for mobile communication nodes using cooperative concepts. In: 2013 IEEE 4th International Conference on Cognitive Infocommunications (CogInfoCom), pp. 693–698 (2013)

149. Csorba, K., Ekler, P., Bori, A., Charaf, H.: Analysis of mobile bittorrent client behavior. In: 2013 IEEE 4th International Conference on Cognitive Infocommunications (CogInfoCom), pp. 613–618 (2013)
150. Péntek, I., Adamkó, A.: Monitoring pandemics, using sensor data from smart ecosystems. Acta Polytech. Hung. **20**(5), 75–90 (2023)
151. Sallai, G.: Evolution of digitization toward the internet of digital & cognitive realities and smart ecosystems. Infocommunications J. **15**, no. Special Issue, 2–8 (2023)
152. Török, M., Tóth, M.J., Szöllősi, A.: Foundations and perspectives of mathability in relation to the coginfocom domain. In: 2013 IEEE 4th International Conference on Cognitive Infocommunications (CogInfoCom), pp. 869–872. IEEE (2013)
153. Chmielewska, K.: From mathability to learnability. In: 2019 10th IEEE International Conference on Cognitive Infocommunications (CogInfoCom). IEEE (2019)
154. Chmielewska, K., Gilnyi, A.: Educational context of mathability. Acta Polytech. Hung. **15**(5) (2018)
155. Bubnó, K., Takács, V.L.: Cognitive aspects of mathematics-aided computer science teaching. Acta Polytech. Hung. **16**(5) (2019)
156. Kvasznicza, Z., Gyurcsek, I., Elmer, G., Bagdán, V., Horváth, I.: Mathability of EMC emission testing for mission crucial devices in GTEM waveguide. Acta Polytech. Hung. **18**(1), 159–173 (2021)
157. Chmielewska, K., Lukasiewicz, A., Mroczynska, K., Gilányi, A.: Mathematical and technical education - experiences gained during online teaching. In: 2022 13th IEEE International Conference on Cognitive Infocommunications (CogInfoCom). IEEE (2022)
158. Bubnó, K., Takács, V.L.: Mathematical and computational awareness before and after the pandemic. Front. Educ. **7** (2022)
159. Zöldy, M., Baranyi, P.: Cognitive mobility – CogMob. In: 12th IEEE International Conference on Cognitive Infocommunications (CogInfoCom), pp. 921–925 (2021)
160. Zöldy, M., Baranyi, P.: The cognitive mobility concept. Infocommunications J. **15**, no. Special Issue, 35–40 (2023)
161. Kampis, G., Gulyas, L., Soos, S.: Do institutions think? comparing past success and future investment. In: 2013 IEEE 4th International Conference on Cognitive Infocommunications (CogInfoCom). IEEE (2013)
162. Susniene, D., Torma, A., Buruzs, A., Hatwagner, M.F., Koczy, L.T.: Using fuzzy cognitive map approach to model the casual relationships in stakeholder management at companies. In: 2014 5th IEEE Conference on Cognitive Infocommunications (CogInfoCom). IEEE (2014)
163. Waldbuesser, P., Komlosi, L.I.: Empirical findings in cognitive entity management: a challenge in the digital era. In: 2015 6th IEEE International Conference on Cognitive Infocommunications (CogInfoCom). IEEE (2015)
164. Szmodics, P.: Knowledge-based process management. In: 2015 6th IEEE International Conference on Cognitive Infocommunications (CogInfoCom). IEEE (2015)
165. Rumiński, D., Maik, M., Walczak, K.: Visualizing financial stock data within an augmented reality trading environment. Acta Polytech. Hung. **16**(6), 223–239 (2019)
166. Korecko, S., Hudak, M., Sobota, B., Marko, M., Cimrova, B., Farkas, I., Rosipal, R.: Assessment and training of visuospatial cognitive functions in virtual reality: proposal and perspective. In: 2018 9th IEEE International Conference on Cognitive Infocommunications (CogInfoCom). IEEE (2018)
167. Török, Z.G., Török, Á.: Looking at the map – or navigating in a virtual city: interaction of visuospatial display and spatial strategies in VR. In: 2018 9th IEEE International Conference on Cognitive Infocommunications (CogInfoCom), pp. 327–332. IEEE (2018)
168. Sudár, A., Csapó, Á.: Interaction patterns of spatial navigation and smartboard use in VR workspaces. In: Accentuated Innovations in Cognitive Info-Communication, pp. 149–166. Springer International Publishing (2022)
169. Berki, B.: Role of presence, memory and spatial ability in a desktop virtual reality. In: Accentuated Innovations in Cognitive Info-Communication, pp. 79–97. International Publishing (2022)

170. Sudár, A., Csapó, Á.B.: Descriptive markers for the cognitive profiling of desktop 3D spaces. Electronics **12**(2), 448 (2023)
171. Sudár, A., Csapó, Á.B.: Elicitation of content layout preferences in virtual 3D spaces based on a free layout creation task. Electronics **12**(9), 2078 (2023). https://www.mdpi.com/2079-9292/12/9/2078
172. Erdős, F., Németh, R., Firuza, B.: Virtual teamwork in gamified 3D environment. Infocommunications J. **15**, no. Special Issue, 15–20 (2023)
173. Zsebi, S., Szabó, P., Filotás, P., Sik-Lányi, C., Cserjési, R.: Enhancing neuropsychological assesment through virtual reality: a pilot study of the corsi-block tapping task. In: 2023 IEEE 2nd International Conference on Cognitive Aspects of Virtual Reality (CVR). IEEE (2023)
174. Horváth, I., Berki, B.: Unlocking the potential of immersive VR systems: the role of gesture and natural controls. In: 2023 IEEE 2nd International Conference on Cognitive Aspects of Virtual Reality (CVR). IEEE (2023)
175. Kovari, A.: CogInfoCom supported education: a review of CogInfoCom based conference papers. In: 2018 9th IEEE International Conference on Cognitive Infocommunications (CogInfoCom), pp. 000 233–000 236. IEEE (2018)
176. Dergham, M., Gilányi, A.: Application of virtual reality in kinematics education. In: Proceedings of the 10th IEEE International Conference on Cognitive Infocommunications: CogInfoCom 2019, pp. 107–112. IEEE (2019)
177. Molnár, G., Sík, D.: Smart devices, smart environments, smart students - a review on educational opportunities in virtual and augmented reality learning environments. In: 2019 10th IEEE International Conference on Cognitive Infocommunications (CogInfoCom), pp. 495–498. IEEE (2019)
178. Mikropoulos, T.A., Bellou, J.: The unique features of educational virtual environments. In: Isaías, P., McPherson, M., Bannister, F. (eds.) Proceedings of the IADIS International Conference on e-Society, pp. 122–128. IADIS Press, Dublin, Ireland (2006)
179. Budai, T., Kuczmann, M.: A multi-purpose virtual laboratory with interactive knowledge integration. In: Baranyi, P. (eds.) Proceedings of the 10th IEEE International Conference on Cognitive Infocommunications: CogInfoCom 2019, pp. 529–532. IEEE (2019)
180. Wijkmark, C.H., Heldal, I., Fankvist, S., Metallinou, M.-M.: Remote virtual simulation for incident commanders: opportunities and possibilities. In: 2020 11th IEEE International Conference on Cognitive Infocommunications (CogInfoCom). IEEE (2020)
181. Boda, I.K., Tóth, E.: Text-based second language learning in the three-dimensional space. In: Accentuated Innovations in Cognitive Info-Communication, pp. 125–148. Springer International Publishing (2022)
182. Guzsvinecz, T., Orbán-Mihálykó, É., Perge, E., Sik-Lányi, C.: The effect of engineering education, on spatial ability, in virtual environments. Acta Polytech. Hung. **20**(5), 11–30 (2023)
183. Sobota, B., Korečko, Š, Mattová, M.: Web-based 3D virtual environments utilization in primary and secondary education of children with multiple impairments. Electronics **12**(13), 2792 (2023)
184. Froland, T.H., Ravyse, W., Luimula, M., Heldal, I.: Single- and multiuser virtual reality training for forklift driving. In: 2023 IEEE 2nd International Conference on Cognitive Aspects of Virtual Reality (CVR). IEEE (2023)
185. Bólya, A.M., Csáji, L.K.: Benefits of virtual reality in ballet didactics. In: 2023 IEEE 2nd International Conference on Cognitive Aspects of Virtual Reality (CVR). IEEE (2023)
186. Alhamad, A., Gilányi, A., Al-Fayyadh, F.: Effectiveness and usability of virtual reality in physical education: a case study on dance teaching. In: 2023 IEEE 2nd International Conference on Cognitive Aspects of Virtual Reality (CVR). IEEE (2023)
187. Kerrebroeck, H.V., Brengman, M., Willems, K.: When brands come to life: experimental research on the vividness effect of virtual reality in transformational marketing communications. Virtual Reality **21**(4), 177–191 (2017)
188. Berki, B.: 2D advertising in 3D virtual spaces. Acta Polytech. Hung. **15**(3), 175–190 (2018)

189. Plancher, G., Tirard, A., Gyselinck, V., Nicolas, S., Piolino, P.: Using virtual reality to characterize episodic memory profiles in amnestic mild cognitive impairment and Alzheimer's disease: influence of active and passive encoding. Neuropsychologia **50**(5), 592–602 (2012)
190. Nascivera, N., Alfano, Y.M., Annunziato, T., Messina, M., Iorio, V.S., Cioffi, V., Sperandeo, R., Rosato, M., Longobardi, T., Maldonato, N.M.: Virtual empathy: the added value of virtual reality in psychotherapy. In: 2018 9th IEEE International Conference on Cognitive Infocommunications (CogInfoCom). IEEE (2018)
191. Boros, K., Bujdoso, G., Novac, C.M., Novac, O.C.: E-health promotion virtual reality services in MaxWhere VR spaces - design and development. In: 2019 10th IEEE International Conference on Cognitive Infocommunications (CogInfoCom). IEEE (2019)
192. van der Kruk, S.R., Zielinski, R., MacDougall, H., Hughes-Barton, D., Gunn, K.M.: Virtual reality as a patient education tool in healthcare: a scoping review. Patient Educ, Counseling (2022)
193. Tafuro, A., Masi, G., Terracciano, M., Maldonato, N.M., Bottone, M., Sperandeo, R., Scandurra, C., Muzii, B.: Virtual reality in the treatment of social anxiety disorder: a systematic review. In: 2022 13th IEEE International Conference on Cognitive Infocommunications (CogInfoCom). IEEE (2022)
194. Bellalouna, F.: Industrial use cases for augmented reality application. In: 2020 11th IEEE International Conference on Cognitive Infocommunications (CogInfoCom). IEEE (2020)
195. Luimula, M., Linder, M., Pieska, S., Laimio, E., Lahde, T., Porramo, P.: Unlimited safety productivity - a finnish perspective using virtual learning methods to improve quality and productivity in the construction industry. In: 2020 11th IEEE International Conference on Cognitive Infocommunications (CogInfoCom). IEEE (2020)
196. Luimula, M., Markopoulos, E., Kaakinen, J.K., Markopoulos, P., Laivuori, N., Ravyse, W.: Eye tracking in maritime immersive safe oceans technology. In: 2020 11th IEEE International Conference on Cognitive Infocommunications (CogInfoCom). IEEE (2020)

Chapter 2
Internet of Digital Reality

Abstract With the ever-increasing prevalence and widespread adoption of Virtual Reality, Augmented Reality, Mixed Reality, and Digital Twins in conjunction with artificial intelligence, 5G networks, and the omnipresent 2D Web, we find ourselves on the verge of a new era. This era is characterized by a multi-modal entangled combination of previously separate domains of information technology with human and social cognitive systems. The complex process of entanglement is bringing us towards a new and unique form of reality. The boundaries between the physical and digital worlds, as well as the digital representations and simulations of them, are becoming increasingly unclear and uncertain. This chapter aims to review and reinterpret well-established concepts such as virtual reality, augmented reality, mixed reality, and digital simulations and twins, building upon recent advancements. Based on innovative interpretations, the chapter presents the concept of higher-level structures, namely Digital Reality (DR) and Internet of Digital Reality (IoD). These structures have the potential to improve our understanding of the unparalleled opportunities provided to humanity by the pervasive nature of digital technologies.

Keywords Digital reality · Future internet · Internet of Things · Internet of Everything · Internet of Digital Reality

2.1 Introduction

The presentation of the concept of the Internet of Digital Reality is primarily based on [1].

The rapid advancement of modern information technology has arguably caused a significant transformation in human cognition [2, 3]. Whether this transformation is positive or negative remains a subject of debate. However, the process of digitization has become a long-lasting phenomenon that is anticipated to further infiltrate every aspect of human life [4–8]. Consequently, new conceptualizations have emerged to explain the coevolution of humans, machines, and computer networks. These include the notions of human-machine entanglement, cognitive merging/co-evolution, singularity, and post-humanism. All these concepts emphasize the blurring boundary

I. Horváth et al., *Cognitive Aspects of Virtual Reality*, Studies in Big Data 156,
https://doi.org/10.1007/978-3-031-68130-1_2

between humans and technology and are extensively explored within the field of Cognitive Infocommunications (CogInfoCom) [7–9].

Based on the understanding that society and technology have reached a turning point, this chapter delves into the deepening entanglement between humans and information and communication technology (ICT). It explains how this deepening entanglement is expected to give rise to a qualitatively new form of reality known as digital reality. Moreover, the chapter presents the concept of the Internet of Digital Reality (IoD), incorporating the aspects of communication, management, and harmonization among different segments within this digital reality.

It is important to note that the term 'digital reality' is not completely new, as Deloitte Consulting LLP and the Consumer Technology Association have trademarked it to refer to various technologies and capabilities associated with augmented reality (AR), virtual reality (VR), mixed reality (MR), 360° video, and immersive experiences that simulate reality [10–12]. However, In this chapter, similarly as in [1], we aim to provide a broader perspective and context to the term, focusing not only on the visualization of information but also on artificial intelligence, capabilities, and social cognition aspects. Additionally, we explore the implications of digital reality in the emergence of a new interconnectedness referred to as the Internet of Digital Reality (IoD). A detailed discussion of the infrastructural background of IoD can be found in [13].

The next section (Sect. 2.2) explores the concepts in the background of Digital Reality, incorporating the concepts of reality (Sect. 2.2.1) and virtuality (Sect. 2.2.2). Section 2.2.3 explores the different levels of physicality in VR, AR, and XR, while Sect. 2.2.4 examines the differences and relationships of virtual simulation, digital simulation, virtual twins, and digital twins, to better comprehend computer-generated realities. Building upon the concepts discussed in the previous section, Sect. 2.3 defines Digital Reality (DR) in the context of contemporary technological advancements, focusing on the prevalence of 2D digital environments (Sect. 2.3.1) and cognitive entities (Sect. 2.3.2). Then, to examine the technologies for managing, transmitting, and synchronizing digital realities in networked environments, Sect. 2.4 presents the concept of Internet of Digital Reality, alongside its historical context (Sect. 2.4.1), then emphasizing its main pillars (Sect. 2.4.2). The chapter is briefly summarized in Sect. 2.5.

2.2 Key Concepts

To establish a clear understanding of the concept known as Digital Reality, first, we begin by examining various technological and social concepts that are closely related. Our inquiry aims to clarify commonly held assumptions about virtual reality, augmented reality, mixed reality, and related fields. Through these deliberations, a fresh perspective emerges, which ultimately aids us in defining Digital Reality by the conclusion of this section. Subsequently, the term "Internet of Digital Reality" is further explored in subsequent sections of this chapter.

2.2.1 Reality Concept

Before we proceed, it is crucial to define what we mean by the term 'reality'. In this chapter, we will be using the working definition of reality from [1]: ***a set of conceptions and perceptions that form an integrated unit of comprehension, and create an understanding of what is possible, desirable and actual***.

As we will discover later, this definition encompasses many different facets of human experience, both in the present moment and throughout history. When we refer to an 'integrated unit of comprehension', we are describing the idea that even though reality may include different distinct parts, they can come together to create a new meaning that wouldn't exist if you were to analyze each part separately. By highlighting what is possible, desirable, and actual, we suggest that reality shapes our fundamental experiences.

Moreover, if a set of options, desires, or outcomes are mutually exclusive or centered on entities that have no meaningful influence over each other, they can be identified as distinct realities. For instance, the files stored on one person's computer and those on another person's computer belong to different realities, at least in a technical sense, as long as the two individuals are strangers and lead independent lives. Conversely, the files on one person's computer and the filing cabinet belonging to the same individual likely belong to the same reality, irrespective of the former existing in digital form and the latter in physical form. These examples illustrate that the concept of reality is independent of the physical/digital and real/virtual dimensions, which may initially seem counterintuitive given the common associations evoked by terms like 'virtual reality' often used to describe computer-generated 3D applications.

To put it in simpler terms, the reason why we consider virtual reality, augmented reality, and physical reality as separate concepts is due to the limited availability of technology which restricts our access, manipulation, and organization of information through different mediums. If the technology were more advanced, virtual and physical realities would have a significant overlap.

2.2.2 Layers of Virtuality

In informal usage, 'virtual' refers to something intangible but resembling its physical counterpart. However, we argue that this interpretation is limited, particularly when considering the definition of reality provided in Sect. 2.2.1. Thus, ***a broader definition of 'virtual' is proposed that encompasses any manifestation with a referential aspect, whether it exists solely in one's imagination or a specific physical or digital form. Furthermore, it includes manifestations that point to either a real or purely imaginary concept or to a specific concrete object.***

Upon closer examination of this term, we observe the following:

- 'Virtual' is not an absolute term and is not solely confined to the digital realm. For instance, both a card game played on a table and a 3D digital game can be considered virtual, depending on the realities they refer to. Similarly, a tangible user interface (TUI) can be considered virtual if it is connected to and synchronized with a backend system, whether digital or physical [14, 15]. Furthermore, open-air museums that depict traditional villages, such as Skansen, can be deemed 'virtual' even though they are physical and unrelated to anything digital, as they represent an architectural and cultural heritage from the past. Therefore, the term 'virtual' can be seen as a relative concept that delineates the boundary between artificial and natural, or constructed and base realities.
- 'Virtual' does not exclusively pertain to visual manifestations. While many people associate the term, especially in the context of 'virtual reality', with visual stimuli, the degree of resemblance to the referent counterpart depends on various factors, such as the counterpart's visual appearance, affordances in other modalities, or its purely abstract nature. It should be noted that certain theoretical frameworks, such as embodied cognition, argue that there is no purely abstract existence divorced from the spatial experiences and interactions of the human mind-body [16]. It is possible to differentiate between concepts that are closely related to physical reality and those that are more abstract [17, 18].

In summary, it is feasible to classify both tangible and intangible entities that are perceived by individuals as virtual to varying degrees, contingent upon their proximity to the reality they pertain to. A digital representation of a tangible entity that lacks physical existence can appear more virtual (less real) than a digital representation of an entity (an embodiment of the idea) that truly exists in the physical realm. On the other hand, tangible objects, such as banknotes, may seem even less virtual, despite being entirely virtual since they represent an abstraction that corresponds to an individual's societal contributions, as recognized by society.

2.2.3 Degree of Physicality: VR, AR, XR

According to the prevailing perspective, **virtual reality (VR)** pertains to a more advanced configuration of computer-generated and visually displayed objects that bear resemblance to objects in the physical realm.

Nonetheless, as previously explicated, the notions of 'reality' and 'virtual' are detached from the medium or particular technology employed. Consequently, akin to the term 'virtual', ***"virtual reality" can be perceived as any artificially constructed setting, be it physical or digital, that encompasses virtual objects.*** It is imperative to emphasize that VR can resemble or diverge from the tangible physical world, and it can be experienced through both physical and digital tools. In other words, the elements of virtual reality are the sign/symbol and the object being referred to which can be physical, digital, or a combination of both.

VR is mainly implemented on digital devices with 3D displays and is associated with the use of 3D glasses to create a simulated virtual environment. In this context, VR is generally regarded as a collection of 3D visualizations that depict either an artificial or real-world scenario (although other sensory modalities such as auditory, haptic, somatosensory, and olfactory cues can also be relevant). It is important to note that just because an environment is 2D (as opposed to 3D), not primarily visual (as opposed to visual), augmentative (in the sense that it does not exclude the physical environment, as opposed to being completely immersive), or physical (as opposed to digital), it does not mean that the environment cannot be artificially constructed or that it cannot refer to other concepts or objects. The combination of these characteristics does not necessarily limit the functionality of the environment.

However, the mere fact that an environment is 2D (as opposed to 3D), not primarily visual (as opposed to visual), augmentative (in the sense that it does not exclude the physical environment, as opposed to being completely immersive), or physical (as opposed to digital)—indeed, any combination of these characteristics—does not signify that the environment cannot be artificially constructed or that it cannot refer to other concepts or objects. Furthermore, this does not affect the extent to which the environment represents a blend of the physical and the digital—one could be situated in an insulated room while wearing a fully immersive headset, and yet one's thoughts and emotions could pertain to concepts and objects existing in any medium, be it physical or digital—thereby relating to the same "reality". One could even argue that VR represents a novel type of infocommunication tool [19], with direct implications for everyday life. As we will explore later in the following Sect. 2.2.4, this aspect holds significant importance for all computer-generated realities, particularly when considering their capacity to refer to realities at various levels of virtuality.

Augmented reality (AR) generally refers to a computer-generated version of reality that involves the presentation of virtual objects within a physical environment. It is worth observing that, similar to virtual reality (VR), augmented reality can be made by means other than a computer. Nevertheless, the defining characteristic of AR, as opposed to VR, is that the user is primarily situated within a "physical base reality" that is in some way augmented.

In general, the physical objects in the user's surroundings can serve as platforms capable of accommodating various forms of data representation. For example, a big dashboard or video display can be projected onto a vacant vertical surface, such as a wall. On the other hand, a more structured and purpose-specific display, like a clock, can be projected onto a smaller, flat surface like the top of a nightstand. Augmented reality also includes wearable displays and bionic devices as they augment the physical human body.

Augmented reality (AR) is often referred to as **extended reality (XR)** because it complements physical reality. **Mixed reality (MR)** is also related to AR. MR is defined as an environment in which "real world and virtual world objects are presented together within a single display", encompassing both augmented reality and augmented virtuality, and falling "anywhere between the extrema of the Virtuality Continuum" [20]. The Virtuality Continuum is a hypothetical continuum that spans from purely physical to purely virtual, with VR being seen as a singular limiting case where everything is purely virtual. All other realities necessarily combine elements of both. When we examine the concepts of reality and virtuality described in Sects. 2.2.1 and 2.2.2, we can see that the difference between them does not rely on a particular medium, whether it is digital or physical. In fact, a single reality can exist beyond the boundary between the digital and the physical. The terms virtual reality and augmented reality are often used interchangeably, which is a clear indication of this phenomenon [21].[1]

As explained in the subsequent subsection, computer-generated realities can be characterized not only based on the levels of virtuality in which they operate but also based on the qualities of the "referent realities" to which the objects and processes within those computer-generated realities refer.

2.2.4 Linking Together Realities via Representation and Simulation

The considerable potential of computer-generated realities does not solely derive from their ability to depict physical objects in an abstract or entertainment-oriented manner, nor does it solely come from their capacity to overlay information about fictional entities onto physical objects. Furthermore, their effectiveness is not inherently rooted in their integration with our physical reality, whether through portable displays or computer screens.

Instead, the true power of computer-generated realities lies in their ability to communicate with humans not only about abstract concepts but also about objects that exist in different realities. This is achieved through a diverse language that encompasses familiar entities as well as abstract representations. This section explores the referential capabilities of computer-generated realities, specifically focusing on the concepts of simulation and twin representation.

A **virtual simulation (VS)** refers to the emulation of three-dimensional motion and changes in the geometric representation of an object, whether it is tangible or abstract, within a computer-generated or physical environment. It is important to note that while virtual simulations typically involve the portrayal of physically existing objects in a computer-generated environment, the definition of a VS allows for alternative combinations between the physical and non-physical realms to reference each other. This is possible as long as both the sign/symbol and the referent have temporal geometric attributes. Therefore, even a purely abstract process like the training procedure of a deep neural network can be the subject of a virtual simulation, especially when the goal is to illustrate certain aspects of the training process to a group of students or experts using graphical concepts. In this context, any graphical model with a temporal dimension can be considered a virtual simulation. Regardless, a VS places emphasis on visualizing geometric properties and/or motion in three dimensions.

[1] For a concise historical overview of both VR and AR applications, see [22].

In a more expansive context and at the same time with greater precision, a **digital simulation (DS)** refers to a computer-generated model of a temporal process, regardless of whether it is actual, tangible, purely imaginary, or digital. Unlike virtual simulations (VS), which focus primarily on the three-dimensional virtual representation of various objects, DS emphasizes the portrayal of processes that manifest as changes over time, utilizing digital technologies. Consequently, the concept of DS encompasses elements that may deviate from the goal of achieving a 'high-fidelity visual replica' characteristic of virtual reality and can be seen as a digital portrayal of a given physical process, whether it is ongoing or envisioned. In certain situations, even potential future states of the system can be estimated or simulated and presented to users, thereby providing a 'multi-state' representation. Returning to the previous example of simulating deep neural networks, if the objective is not to explain the model's operations to humans but rather to execute it to obtain viable estimates for specific hyper-parameters, which are subsequently used in the actual training process, then this hyper-parameter search qualifies as a digital simulation.

It is important to note that both VS and DS can be utilized to simulate purely mental or digital phenomena, where objects and processes may refer to cognitive objects and cognitive representations of processes, or to processes that are already being executed on computers. Due to its process-oriented nature, DS, in its most extreme form, does not require the use of any specific three-dimensional representation from the physical world, whereas, in the case of VS, three-dimensional representations are inherently object-based, involving the conveyance of geometric properties.

Exploring the concepts of virtual twins and digital twins can help us better understand the connections between physical and mental/computer-generated realities.

A **virtual twin (VT)** refers to an object that denotes another object (known as the referent) in a separate reality. It constantly updates its geometric appearance to mirror that of the referent object. VT primarily represents a highly accurate geometric replica that showcases photorealistic visualization of various physical characteristics like color, materials, and textures, particularly within VR environments. The three-dimensional movement of a virtual twin can also be synchronized in real-time with its counterpart, especially in situations where the VT is utilized to manipulate the real object or vice versa (control/monitoring). Practically, the movement of the three-dimensional representation of the VT typically combines information from simulations (VS) and the real object. In a broader sense, when two objects—one tangible and the other represented in VR—are intentionally similar in visual appearance and three-dimensional movement, the application can be recognized as an example of a virtual twin.

A **digital twin (DT)** builds upon the virtual twin (VT) concept by incorporating digital simulation (DS), thereby not only focusing on the three-dimensional visualization and movement of an object but also encompassing all process-based functionalities and interactions involving the object(s), even without a direct three-dimensional geometric representation for the given phenomenon. DT enhances the VT concept with user interactions and various additional functionalities derived from digital simulation or obtained through a connection to the referent object. Essentially, when two objects—one physical and the other represented in VR—not only demonstrate visual similarity but also allow interaction in similar ways and synchronously, the application can be categorized as an instance of digital twins.

2.2.5 Summarizing the Key Concepts

To summarize, virtual, augmented, and other mixed realities consist of arrangements of virtual objects that bear some resemblance to counterparts from different realities, whether physical, digital, real, or imagined. The purpose of these realities may involve providing an aesthetically pleasing experience or serving as a means of communication to convey related concepts more effectively [19]. It is important to note that the key distinction between virtual and augmented reality is that AR is grounded in physical reality, with added extensions, whether they are physical or digital. Even augmentations to the human body can be seen as augmented reality if they possess the qualities outlined in Sect. 2.2.1.

Virtual and digital simulations, as well as virtual and digital twins, can be seen as components that can be integrated into virtual or augmented reality environments. Conceptually, virtual simulations and digital simulations represent temporal changes in the physical appearance or state of certain objects. It is worth mentioning that these objects do not necessarily exist in another reality and can be purely imagined. The main focus is on the temporal changes in the states of these objects that may potentially exist.

Lastly, virtual twins and digital twins are defined as representations that evolve in synchrony with the states of corresponding objects that do exist in a separate reality. These representations are designed to facilitate the control and monitoring of systems that exist in a base reality of great significance. A summary of these concepts can be found in Table 2.1.

2.3 Definition of Digital Reality

2.3.1 The Increasing Prevalence of 2D Digital Environments

The phrase **2D digital environment (2DE)** contains all software applications accessible to individuals on their infocommunication devices, encompassing any digital feature that presents two-dimensional visual and interaction surfaces. This includes web applications, collaborative solutions, and cloud-based platforms utilized for work, entertainment, and digital socialization. For the past decade, 2D digital environments have played a crucial role in the lives of people living in developed countries. These environments are now becoming even more widespread and influential due to multiple factors that make automation possible at different conceptual levels.

Table 2.1 Summary of concepts defined in Sect. 2.2. The subheaders of the columns denote the letters P, D, G, A, T, R, and I, which represent Physical, Digital, Geometric, Abstract, Static, Temporal, Real, and Imagined, respectively

	Environment	Symbols	References		
	Physical/Digital	*Physical/Digital*	*Geometric/Abstract*	*Static/Temporal*	*Real/Imagined*
Virtual Reality	P/D	P/D	G/A	S/T	R/I
Augmented Reality	P	P/D	G/A	S/T	R/I
Include as components					
Virtual Simulation (VS)	P/D	P/D	G	T	R/I
Digital Simulation (DS)	D	D	A	T	R/I
Virtual Twin (VT)	P/D	P/D	G	T	R
Digital Twin (DT)	D	D	A	T	R

Firstly, enhanced network automation has enabled faster data transmission rates and wider content availability even in remote areas. Secondly, handling big data solutions through automation enables service providers to gain a better understanding of their customers' habits and preferences. Lastly, the rise of AI now allows for efficient context-sensitive filtering and curation of digital content amidst information overload.

These trends initiate a transformative process that blends physical, digital, and mental concepts, affecting our daily lives and the perception of what is possible, desirable, and achievable.

Presently, the World Wide Web extends beyond a mere collection of internet-based technologies used for information retrieval and communication through text, voice, or video messages. Instead, communication between humans and machines is increasingly direct, and machines can also interact with each other. Algorithms vigilantly monitor our activities, while prominent technology companies acquire insights about us that we may not be aware of. Furthermore, they can access information we willingly or unknowingly provide through various modalities. As non-human entities become capable of communicating and making decisions based on adaptive methods on the public internet, the resulting cognitive burden on humans is expected to be immense [23].

2.3.2 The Rise of Cognitive Entities in Digital Reality

As presented in Chap. 1, during their formative years, individuals' neural activity, cognitive processes, and overall psychological framework undergo significant and often irreversible influences as they navigate the complexities of an ever-evolving digital landscape. Simultaneously, the vast amount of information accumulated and processed by artificial intelligence algorithms marks a substantial departure from previous decades. Essentially, the mental realms of humans and the digital domain are intricately intertwined, engaging in a constant cycle of reciprocal interaction. Consequently, a distinctively hybrid system is unfolding wherein human and digital components are inseparably interconnected. This phenomenon can be conceived as the emergence of the Cognitive Entity, characterized by the co-evolution of artificial and organic cognitive systems to yield qualitatively unprecedented capabilities [8, 9, 24, 25].

This perspective gains further significance when contemplating future generations. Expanding upon this conceptual framework, informed by notions such as Virtual Reality (VR), Augmented Reality (AR), Virtual Simulation (VS), Data Simulation (DS), Virtual Twin (VT), Digital Twin (DT), Artificial Intelligence (AI), and others within the domain of cognitive infocommunications (CogInfoCom) [7, 8, 26, 27], the notion of Digital Reality was outlined in [1] as follows:

> **Digital Reality (DR)** is a high-level integration of virtual reality (including augmented reality, virtual and digital simulations and twins), artificial intelligence and 2D digital environments which creates a highly contextual reality for humans in which previously disparate realms of human experience are brought together. DR encompasses not only industrial applications but also helps increase productivity in all corners of life (both physical and digital), thereby enabling the development of new social entities and structures, such as 3D digital universities, 3D businesses, 3D governance, 3D web-based digital entertainment, 3D collaborative sites and marketplaces.

A fundamental distinction exists between the Digital Reality (DR) concept in comparison to *all digital solutions* or *everything that is digital*. From our perspective, the term DR holds a more precise meaning, referring to the integration of digital solutions that aim to define and consequently promote the creation of a new reality characterized by human immersion and contextual relevance, as discussed within the field of Cognitive Infocommunications.

To illustrate, let us consider an example of integration involving physical appliances such as a refrigerator, microwave oven, smart stove, and deep fryer, each accompanied by their corresponding digital counterparts. These physical and digital elements are further enhanced by various virtual objects within a shared virtual reality space. Moreover, this environment incorporates a 2D digital landscape that includes collaborative cookbooks, cooking notes, blogs, and video chats, which facilitate interaction with renowned culinary experts. Through the integration of these physical devices, digital replicas, documentation, and access to expert knowledge, a mixed reality is created—a Digital Reality specifically designed for cooking. Such a digital reality may also incorporate AI algorithms that provide suggestions during cooking, track nutritional intake over time, or predict ingredient replenishments.

This example emphasizes that a digital reality goes beyond encompassing all digital elements. Instead, it signifies a level of cohesion or concentration of components—whether they are physical, digital, static, dynamic, pre-programmed, or AI-driven—that enables the formulation and implementation of higher-level goals and functionalities. The capacity of DR to push the limits of our cognitive concepts and expand our human capabilities in unprecedented ways sets it apart from mere collections of digital tools. While digital realities certainly include digital tools, these tools are organized and integrated in a specific, topic-oriented manner. Essentially, DRs have the potential to create new realities that resemble our understanding of reality as a combination of concepts and experiences, encompassing aspirations and actualization in real life.

2.4 Internet of Digital Reality (IoD)

The Internet of Digital Reality (IoD) includes technologies for managing, transmitting, and synchronizing digital realities in networked environments, whether public or private. IoD focuses on enhancing user accessibility, immersiveness, and experience by integrating virtual reality and artificial intelligence. It is worth noting that IoD requires the consideration of connections among various cognitive entities, not only at the end-user level of virtual reality displays and software, but also across network protocols, network management systems, physical media (wired or wireless), hardware interfaces, and other equipment. Artificial intelligence plays a crucial role in both digital reality and IoD, facilitating the seamless integration of context-driven content and intelligent network routing. This section provides a succinct overview of IoD. For more comprehensive discussions on IoD, its implications, and the underlying technologies, readers are advised to refer to [13].

2.4.1 Historical Context of IoD

The rise of the Internet of Things (IoT) has resulted in a network of interconnected *things* that encompass sensors, actuators, wearables, and digital twins. This integration combines distributed computation with intelligent connections, allowing for the interconnection, interaction, and management of digital representations of physical entities in the real world. Even without constant human supervision, communication among these entities is fostered.

Cisco introduced the concept of the Internet of Everything (IoE) in 2013, building on the foundation of IoT [28, 29]:

> Bringing together people, process, data, and things to make networked connections more relevant and valuable than ever before, turning information into actions that create new capabilities, richer experiences, and unprecedented economic opportunity for businesses, individuals, and countries.

One notable difference between IoE and IoT lies in the intelligence embedded within the connections. While IoT primarily focuses on communication among physical objects and concepts, IoE introduces network intelligence to integrate these concepts into a cohesive system.

The term *'Internet-of-X'* has become increasingly popular, where the 'X' can represent various things such as Internet of Nano Things, Internet of Mission-Critical Things, and Internet of Mobile Things [30]. This expansion reflects the growth observed in more conventional areas that involve different types of interactions, including human-computer interactions, human-machine interactions, and human-robot interactions. This emergence of diverse fields, which share similar concepts and methodologies but are applied to distinct contexts, has led to the development of cognitive infocommunications (CogInfoCom) [7–9]. Similarly, the term *'Internet of Digital Reality (IoD)'* was introduced in [1] to partially integrate, complement, and notably enhance earlier concepts like the *Internet-of-X*. The fundamental motivation driving these technological trajectories has always been to merge, enhance, and disseminate realities, an idea already encompassed by Digital Reality, although excluding the network aspects.

Arguably, the most significant aspect of IoD is its capacity to connect digital realities, which are combinations of technologies and data that result in a higher-level functional integration. For example, IoD serves as the network through which amalgamations of 3D virtual spaces and their real-world counterparts can be shared, along with all relevant data and interactive support, facilitating extended capabilities.

2.4.2 Pillars of IoD

The Internet of Digital Reality (IoD) goes beyond the Internet of Things (IoT) and the Internet of Everything (IoE) by connecting physical objects and integrating complete digital environments through a network, whether it is public or private. In simpler terms, IoD aims to establish a connection between the physical and digital world, to create a seamless experience for users. At present, cognitive entities and virtual/digital twins heavily depend on Internet connections [31].

The pillars of the Internet of Digital Reality include the following, which are also presented in Fig. 2.1:

- **Cognitive Entities** These entities represent qualitatively novel capabilities that arise from the interaction of machines, sensors, artificial intelligence (AI), digital twins, avatars, algorithms, bots, robotic process automation (RPA), and higher-level organizational structures within the digital realm.
- **Information** This encompasses data, web content, and control.
- **Communication Networks** These networks enable intelligent connections, spanning both wired and RF physical layers across public and private networks.

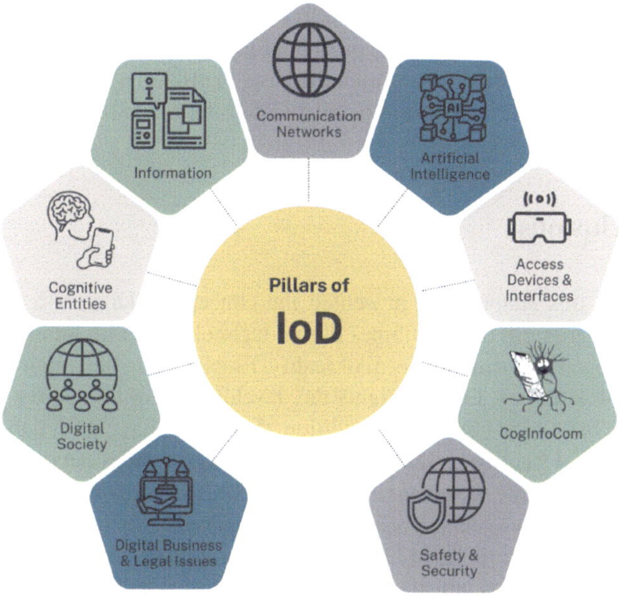

Fig. 2.1 The pillars of Internet of Digital Reality

- **Artificial Intelligence** With the interconnectedness and global sharing of digital realities, artificial intelligence assumes a newfound significance as a network capable of distributed learning and continuous evolution.
- **Access Devices and Interfaces** This category includes headsets, tactile devices, AR/VR/MR spaces, 360-degree immersive scenarios, as well as mobility and navigation tools.
- **Cognitive Infocommunications** This field involves sensation and perception, human factors, and the co-evolution of humans with infocommunication technologies.
- **Safety and Security** Addressing concerns related to data security, information security, and the physical safety of users.
- **Digital Business and Legal Issues** This refers to the rules and regulations as well as the standard procedures that govern the legal and business operations in the digital world.
- **Digital Society** This includes aspects such as education, technology acceptance, digital work, e-government, and digital arts and gaming.

These pillars collectively lay the groundwork for IoD. However, they also expand in scope as IoD offers new perspectives. For example, artificial intelligence evolves into a global network (networked AI) that can dynamically learn from user activities and data. Naturally, the dissolution of barriers between different contexts and use cases introduces new challenges related to data security and legal considerations. These examples demonstrate how the pillars of IoD both contribute to its development and are influenced by it. For a comprehensive exploration of IoD, including its implications and underlying technologies, please refer to [13].

2.5 Conclusions

In conclusion, this chapter has presented the concept of Digital Reality (DR) as an innovative paradigm that combines digital technologies with human interaction, thus leading to a transformative form of reality. Emphasizing the importance of integrating digital solutions, DR highlights the development of immersive experiences deeply rooted in specific contexts. In addition, the chapter has clarified the idea of the Internet of Digital Reality (IoD) as a framework that includes network and related technologies to facilitate the management, transition, and alignment of Digital Realities. By elucidating these concepts, the chapter argues that the structural frameworks of Digital Reality and Internet of Digital Reality provide significant insights into the emerging possibilities brought about by ubiquitous digital technologies. This investigation sets the stage for further exploration and advancement in comprehending and harnessing the potential of Digital Reality and its interconnected landscape within the wider digital ecosystem.

References

1. Baranyi, P., Csapó, Á., Budai, T., Wersényi, G.: Introducing the concept of internet of digital reality – part I. Acta Polytech. Hungarica **18**(7), 225–240 (2021)
2. Loh, K.K., Kanai, R.: How has the internet reshaped human cognition? Neuroscientist **22**(5), 506–520 (2016)
3. Prensky, M.: Homo sapiens digital: from digital immigrants and digital natives to digital wisdom. Innovate: J. Online Educ. **5**(3) (2009)
4. Frauenberger, C.: Entanglement HCI the next wave? ACM Trans. Comput.-Hum. Inter. (TOCHI) **27**(1), 1–27 (2019)
5. Kurzweil, R.: The Singularity is Near. Penguin Books, New York, NY [u.a.] (2006)
6. Pilsch, A.: Transhumanism: Evolutionary Futurism and the Human Technologies of Utopia. University of Minnesota Press (2017)
7. Baranyi, P., Csapó, Á.: Definition and synergies of cognitive infocommunications. Acta Polytech. Hungarica **9**(1), 67–83 (2012)
8. Baranyi, P., Csapo, A., Sallai, G.: Cognitive Infocommunications (coginfocom). Springer (2015)
9. Baranyi, P., Csapó, Á.B., Berki, B.: Concepts of cognitive infocommunications. Infocommunications J. (2024). Accepted, in press
10. Cook, A., Cotteleer, M., Holdowsky, J.: Digital reality – a technical primer (2021). https://www2.deloitte.com/insights/us/en/topics/emerging-technologies/digital-reality-technical-primer.html
11. Schwartz, J., Hatfield, S., Jones, R., Anderson, S.: What is the future of work? redefining work, workforces, and workplaces. Part of a Deloitte Series on the Future of Work (2019)

12. Kang, Y., Yang, K.C.C.: Employing digital reality technologies in art exhibitions and museums: a global survey of best practices and implications. IGI Global 139–161 (2020)
13. Wersényi, G., Csapó, Á., Budai, T., Baranyi, P.: Internet of digital reality: Infrastructuralbackground – part ii. Acta Polytech. Hungarica **18**(8), 91–104 (2021)
14. Ishii, H.: The tangible user interface and its evolution. Commun. ACM **51**(6), 32–36 (2008)
15. Zhao, Y., Qin, Y., Liu, Y., Liu, S., Shi, Y.: Qook: a new physical-virtual coupling experience for active reading. In: Proceedings of the Adjunct Publication of the 26th Annual ACM Symposium on User Interface Software and Technology - UIST –13 Adjunct. UIST – 13 Adjunct, pp. 5–6. ACM Press (2013)
16. Glenberg, A.M.: Few believe the world is flat: how embodiment is changing the scientific understanding of cognition. Can. J. Exper. Psychol./Revue canadienne de psychologie expérimentale **69**(2), 165–171 (2015)
17. Barsalou, L.W.: Perceptual symbol systems. Behav. Brain Sci. **22**(4), 577–660 (1999)
18. Mahon, B.Z., Caramazza, A.: A critical look at the embodied cognition hypothesis and a new proposal for grounding conceptual content. J. Physiol.-Paris **102**(1–3), 59–70 (2008)
19. Csapó, Á.B., Horváth, I., Galambos, P., Baranyi, P.: VR as a medium of communication: from memory palaces to comprehensive memory management. In: 2018 IEEE 9th International Conference on Cognitive Infocommunications (CogInfoCom), pp. 000 389–000 394. IEEE (2018)
20. Milgram, P., Takemura, H., Utsumi, A., Kishino, F.: Augmented reality: a class of displays on the reality-virtuality continuum. In: Das, H. (ed.) Telemanipulator and Telepresence Technologies, vol. 2351, pp. 282–292. International Society for Optics and Photonics. SPIE (1995)
21. Grossmann, R.J.: Virtual reality, augmented reality. . . i call it i-reality. mHealth **1**(4) (2015). https://mhealth.amegroups.org/article/view/6573
22. Williams, D.: The history of augmented reality (infographic) (2017). https://www.huffpost.com/entry/the-history-of-augmented-_b_9955048
23. Pasquale, F.: The Alien Intelligence of Automated Media, pp. 89–118. Harvard University Press (2020)
24. Baranyi, P., Csapo, A.B.: Revisiting the concept of generation CE - generation of cognitive entities. In: 2015 6th IEEE International Conference on Cognitive Infocommunications (CogInfoCom), pp. 583–586. IEEE (2015)
25. Komlosi, L.I., Waldbuesser, P.: The cognitive entity generation: emergent properties in social cognition. In: 2015 6th IEEE International Conference on Cognitive Infocommunications (CogInfoCom), pp. 439–442. IEEE (2015)
26. Csapo, A., Baranyi, P.: A unified terminology for coginfocom applications. In: 2011 2nd International Conference on Cognitive Infocommunications (CogInfoCom), pp. 1–6 (2011)
27. Baranyi, P., Persa, G., Csapo, A.: Definition of cognitive infocommunications and an architectural implementation of cognitive infocommunications systems. Int. J. Electr. Commun. Eng. (2011)
28. Cisco: The Internet of Everything - Global Private Sector Economic Analysis (2013). https://www.cisco.com/c/dam/en_us/about/ac79/docs/innov/IoE_Economy_FAQ.pdf
29. Cisco: The Internet of Everything - Cisco IoE Value Index Study (2013). https://www.cisco.com/c/dam/en_us/about/business-insights/docs/ioe-value-index-faq.pdf
30. Srinivasan, C., Rajesh, B., Saikalyan, P., Premsagar, K., Yadav, E.S.: A review on the different types of internet of things (IoT). J. Adv. Res. Dyn. Control Syst. **11**(1), 154–158 (2019)
31. Jones, D., Snider, C., Nassehi, A., Yon, J., Hicks, B.: Characterising the digital twin: a systematic literature review. CIRP J. Manuf. Sci. Technol. **29**, 36–52 (2020)

Chapter 3
Cognitive Mobility

Abstract The rapid advancement of technology has had a significant impact on mobility, which is no longer limited to transportation but encompasses digital realms and other dimensions, leading to the enhancement of cognitive abilities in various domains. This chapter presents the concept of cognitive mobility (CogMob), which represents a holistic understanding of mobility through cognitive frameworks, integrating natural and artificial cognitive systems. Drawing on research from diverse disciplines such as mobility science, transportation engineering, artificial intelligence, and social sciences, CogMob aims to provide a comprehensive perspective on mobility and its optimization. The hierarchical structure of mobility, ranging from intracognitive to intercognitive mobility, is elucidated, along with its constituent elements such as necessity, decision-making, devices/vehicles, infrastructure/resources, and human-machine interfaces. Through historical evolution and interdisciplinary synergy, the chapter demonstrates how cognitive processes have increasingly become integral to various aspects of mobility, including vehicular cybersecurity, infocommunications, decision-making, sustainable transportation, and human-machine communication. Practical examples illustrate the seamless integration of cognitive levels, mobility modes, and elements in domains such as vehicular cybersecurity, hybrid propulsion systems, infrastructure digitalization, environment perception, and intermodality decision-making. By providing a unified framework for understanding and optimizing mobility systems, CogMob aims to address contemporary challenges and contribute to the development of efficient, sustainable, and user-centric mobility solutions.

Keywords Cognitive mobility · Vehicles · Infrastructure · Digital mobility · Human-machine communication · Infocommunications

I. Horváth et al., *Cognitive Aspects of Virtual Reality*, Studies in Big Data 156,
https://doi.org/10.1007/978-3-031-68130-1_3

3.1 Introduction

The presentation of the concept of cognitive mobility is based on [1, 2].

Mobility is essential to human existence due to our constant movement. While commonly associated with physical transportation, whereby individuals relocate with the aid of various conveyance methods, the concept of mobility encompasses a broader spectrum. Social mobility denotes shifts in an individual or family's societal standing, while labor mobility establishes connections between employment and residential preferences.

Cognitive mobility refers to gaining a deeper understanding of mobility through cognitive frameworks. The process of transport mobility is complex and involves recognizing needs, making decisions, and fulfilling those needs. The key components of transport mobility include the mode of transportation used, infrastructure, environmental impact, time and energy requirements, and the resources needed in both the short and long term. Mobility helps fulfill identified needs, creating value for individuals and societies. Analyzing historical mobility patterns is a useful tool for gaining insight into current and future mobility dynamics

The process of cognition and its mapping have been crucial to human existence for a long time. With the rapid advancements in digital technology in the 21st century, we have been able to improve our experiential learning by using smaller, more cost-effective, and increasingly sensitive sensors. When combined with sophisticated data fusion techniques and artificial intelligence algorithms, these sensors have resulted in an unprecedented abundance of data [3–6]. Technological innovations have made it easier to understand and map the world, including its mobility dynamics [5, 7]. Moreover, they have opened up new avenues for comprehension and advancement. While many people have traditionally focused on analyzing existing system components, there is a growing recognition of the importance of detailed empirical insights into initial needs and the resulting value. This approach can potentially lead to innovative solutions.

Enhancing our understanding of the world, our decision-making processes, and our opportunities can, in extreme circumstances, entail reducing or abandoning mobility, as evidenced by the impact of the COVID-19 pandemic waves and the emergence of new compact urban planning objectives. Traditionally, cognitive mobility has been narrowly construed as measuring physical transitions between locations [8, 9]. However, our approach considers this form of mobility a subset of cognitive mobility, and it can be comprehensively analyzed within this framework.

The second decade of the 21st century is characterized by unique challenges, such as the COVID-19 pandemic and geopolitical conflicts like the Russo-Ukrainian war in the EU. These challenges create unprecedented mobility issues that go beyond energy crises, logistics chain disruptions, and vehicle production slowdowns. Additionally, sensor system developments impact both new and existing mobility systems. As artificial intelligence continues to evolve, the integration of human cognitive abilities and machine knowledge is increasingly influencing mobility paradigms. In response, the field of Cognitive Mobility has been proposed by Zöldy and Baranyi [1, 2], which emphasizes the integration of human and machine capabilities within mobility systems.

This chapter utilizes the scientific perspective, mindset, and paradigm that have been derived and established from the various perspectives within the following foundational articles: [10–17].

Section 3.2 defines cognitive mobility and explains its hierarchical structure and constituent elements. The following Sect. 3.3 argues for this holistic approach of cognitive mobility, which considers both natural and artificial cognitive processes within the complex system of vehicles, infrastructure, decision-making, transport, sustainability, and human-machine communication. Section 3.4 presents examples of integrating cognitive levels, mobility modes, and elements in practical contexts, thereby demonstrating the effectiveness of this integration. It emphasizes the growing significance of cybersecurity in the automotive industry, the influence of digitization on transportation, advancements in vehicle propulsion systems and energy management, environment perception in automotive advancements, the role of digital twins in mobility infrastructure development, and the use of cognitive methodologies in transportation decision-making. The chapter ends with conclusions (Sect. 3.5).

3.2 Definition of Cognitive Mobility

This section presents the definition of cognitive mobility, aiming to provide a holistic understanding of mobility and its optimization, particularly emphasizing its engineering applications, while identifying its hierarchical structure and constituent elements. The following definition was proposed for Cognitive Mobility (CogMob) in [2]:

> ***Cognitive mobility (CogMob)*** examines the intertwined combination of research areas such as mobility, transport, and its management, vehicle manufacturing, related social sciences, artificial intelligence and its applications, and cognitive info-communications. The main goal of CogMob is to give a holistic picture of mobility and its broad understanding. It thus describes, models, and optimizes as a mixed combination of artificial and natural/human cognitive systems. It sees the whole combination as an inseparable CogMob system and explores what new cognitive abilities come from this CogMob system. One of CogMob's focus areas is, of course, engineering applications in the mobility sector.

Within the domain of cognitive mobility, it is crucial to identify two key dimensions: the hierarchical structure of mobility (i.e., its ranking or levels) and its constituent elements.

Ranking of Mobility

Within the theoretical framework of mobility classification, ranking emerges as a hierarchical structure defined by two conceptual end-points. These theoretical end-points signify the diverse cognitive level of entities involved in mobility.

- *Intracognitive mobility*—Entities engaged in mobility exhibit similar cognitive capacities (e.g.: pedestrians navigating a pedestrian area, or warehouse workers transporting goods).
- *Intercognitive mobility*—Entities engaged in mobility exhibit different cognitive capacities (e.g.: highly automated vehicles and bicycles travel in a shared urban environment).

Elements of Mobility

Elements of mobility are intricately linked to its fundamental nature:

- *Necessity*—Embodies the motivation that drives mobility.
- *Decision*—Determines the manner and frequency of mobility occurrences, influencing its realization. It can happen multiple times during an activity.
- *Device/Vehicle/Quality*—Encompasses diverse quality metrics impacting decisions, resource utilization, and mobility outcomes.
- *Infrastructure/Resources*—Apart from physical infrastructure, mobility necessitates substantial resource expenditure, including financial, temporal, and energetic investments.
- *Human-machine interface*—Encompasses an array of interfaces ranging from smartphones and interactive maps to web-tracking interfaces and simulation software, facilitating seamless interaction between humans and technology.

3.3 Areas of Mobility

In this section, we will explore the historical evolution of research domains related to CogMob, presenting the increasing importance of both natural and artificially induced cognitive processes in the field of mobility science. It explains how human agents and environmental factors are progressively converging within mobility contexts, providing a unified modeling framework that leads to a better understanding of mobility phenomena.

Traditionally, research on mobility was separated into different areas and didn't have much collaboration between them. Investigations usually focused on specific areas, leaving other domains in a secondary position. The pace of mobility advancement has mirrored the rapid acceleration observed across various spheres throughout the 20th century. Predominant domains encompassed automotive technologies, infrastructure development, road network planning, decision-making processes, transportation studies, human-machine interfaces, and socio-scientific inquiries, with sustainability considerations emerging as a salient thematic concern in recent years. Disciplinary frontiers have converged, facilitating interdisciplinary synergy (see Fig. 3.1).

Interdisciplinary synergy between vehicles and infrastructure is flourishing. Previously viewed as passive recipients, contemporary vehicles now provide valuable input for informed decision-making and route optimization.

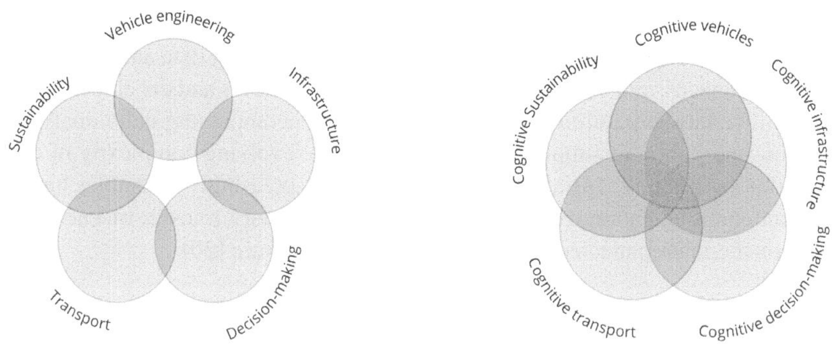

Fig. 3.1 The main areas of mobility demonstrate a heightened synergy, promoting enhanced collaboration among their constituent parts for more effective cooperation

3.3.1 Cognitive Vehicles

The growing importance given to safety considerations has become the primary driving force behind the development of cognitive features in vehicular systems. Initially, safety features were mechanical aids to help driver capabilities. However, in recent decades, driver assistance systems have expanded in scope and efficacy [18]. In 2019, a workshop was organized to explore and understand human-inspired paradigms related to perception, learning, and decision-making processes in the field of cognitive vehicle engineering.[1] The workshop aimed to improve the cognitive abilities of future vehicles and other autonomous entities. The workshop provided a platform to discuss the advantages and disadvantages of using artificial and naturalistic approaches in designing intelligent systems. These systems are responsible for carrying out tasks such as perception, interaction, learning, and decision-making.

3.3.2 Cognitive Infrastructure

Throughout history, the establishment of transportation infrastructure has been primarily motivated by commercial interests. Examples such as the ancient Silk Road or the network of Roman military roads highlight the importance of commercial intent. This trend has continued over time with projects like the Panama Canal and early railway systems. Similarly, the inception of the electricity network in the early 20th century was also driven by analogous commercial considerations. Despite the inherent complexity of these infrastructural systems, they operate as distinct tools that facilitate the movement of goods.

In contrast, cognitive transport infrastructure represents a meta-infrastructure characterized by its complex interconnectivity, wherein numerous factors interact

[1] Cognitive vehicles/vehicle engineering workshop (https://cogvehicles2019.github.io/).

continuously. This interconnectivity is designed to improve capacity utilization using emerging technologies such as 5G or 6G networks, AI and big data analytics, social media, IoT devices, and cloud storage. The ecosystem of users and developers within this cognitive infrastructure is notably multifaceted, encompassing individuals, corporations, and public institutions [19]. Moreover, the evolving complexity of communication between the infrastructure and its users increasingly resembles human-to-human communication dynamics. The cross-disciplinary transfer of knowledge highlights the interconnectedness of cognitive infrastructure [20].

3.3.3 Cognitive Decision-Making

Automated vehicles and driver assistance systems require effective decision-making, which has become a significant area of research. Within the domain of transportation decision-making, cognitive memory assumes a central role, demonstrated by the continuous and dynamic recalibration of routes through the incorporation of cognitive variables. The refinement of decision outcomes, characterized by error reduction, signifies the vehicle's capacity to assimilate cognitive elements, thereby fostering the potential for unsupervised learning in the domain of transportation decision-making. Vehicle decision-making mimics human cognitive processes, where event classification is mapped onto processed patterns.

Further exploration into other characteristics of human cognition, including emotions and memory structures, promises novel avenues for future inquiry. Unsupervised learning is crucial for the development of self-driving cars, while optimized data mining can improve routing efficiency through optimized data selection strategies [21].

3.3.4 Cognitive Transport

The implementation of intelligent transportation systems is a significant challenge in contemporary traffic planning. Fostering collaboration between conventional and self-driving vehicles at all levels of the global transportation system is an urgent necessity. In the following years, there is going to be a noticeable increase in the use of self-driving vehicles, as well as a wider range of traffic patterns and transportation infrastructure. These developments are intended to improve the management of transportation systems and provide better quality services to users.

The development of a cognitive multimodal transport system involves the integration of vehicular intelligence, transport infrastructure, and interconnected systems and networks. Artificial intelligence plays a crucial role in advancing transportation systems. A well-designed transportation system can significantly improve the safety of passengers and cargo, reduce transportation time for both passengers and freight and optimize the use of national transportation resources. This, in turn, can elevate environmental safety standards and positively impact the economy. It can be inferred that a smart transportation system would greatly contribute to the overall economy [22].

3.3.5 Human-Machine Communication

The evolution of the human-machine interface (HMI) is intricately intertwined with the historical narrative of driving. During the 1949 Geneva Convention on Road Traffic, an early international agreement on vehicle HMI was formalized [23]. This agreement established regulations and recommendations for internal vehicle HMI components, such as the mandatory installation of a speedometer and steering wheel.

A new era of driver assistance began in the 1970s with the development of routing assistance systems. Subsequent iterations of these systems, marked by increasing complexity, engendered enhanced communication channels between drivers and vehicles. Noteworthy examples of early driving support systems, including ABS, ASR, and ESP, were already in operation by the conclusion of the 20th century. However, the introduction of 'cruise control' at the turn of the millennium marked a significant milestone in managing tracking distance.

The early 2010s witnessed a substantial leap forward in management support systems with the introduction of complex systems [24], as evidenced by the proliferation of enlarged screens and onboard displays. This era also witnessed a heightened level of communication between infrastructure elements and drivers. Despite the evolution towards partially autonomous driving functionalities in various traffic sectors, the indispensable role of the driver in mobility remains largely unassailable. Nonetheless, the increasing complexity of systems facilitating autonomous driving necessitates the development of sophisticated communication interfaces.

3.3.6 Cognitive Sustainability

Sustainability emerges as a vital aspect of existence in the early stages of the third millennium. Our society is changing faster and more intensely than ever before. Mobility, often serving as a catalyst, is deeply intertwined with this rapid evolution, thus rendering sustainability considerations imperative, given society's pronounced reliance on mobility [25].

CogMob integrates insights from the social sciences to create a comprehensive approach that surpasses the examination of mobility based solely on artificial and natural cognitive capabilities. The link between mobility and economic processes is evident, highlighting the importance of using social science methodologies to understand and interpret mobility dynamics. Exploring the social implications of mobility can provide insights for future development efforts.

The main objective of CogMob is to promote a comprehensive mobility paradigm, facilitated significantly by information and communication technologies. It establishes connections across various system aspects, including decision-making processes, infrastructure, and tools. CogMob provides a collaborative workspace for optimizing through the conscious and purposeful use of IT tools in a cognitive framework [1].

3.4 Cognitive Mobility in Practice

This section presents a series of examples that demonstrate the integration of cognitive levels, mobility modes, and mobility elements. These examples serve to illustrate the seamless and efficient integration of these three components.

3.4.1 Vehicular Cybersecurity

The field of cybersecurity has become a crucial area of research in the automotive industry in recent years and has attracted the attention of scholars from various perspectives. Although competitors have different safety measures for their vehicles, research on artificial intelligence and defense mechanisms are lacking, as indicated by the results of [26]. Furthermore, vehicle communication is a crucial area of focus in contemporary developments within the field.

3.4.2 Infocommunications and Mobility

Digitization has an impact on various aspects of transportation, including vehicles, drivers, infrastructure, and the individuals occupying the vehicles. A recent study [27], conducted in Hungary, has shed light on the specific activities carried out during non-local journeys, particularly on public transportation modes such as trains, city buses, and suburban buses. The findings of this study highlight that engagement in activities on smartphones or tablets during non-local trips starts at a relatively low level but steadily increases. This trend is closely linked to the ownership of smartphones and access to mobile internet. Furthermore, the study reveals that the utilization of electronic devices decreases with age.

In-vehicle information systems are another characteristic of infocommunication technology in mobility. Turn-by-turn navigational assistance provided by GPS devices can have negative impacts on human spatial abilities. However, this effect can be reduced by increasing driver awareness, improving the interaction design of the devices to encourage active encoding, and incorporating landmarks as reference points [28].

The actual configuration and interaction with a GPS and air conditioning system impact human cognitive load. Traditional interfaces such as buttons and rotary dials are found to have a lower cognitive load for the air conditioning system, while innovative interfaces like touchscreens and voice control offer advantages for GPS configuration in terms of user-friendliness and safety. Effective management of cognitive load is critical for drivers. Prioritizing information, ensuring simplicity and consistency, utilizing multimodal displays, monitoring driver states, and incorporating voice control are effective strategies [29].

3.4.3 Hybrid and Electric Propulsion Systems

In addition to the control of vehicles, the utilization of artificial intelligence has also progressed in the driveline domain. The implementation of internal combustion engines is increasingly confronted with more rigorous standards [30]. The advancement of in-vehicle electronics has enabled the control of components such as brake assist, exhaust gas recirculation, and steering, thereby making previously inaccessible operating conditions attainable. A notable example of this is the intelligent engine control system [31], which is capable of achieving optimal performance even under extreme circumstances. Furthermore, a forecasting model based on support vector machines has been developed to anticipate engine performance under varying operating conditions. The developed model exhibited highly accurate predictions of engine performance.

The use of statistical regression models has also been observed in the field of vehicle development. A three-step statistical analysis algorithm employing vibration and sound pressure data as covariates has become demonstrable in predicting exhaust gas composition [32].

Machine learning outcomes can also be implemented in vehicle control, particularly in the management of autonomous vehicles. The road tracking function is imperative to ensure the safe speed and movement profile of vehicles, taking into account variable tire-road contact. The solution for this scenario combines the integration of Linear Parameter-Varying (LPV) control and machine learning-based analysis. The integration occurs in two steps. The estimation method produces the result utilizing the coefficient of adhesion decision trees. This estimation result is then incorporated into the robust LPV controller through a scheduling variable. An inherent error in the machine learning algorithm is incorporated into the control design. In a second step, optimization of longitudinal velocity is proposed over a projected timeframe, with a defined approximation of the feasibility of steering intervention based on machine learning [33].

3.4.4 Infrastructure Digitalization

The cognitive approach is increasingly influential in water transportation. Alongside the establishment of navigation infrastructure, the development of ports holds tangible significance. Enhancing digitization leads to increased efficiency, thus serving as the primary path of progress. Ports exhibit varying degrees of digitization, a phenomenon influenced by factors such as port size, historical background, traditions, as well as traffic volume and type. An alternative methodology endeavors to unify these distinct levels and their associated advancements [34]. Ports were categorized into small, medium, and large groups for comparison purposes. It is estimated that the digitalization level in small and medium-sized harbors is roughly 30% lower compared to large seaports. The research findings may prove valuable to ports aiming to evaluate their digitization level and determine optimal digital development solutions.

3.4.5 Onboard Energy Management

Vehicle recharge management is an intricate manifestation of cognitive vehicle technology. The significance of this domain is on the rise as vehicles continue to advance in their self-driving capabilities. The primary means of vehicle propulsion in the future remains uncertain. Among the array of available technologies, plug-in hybrid technology stands out as the most intricate. This technology amalgamates an electric powertrain with a conventional internal combustion engine, potentially utilizing renewable [35–38] or synthetic [39] propellants.

Regarding fossil fuels, the communication between filling stations and vehicles is unidirectional and limited to providing information solely on fuel prices. In stark contrast, bidirectional communication allows for collaborative optimization concerning route, fuel, charging time, and even price [40, 41]. For instance, the duration of charging for electric vehicles and the subsequent waiting time caused by other vehicles' charging needs can be considered as a decision criterion, but only with enhanced communication systems that offer superior quality and quantity. Additionally, the domains of waterborne transport and the utilization of agrofuels present further opportunities for enhancing the cognitive tools that facilitate optimal energy management in vehicles [42].

3.4.6 Environment Perception

Perceiving and comprehending the surrounding environment constitutes a central domain within the automotive advancements in the 21st century. This solution holds utmost significance not only in autonomous vehicles but also in the driver assistance mechanisms of conventional automobiles. These systems are adept at processing and managing dynamic relay systems such as adjustable signals, adaptable traffic control with traffic lights, and even traffic updates. By establishing a connection between in-vehicle systems and intelligent infrastructure, the scope of potential decision options expands. Given that perception serves as the pivotal input parameter for both self-driving and driving assistance functions, as driving is predicated on the outcomes of diverse sensing algorithms, perception can be dissected into subproblems. Common subproblems include lanes, traffic signals, objects, and the environment. These solutions, which rely on artificial intelligence, particularly neural networks, are already on the verge of practicality today and are anticipated to become increasingly prevalent in the future [43–45].

3.4.7 Digital Twins

The emergence of intermodal transport infrastructure serves as a prime illustration of the escalating and increasingly significant role played by cognitive elements. The design, optimization, and configuration of mobility have become an increasingly crucial component of models of urban ecosystems. By employing cognitive tools, one can ascertain the correlation between urban structure, mobility, and urban development [46]. Moreover, the simulation of mobility infrastructure assumes a pivotal position in the advancement of vehicle development.

The development of transport simulations necessitates this for augmenting mobility safety. The continuous updating of conventional static maps proves to be resource-intensive and, concurrently, presents an opportunity for computer simulations [47]. The integration of self-driving vehicles into the arduous homologation procedures of conventional vehicles is nearly unfeasible due to their technological solutions [48]. However, the implementation of digital-twin solutions, such as the utilization of simulation and the simultaneous operation of reality, has the potential to drastically reduce development time [49]. As a result, validation can be conducted at any given moment.

3.4.8 Intermodality Decision-Making

In the domain of transportation, cognitive methodologies have the potential to facilitate prompt decision-making. The implementation of roundabouts has long been employed as a solution for intersections due to their adaptable nature and high degree of safety. However, when it comes to autonomous vehicles, the focus of development often leans towards scenarios involving highways, as navigating roundabouts necessitates more time and effort. Extensive research has illustrated the effectiveness of various decision-support applications, highlighting the fact that cognitive techniques enhance operational efficiency [50]. The primary objectives of driving are to avoid collisions and minimize energy consumption. These optimization tasks are interconnected, meaning that quadratic optimization is employed alongside the vehicle model during training [51]. The application of novel cognitive approaches is imperative not only in terrestrial transportation but also in air mobility. Unmanned aerial vehicles exhibit similarities to self-driving cars, thus requiring similar cognitive applications [52].

In the field of intermodal transportation, achieving synergy among various subsectors can be seen as a comprehensive approach to address the aforementioned concerns. Implementing these upgrades will not only improve transport efficiency but also contribute towards sustainability goals. Research [50] collects and examines the factors that have been identified as influential in determining the efficiency and sustainability of intermodal transport.

3.5 Conclusions

The concept of mobility is an essential pillar of human existence. This chapter presented the CogMob concept, aimed at deepening and expanding the notion of cognitive mobility. The chapter explored the dimensions of mobility, such as the ranking of mobility and its constituent elements, enabling us to comprehend the vast subject of mobility through cognitive lens. Through various examples, the significance of the cognitive approach was highlighted across numerous domains and aspects of mobility. CogMob aims to establish a comprehensive framework for managing natural and artificial cognition within a unified system by uncovering commonalities across different subject areas.

References

1. Zöldy, M., Baranyi, P.: Cognitive mobility – CogMob. In: 12th IEEE International Conference on Cognitive Infocommunications (CogInfoCom), pp. 921–925 (2021)
2. Zöldy, M., Baranyi, P.: The cognitive mobility concept. Infocommunications J. **15**, no. Special Issue, 35–40 (2023)
3. Baranyi, P., Csapo, A.: Cognitive infocommunications: Coginfocom. In: 2010 11th International Symposium on Computational Intelligence and Informatics (CINTI), pp. 141–146 (2010)
4. Baranyi, P., Persa, G., Csapo, A.: Definition of cognitive infocommunications and an architectural implementation of cognitive infocommunications systems. Int. J. Electron. Commun. Eng. (2011)
5. Csapo, A., Baranyi, P.: A unified terminology for coginfocom applications. In: 2011 2nd International Conference on Cognitive Infocommunications (CogInfoCom), pp. 1–6 (2011)
6. Csapo, A., Baranyi, P.: An application-oriented review of coginfocom: the state-of-the-art and future perspectives. In: 2012 IEEE 10th International Symposium on Applied Machine Intelligence and Informatics (SAMI). IEEE (2012)

7. Baranyi, P., Csapo, A., Sallai, G.: Cognitive Capabilities in the Future Internet, pp. 173–185. Springer International Publishing (2015)
8. Borjas, G., Doran, K.: Cognitive mobility: labor market responses to supply shocks in the space of ideas. Natl. Bureau Econ. Res. (2012)
9. Dobler, R.: Cognitive Mobility: A Macroscopic Investigation of Migration of Scientists between Research Fields Studied by Example of Mathematics, vol. 01, pp. 225–245. Allied Publishers (2000)
10. Baranyi, P., Csapó, Á.: Definition and synergies of cognitive infocommunications. Acta Polytech. Hungarica **9**(1), 67–83 (2012)
11. Baranyi, P., Csapo, A., Sallai, G.: Cognitive Infocommunications (coginfocom). Springer (2015)
12. Baranyi, P., Csapó, Á.B., Berki, B.: Concepts of cognitive infocommunications. Infocommunications J. (2024). Accepted, in press
13. Horváth, I., Csapó, Á.B., Berki, B., Sudár, A., Baranyi, P.: Definition, background and research perspectives behind 'Cognitive Aspects of Virtual Reality' (cVR). Infocommunications J. **15**, no. Special Issue, 9–14 (2023)
14. Kő, A., Szabó, I., Csapó, Á.B., Kovács, T., Lőrincz, L., Baranyi, P.: Digital & cognitive corporate reality. Infocommunications J. **15**, no. Special Issue, 2–10 (2023)
15. Baranyi, P., Csapó, Á., Budai, T., Wersényi, G.: Introducing the concept of internet of digital reality - part I. Acta Polytech. Hungarica **18**(7), 225–240 (2021)
16. Wersényi, G., Csapó, Á., Budai, T., Baranyi, P.: Internet of digital reality: infrastructuralbackground - part ii. Acta Polytech. Hungarica **18**(8), 91–104 (2021)
17. Baranyi, P., Gilányi, A.: Mathability: emulating and enhancing human mathematical capabilities. In: 2013 IEEE 4th International Conference on Cognitive Infocommunications (CogInfoCom), pp. 555–558. IEEE (2013)
18. Torok, A., Pauer, G.: Safety aspects of critical scenario identification for autonomous transport. Cognit. Sustainab. **1**(3) (2022)
19. Allenby, B.: 5g, ai, and big data: we're building a new cognitive infrastructure and don't even know it. Bull. Atom. Sci. **75** (2019)
20. de Ruiter, J.P., Noordzij, M.L., Newman-Norlund, S., Newman-Norlund, R., Hagoort, P., Levinson, S.C., Toni, I.: Exploring the cognitive infrastructure of communication. Inter. Stud./Soc. Behav. Commun. Biol. Artif. Syst. **11**(1), 51–77 (2010)
21. Saeed, Y., Ahmed, K., Zareei, M., Zeb, A., Vargas-Rosales, C., Awan, K.M.: In-vehicle cognitive route decision using fuzzy modeling and artificial neural network. IEEE Access **7**, 20 262–20 272 (2019)
22. Komashinskiy, V., Malygin, I., Korolev, O.: Introduction into cognitive multimodal transportation systems. Transp. Res. Procedia **50**, 273–279 (2020)
23. United Nations Conference on Road and Motor Transport. Geneva convention on road traffic (1949). Accessed 8 Feb 2024. https://unece.org/fileadmin/DAM/trans/conventn/Convention_on_Road_Traffic_of_1949.pdf
24. Szabó, Z., Török, Á., Sipos, T.: Order of the cities: usage as a transportation economic parameter. Period. Polytech. Transp. Eng. **49**(2), 164–169 (2019)
25. Zoldy, M., Szalmane Csete, M., Kolozsi, P.P., Bordas, P., Torok, A.: Cognitive sustainability. Cognit. Sustainabil. **1**(1) (2022)
26. Pethő, Z., Török, Á., Szalay, Z.: A survey of new orientations in the field of vehicular cybersecurity, applying artificial intelligence based methods. Trans. Emerg. Telecommun. Technol. **32**(10) (2021)
27. Munkácsy, A., Keserű, I., Siska, M.: Travel-based multitasking on public transport: an empirical research in hungary. Periodica Polytechn. Transp. Eng. **50**(1), 43–48 (2021)
28. Berki, B.: Overview of the relationship between human spatial abilities and GPS usage. In: 2022 IEEE 1st International Conference on Cognitive Mobility (CogMob), pp. 21–24. IEEE (2022)
29. Horváth, I., Berki, B.: Analysis of in-vehicle information systems from a cognitive perspective. In: 2023 IEEE 2nd International Conference on Cognitive Mobility (CogMob), pp. 95–100 (2023)

30. Lekić, M., Rogić, K., Boldizsár, A., Zöldy, M., Török, Á.: Big data in logistics. Periodica Polytech. Transp. Eng. **49**(1), 60–65 (2019)
31. Naradasu, K., Jyothirmai, S., Ramesh, R.: Towards artificial intelligence based diesel engine performance control under varying operating conditions using support vector regression. Thermal Sci. **17**(1), 167–178 (2013)
32. Žvirblis, T., Vainorius, D., Matijošius, J., Kilikevičienė, K., Rimkus, A., Bereczky, Á., Lukács, K., Kilikevičius, A.: Engine vibration data increases prognosis accuracy on emission loads: a novel statistical regressions algorithm approach for vibration analysis in time domain. Symmetry **13**(7), 1234 (2021)
33. Fényes, D., Németh, B., Gáspár, P.: Design of LPV control for autonomous vehicles using the contributions of big data analysis. Int. J. Control **95**(7), 1802–1813 (2021)
34. Paulauskas, V., Filina-Dawidowicz, L., Paulauskas, D.: Ports digitalization level evaluation. Sensors **21**(18), 6134 (2021)
35. Emőd, I., Füle, M., Tánczos, K., Zöldy, M.: A bioetanol magyarországi bevezetésének műszaki, gazdasági és környezetvédelmi feltételei. Magyar tudomány **50**, 278–286 (2005)
36. Goldbach, A., Meier, H., Wiggers, V., Chiarello, L., Barros, A.: Combustion performance of bio-gasoline produced by waste fish oil pyrolysis. Chem. Ind. Chem. Eng. Q. **28**(1), 1–8 (2022)
37. Alahmer, A., Rezk, H., Aladayleh, W., Mostafa, A.O., Abu-Zaid, M., Alahmer, H., Gomaa, M.R., Alhussan, A.A., Ghoniem, R.M.: Modeling and optimization of a compression ignition engine fueled with biodiesel blends for performance improvement. Mathematics **10**(3), 420 (2022)
38. Annamalai, M., Dhinesh, B., Nanthagopal, K., SivaramaKrishnan, P., Isaac JoshuaRamesh Lalvani, J., Parthasarathy, M., Annamalai, K.: An assessment on performance, combustion and emission behavior of a diesel engine powered by ceria nanoparticle blended emulsified biofuel. Energy Convers. Manag. **123**, 372–380 (2016)
39. Virt, M., Arnold, U.: Effects of oxymethylene ether in a commercial diesel engine. Cognit. Sustainab. **1**(3) (2022)
40. Mrówczyńska, M., Alsabry, A., Skiba, M., Bazan Krzywoszańska, A.: Research on energy consumption in the city. A system of modelling potential for energy effectiveness using artificial intelligence. Inter. J. Eng. Sci. Innov. Technol. (IJESIT) **3**, 373–384 (2014)
41. Zöldy, M., Zsombók, I.: Modelling fuel consumption and refuelling of autonomous vehicles. MATEC Web Conf. **235**, 00037 (2018)
42. Zalacko, R., Zöldy, M., SimongÃÂ¡ti, G.: Comparison of alternative propulsion systems - a case study of a passenger ship used in public transport. Brodogradnja **72**(2), 1–18 (2021)
43. Rövid, A., Remeli, V., Paufler, N., Lengyel, H., Zöldy, M., Szalay, Z.: Towards reliable multisensory perception and its automotive applications. Periodica Polytech. Transp. Eng. **48**(4), 334–340 (2020)
44. Varga, B., Tettamanti, T., Szalay, Z.: System architecture for scenario-in-the-loop automotive testing. Transport Telecommun. J. **22**(2), 141–151 (2021)
45. Varga, I., Bokor, J.: A new approach in urban traffic control systems. Periodica Polytech. Transp. Eng. **35**, 3–13 (2007)
46. Barsi, Á.: Performing coordinate transformation by artificial neural network. Allgemeine Vermessungs-Nachrichten **108**(4), 134–137 (2001)
47. Zöldy, M.: Investigation of autonomous vehicles fit into traditional type approval process. In: Proceedings of the Fourth International Conference on Traffic and Transport Engineering (2018)
48. Zöldy, M., Szalay, Z., Tihanyi, V.: Challenges in homologation process of vehicles with artificial intelligence. Transport **35**(4), 435–446 (2020)
49. Cao, H., Zoldy, M.: Implementing b-spline path planning method based on roundabout geometry elements. IEEE Access **10**, 81 434–81 446 (2022)

50. Németh, B., Géspár, P.: Design of learning-based control with guarantees for autonomous vehicles in intersections. IFAC-PapersOnLine **54**(2), 210–215 (2021)
51. Bokor, J., Vanek, B.: Research directions of unmanned aerial vehicles. In: Security Challenges in the 21st Century, pp. 411–421. Dialóg Campus Kiadó, Budapest (2018). https://eprints.sztaki.hu/9633/
52. Filina-Dawidowicz, L., Stankiewicz, S., Čižiūnienė, K., Matijošius, J.: Factors influencing intermodal transport efficiency and sustainability. Cognit. Sustainab. **1**(1) (2022)

Chapter 4
Cognitive Aspects of Virtual Reality

Abstract The recent surge in VR technology blurs the lines between physical and digital worlds. While initially limited to gaming, VR is now expanding into education, healthcare, and engineering, fueled by AI and sensor advancements, paving the way for a future "cognitive triad" of humans, AI, and digital twins. It explores VR through the lens of CogInfoCom, highlighting its potential as a revolutionary new communication and information organization tool. VR's unique spatial and temporal perspective enables intuitive information sharing and organization, surpassing traditional 2D interfaces. Furthermore, its future integration with AI and sensor technologies promises a powerful "CogInfoCom platform" that adapts to user needs and supports advanced cognitive functions. The definition of Cognitive Aspects of Virtual Reality (cVR) is presented, highlighting its synergies with related fields like Cognitive Infocommunications, AI, and the Internet of Things. cVR aims to understand how humans interact with and utilize information in 3D virtual environments, drawing insights from AI to enhance these interactions and empower users' cognitive capabilities. This leads to the concept of a cognitive triad, where humans, AI, and digital twins collaborate within VR, potentially paving the way for a future Digital Reality that seamlessly blends the physical and digital worlds.

Keywords Virtual reality · Mixed reality · Augmented reality · Cognitive infocommunications · Metaverse · Cognitive aspects of virtual reality

4.1 Introduction

This chapter presents the recent surge in VR technology, blurring the lines between physical and digital worlds. While initially limited to gaming, VR is now expanding into education, healthcare, and engineering, fueled by AI and sensor advancements, paving the way for a future "cognitive triad" of humans, AI, and digital twins. The key concepts presented in this chapter are based on [1]. The chapter employs the scientific viewpoint, mentality, and framework that have been deduced and established from the diverse standpoints within the subsequent fundamental articles: [2–7].

I. Horváth et al., *Cognitive Aspects of Virtual Reality*, Studies in Big Data 156,
https://doi.org/10.1007/978-3-031-68130-1_4

In the past decade, technology has made significant advancements in computer graphics, making it possible to create applications that blend the physical and digital worlds. These applications are known as extended reality (XR) and include virtual reality (VR), augmented reality (AR), and mixed reality (MR). In this chapter, we will use the term virtual reality (VR) to refer to all of these technologies, as explained in [6, 8].

Within a VR application, the user's digital environment is significantly enriched and can include different digital representations like avatars, objects, digital twins, and simulations. The rising interest in the "*Web 3.0*" concept has led to envisioning a new type of *Metaverse* that complements VR frameworks, fostering seamless interactions between the physical and digital realms. This envisioned Metaverse goes beyond spatial aspects, embracing decentralization and carrying notable social implications [9]. As the line between the real and virtual blurs, and more decentralized and participatory platforms emerge, a heightened level of human-ICT collaboration becomes feasible, integrating the existing digital 2D world while surpassing it in significant ways, thereby extending the capabilities of both humans and ICT for mutual co-evolution.

VR technologies drive significant changes in human-ICT interactions, even without the rise of Web 3.0 and the Metaverse. Leading tech giants, including Apple, Google, Microsoft, Facebook, Samsung, and others, have made significant investments in virtual reality (VR) technologies. This has led to a 12 billion USD VR industry in 2020, with projections exceeding 72 billion USD by 2024 [10]. It is expected that spatial content will become more prevalent, eventually replacing 2D layouts, similar to how graphical user interfaces replaced character-based interfaces in the late 1980s and early 1990s. This growth is anticipated to bring even greater benefits than the previous transition from 2D to 3D interfaces, as highlighted by recent studies [11–15].

Pioneers of virtual reality immediately recognized the importance of this technology, that is, that it was a fundamentally new infocommunications platform [16]. Despite this initial recognition, the prevailing perception often confines VR to gaming, entertainment, and professional domains like simulation and training. Over the last decade, VR has increasingly become a part of basic and higher education [17–21], healthcare [22–25], engineering, and various professional industries [26–30]. With the integration of AI technology, VR has risen to new heights, extending human capabilities and qualitatively enhancing them. At the same time, developments in miniaturized sensors and actuators have allowed for a deeper connection between VR, the digital twin concept, and portable, location-aware informatics devices. As a result, a revolutionary "cognitive triad" of humans, AI, and digital twins has emerged within VR, revolutionizing information access and understanding. This has ultimately led to enhanced efficiency, productivity, and safety in the digital realm.

The field of Cognitive Aspects of Virtual Reality (cVR) is gaining traction thanks to its comprehensive perspective on "reality plus capability" technologies. Advocates believe that these cutting-edge tools have the potential to revolutionize human cognition by introducing the concept of a "spatial cognitive cloud". Various essential factors are driving the advancement of VR technology. Firstly, the an increasing demand for data in VR systems. Secondly, the rise of sophisticated AI solutions capable of managing and interpreting large amounts of data. Thirdly, there is a crucial need for real-time supervision and management of physical processes within VR worlds. Lastly, the critical importance of managing information overload for human users effectively. These advancements can have a profound impact on the future of VR, opening up new and exciting possibilities.

The following Sect. 4.2 explores VR through the lens of CogInfoCom, highlighting its potential as a revolutionary new communication and information organization tool. VR's unique spatial and temporal perspective enables intuitive information sharing and organization, surpassing traditional 2D interfaces. Furthermore, its future integration with AI and sensor technologies promises a powerful "CogInfoCom platform" that adapts to user needs and supports advanced cognitive functions.

Section 4.3 presents the definition for Cognitive Aspects of Virtual Reality (cVR), highlighting its synergies with related fields like Cognitive Infocommunications, AI, and the Internet of Things. cVR aims to understand how humans interact with and utilize information in 3D virtual environments, drawing insights from AI to enhance these interactions and empower users' cognitive capabilities. This leads to the concept of a cognitive triad, where humans, AI, and digital twins collaborate within VR, potentially paving the way for a future Digital Reality that seamlessly blends the physical and digital worlds. The chapter ends with a summary and conclusion (Sect. 4.4).

4.2 Virtual Reality from a CogInfoCom Perspective

This section examines the foundations and promising future of VR through the lens of cognitive infocommunications. This recently established field studies the combined potential of contemporary infocommunications technologies (themselves the product of the fusion of media, communications, and computer science) and cognitive science [2–4, 31], considering also generational differences [32].

In industrial research, development, and innovation processes, the initial stages are often driven by human needs, with subsequent technological advancements generating new technological demands. This cyclical relationship underscores the symbiotic evolution of human requirements and technological development (Fig. 4.1). However, facilitating this progression requires more than conscious human motivation; explicit knowledge integration poses challenges due to granularity limitations. Thus, there is a crucial role for information and communication technologies (ICT) to transform into tools capable of managing our reality effectively.

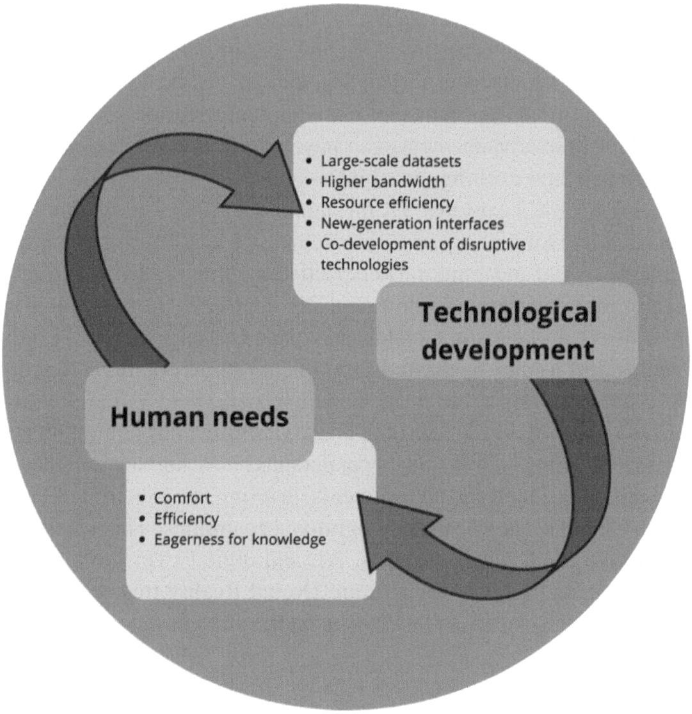

Fig. 4.1 Schematic representation of the cyclical relationship between the evolving human needs and technological developments

4.2.1 VR as an Infocommunications System

Virtual Reality (VR) uniquely expands and reshapes both spatial and temporal perspectives, additionally facilitating novel modes of information sharing, comprehension, and organization among human users.

4.2.1.1 Spatial and Temporal Perspective

A longstanding perspective acknowledges the potential of virtual reality as a fundamentally transformative communication medium, capable of transcending spatiotemporal boundaries. Jaron Lanier, a pioneer in the field and credited with coining the term "virtual reality," posits VR's unique ability to foster a novel cognitive mode exceeding the conceptual level [16]. He argues that within VR, represented objects transmute into self-contained entities, eliminating the need for mental translation from symbol to signified. Furthermore, Lanier envisions VR as a collaborative environment for fostering novel forms of self-expression, unconstrained by physical location or temporal synchronicity.

Irrespective of the ultimate realization of this vision, VR's inherent potential enables users to navigate between past and present seamlessly, and imagined or predicted spatial configurations, regardless of their physical setting. It enables the simultaneous interaction with various spatial and temporal scales within a unique spatiotemporal framework, which mirrors physical reality but with subtle deviations. As a result, VR opens up the possibility for a new, non-physical way of thinking about space (spatial cognition).

4.2.1.2 Information Organization Perspective

Recent VR platforms dedicated to organizing and sharing web-based documents within spatial contexts suggest an equally groundbreaking realization: VR's potential to revolutionize information organization. This perspective necessitates heightened attention to considerations from both informatics and psychology, such as the optimal representation of information through spatial relationships or modeling human comprehension and retention within a spatial framework.

Numerous studies have demonstrated enhanced information sharing and collaborative action at lower cognitive costs when utilizing 3D spaces compared to 2D representations. Such environments allow for direct reflection of semantic, spatial, and temporal relationships through relative size and position in 3D [11, 12, 33–39].

Therefore, VR represents a significant step toward replacing command-based interactions with dynamic, affordance-based interfaces that align with human intuition [40]. In essence, VR acts as an infocommunication tool offering increased effectiveness, efficiency, and an enhanced overall experience. It facilitates the sharing and collaborative utilization of vast information volumes, organized according to their relative importance, workflow position, or other interdependencies, all while transcending spatial and temporal constraints. Notably, spatial interfaces generally offer a more 'life-like' experience, enhancing user comfort and enabling the understanding and manipulation of digital tools across multiple spatiotemporal levels.

4.2.2 VR as a CogInfoCom Platform

In the intermediate to extended future, it is anticipated that virtual reality will further integrate with sensor technologies and artificial intelligence (AI), evolving from a mere infocommunication platform to a cognitive infocommunication (CogInfoCom) platform.

As outlined in [12], VR has the potential to interact with users in new dimensions that surpass traditional platforms. This enhances the utilization of human mental models rooted in space and time. When supplemented with sensor technologies

(IoT) and AI, extended reality technologies are set to permeate all aspects of our daily lives. This includes conventional IT tools (documents, media, 2D interfaces), the physical 3D world (real, imagined, or simulated), the user's mental reality (the organization of information, content relevance to location and time), and the impact of the user's decisions (real or hypothetical) on the aspects mentioned above. With AI, these platforms can gain increasing autonomy, making decisions such as:

- Determining which content should be presented to users, where, and when.
- Identifying suitable concepts to aid users in navigating relevant content, including broader content-group levels that encompass complete spatial layouts.
- Establishing paths, including sequences of locations and viewpoint orientations, to facilitate users' sequential traversal through the displayed information.
- Supporting advanced human cognitive functions such as memory, association, learning, recall, problem-solving, and collaborative effectiveness.

It is important to emphasize that autonomous systems make decisions based on the user's actions, which include movements within the environment, content consumption patterns, and metrics related to knowledge and skills. These metrics can be obtained from gamified scenarios and other sources [41, 42]. Consequently, a symbiotic relationship can evolve between the human user and the AI-facilitated ICT platform, yielding emergent behaviors that would be inconceivable without the mutual involvement of both entities.

Simultaneously, the advancement of such autonomous AI-supported systems presents a downside, as these systems inherently aim to learn and predict human cognition and behavior with utmost fidelity. One approach to achieving this objective involves shaping users' thought patterns to conform to a limited set of predefined categories. Certain scholars contend that this mechanism underpins phenomena such as tribalism on social media platforms or the dominance of a few product categories with high sales volumes on e-commerce platforms. These outcomes are often engineered by efficient learning systems primarily aimed at optimizing singular metrics, such as user engagement measured in terms of time spent on a particular platform [43, 44]. The overarching motivation behind cVR lies in the development of methodologies and solutions that harness the immense potential offered by AI-supported spatial ICT while concurrently addressing these legitimate concerns credibly.

4.3 Definition of Cognitive Aspects of Virtual Reality and Synergies With Related Fields

Based on the above considerations, the following definition for Cognitive Aspects of Virtual Reality (cVR) was proposed in [1]:

> **Cognitive Aspects of Virtual Reality (cVR)** investigates the next phases of IT evolution characterized by a transition from digital environments based on 2D graphical user interfaces (e.g. windows, images, 2D widgets) to 3D spaces which represent a higher-level integration

of VR/AR/MR systems, human spatial cognition, the 2D digital world (i.e. Web 2.0, Web 3.0) and artificial intelligence (AI). A primary focus of cVR is how this transition simultaneously makes use of and augments human capabilities, including psychological, cognitive and social capabilities - especially capabilities linked to a deeper understanding of geometric, temporal and semantic relationships. By extension, cVR further investigates the effects of these changes in human and AI capabilities with respect to a variety of sectors including education, commerce, healthcare, industrial production and others.

cVR is positioned to contribute outcomes and derive insights from interrelated disciplines such as Cognitive Infocommunications, Artificial Intelligence, Internet of Things, and Digital Reality. These relations are presented in the following subsections.

4.3.1 cVR and Cognitive Infocommunications

Cognitive Infocommunications, also known as CogInfoCom, is an interdisciplinary research field that aims to create new synergies between infocommunications and cognitive sciences. The primary objective of CogInfoCom is to enhance the efficacy of human-computer interaction while augmenting human cognitive capabilities by integrating infocommunications devices. Additionally, this discipline seeks to provide a systematic understanding of how infocommunication technologies and cognitive processes evolve together over time [2–4].

McLellan posits virtual reality as a cognitive instrument, conceived to facilitate the management of information [45]. Recent findings elaborated in Sect. 4.2.1 corroborate and broaden this perspective. Notably, this viewpoint aligns with the objectives of CogInfoCom, which endeavor to amplify and enrich human cognitive capacities within the context of spatial technologies. Hence, cVR and CogInfoCom exhibit a close association.

4.3.2 cVR and Artificial Intelligence

Since approximately 2010, there has been a significant advancement in Artificial Intelligence (AI) with the introduction of (deep) neural networks (DNNs), which have proven to be a proficient and highly successful approach to addressing numerous complex challenges. Over the past decade, traditional methodologies such as reinforcement learning [46] and symbolic reasoning [47] have been enriched through the integration of DNN advancements, sparking a noteworthy resurgence in AI research.

Innovative approaches in cVR stand to derive considerable inspiration from contemporary AI methodologies. This is because many of the primary difficulties faced in cVR arise from the attempt to replicate human cognitive abilities, which requires analyzing the patterns of users' interactions within virtual reality environments. Some specific capabilities are particularly relevant in spatial contexts, such as:

- Abilities to mentally organize and navigate a diverse array of information sources, which may concurrently manifest within a single digital 3D environment. Analogous to the scenario where individuals manage numerous open tabs in a web browser, a VR space could present 50 'in-game' browser windows with a simple click [12]. This challenge invokes the concept of ***big big data***, involving the analysis of simultaneous user interactions with an exponentially growing number of information sources.
- Proficiencies in guiding users' attention, as well as facilitating retention and recall of information within VR environments. These capabilities are intricately linked to the intelligent design of environments and content layouts, as well as the semi-supervised semantic modeling of content engaged with by users [12, 48].
- Skills in spatial navigation and avatar control raise questions about the interface between humans and AI, which converts inputs into these forms. This may arise from the homuncular flexibility hypothesis [49, 50].
- A fundamental aspect of human cognition lies in aptitudes for cognitive processes that are grounded in spatial movements and metaphors [51–53]. In this context, we propose the concept of an ***invisible VR***—a virtual reality integrated within the mental models of humans. The aim is to establish a correlation between this hidden VR and the visible VR experienced by users.

4.3.3 cVR and Internet of Things

As outlined in [6, 8], the Internet of Things (IoT) is a field that combines distributed computing, sensors, actuators, wearables, and digital twins with intelligent interconnections. This vision establishes a direct linkage between the dynamic state of the physical realm and the digital domain, precipitating the emergence of various "Internet-of-X" disciplines, including the Internet of Everything, Internet of Nano Things, and Internet of Mobile Things, among others [54–56].

From the perspective of cVR, it's worth noting that the aforementioned disciplines mainly focus on the technological aspects of network infrastructure. Therefore, they do not address questions related to the presentation or representation layers of relevant applications. Moreover, cognitive dimensions, encompassing users' capacities to comprehend and interact with relevant data, are largely overlooked, as they extend beyond the technological framework of efficiently connecting numerous devices and routing data among them. Nevertheless, these cognitive dimensions can often be effectively addressed within the context of 3D spatial interfaces. At the intersection of cVR, IoT, and AI, a novel form of co-evolution emerges, involving humans, digital twins (IoT), and AI, which we denote as the ***cognitive triad*** (Fig. 4.2).

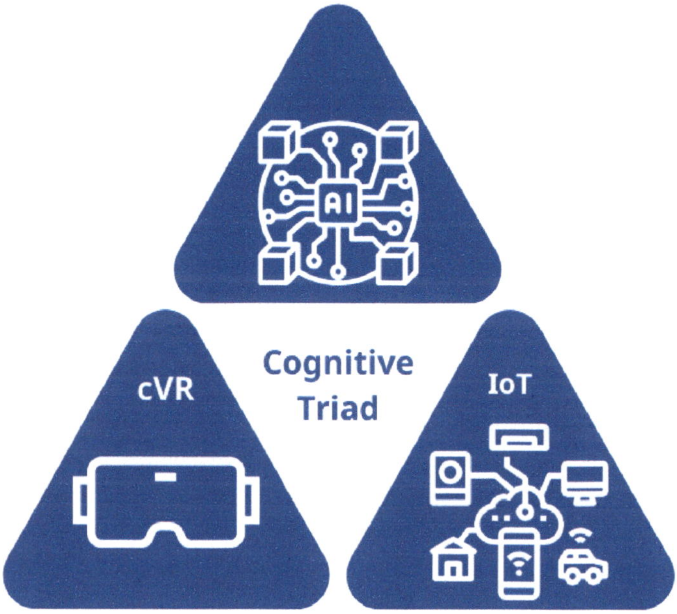

Fig. 4.2 The cognitive triad represents the intersection of humans, digital twins, and AI

4.3.4 cVR and Digital Reality/Internet of Digital Reality

Baranyi, Wersényi, Csapó, and Budai [6, 8] propose that humanity has arrived at a milestone in its socio-technological development, marked by a symbiotic evolution and interdependence between humans and information and communication technologies (ICT). This convergence holds the potential to engender a fundamentally novel reality, termed digital reality.

> A Digital Reality (DR) is a high-level integration of virtual reality (including augmented reality, virtual and digital simulations, and twins), artificial intelligence, and 2D digital environments which creates a highly contextual reality for humans in which previously disparate realms of human experience are brought together. DR encompasses not only industrial applications but also helps increase productivity in all corners of life (both physical and digital), thereby enabling the development of new social entities and structures, such as 3D digital universities, 3D businesses, 3D governance, 3D web-based digital entertainment, 3D collaborative sites, and marketplaces.

The authors [6, 8] emphasize that digital reality (DR) transcends the mere summation of individual digital solutions. Instead, they propose it as a holistic integration of various technologies, prioritizing human immersion and contextuality. In this framework, cVR holds significant potential for realizing DR due to its inherent focus on spatial interactions and its alignment with DR's emphasis on spatial and temporal contextualization. However, it is crucial to acknowledge the partially distinct aims of these fields. While cVR focuses more on understanding the cognitive capabilities of users and ICT platforms within a 3D interface, DR adopts a more technical perspective, aiming to converge disparate aspects of human reality.

4.4 Conclusions

The field of Cognitive Aspects of Virtual Reality (cVR) explores the potential of human-computer interaction in 3D virtual environments to extend and augment cognitive capabilities. This chapter presented the definition of cVR alongside its motivations and background, drawing from diverse perspectives. Building upon this discussion, we hypothesize that virtual realities specifically designed to align with human-ICT cognitive principles can produce simulated environments so lifelike that physical and online activities seamlessly converge into a unified experience.

References

1. Horváth, I., Csapó, Á.B., Berki, B., Sudár, A., Baranyi, P.: Definition, background and research perspectives behind 'Cognitive Aspects of Virtual Reality' (cVR). Infocommunications J. **15**, no. Special Issue, 9–14 (2023)
2. Baranyi, P., Csapó, Á.: Definition and synergies of cognitive infocommunications. Acta Polytech. Hungarica **9**(1), 67–83 (2012)
3. Baranyi, P., Csapo, A., Sallai, G.: Cognitive Infocommunications (coginfocom). Springer (2015)
4. Baranyi, P., Csapó, Á.B., Berki, B.: Concepts of cognitive infocommunications. Infocommunications J. (2024). Accepted, in press
5. Kő, A., Szabó, I., Csapó, Á.B., Kovács, T., Lőrincz, L., Baranyi, P.: Digital & cognitive corporate reality. Infocommunications J. **15**, no. Special Issue, 2–10 (2023).
6. Baranyi, P., Csapó, Á., Budai, T., Wersényi, G.: Introducing the concept of internet of digital reality – part I. Acta Polytech. Hungarica **18**(7), 225–240 (2021)
7. Baranyi, P., Gilányi, A.: Mathability: emulating and enhancing human mathematical capabilities. In: 2013 IEEE 4th International Conference on Cognitive Infocommunications (CogInfoCom), pp. 555–558. IEEE (2013)
8. Wersényi, G., Csapó, Á., Budai, T., Baranyi, P.: Internet of digital reality: Infrastructuralbackground – part ii. Acta Polytech. Hungarica **18**(8), 91–104 (2021)
9. Park, S.-M., Kim, Y.-G.: A metaverse: Taxonomy, components, applications, and open challenges. IEEE Access **10**, 4209–4251 (2022)
10. IDC: Worldwide Spending on Augmented and Virtual Reality Forecast to Deliver Strong Growth Through 2024, According to a New IDC Spending Guide (2020). https://www.idc.com/getdoc.jsp?containerId=prUS47012020
11. Horváth, I.: MaxWhere 3D capabilities contributing to the enhanced efficiency of the Trello 2D management software. Acta Polytech. Hungarica **16**(6), 55–71 (2019)
12. Csapó, Á.B., Horvath, I., Galambos, P., Baranyi, P.: Vr as a medium of communication: from memory palaces to comprehensive memory management. In: 2018 9th IEEE International Conference on Cognitive Infocommunications (CogInfoCom), pp. 000 389–000 394. IEEE (2018)

13. Bellalouna, F.: Virtual-reality-based approach for cognitive design-review and fmea in the industrial and manufacturing engineering. In: Baranyi, P. (ed.) Proceedings of the 10th IEEE International Conference on Cognitive Infocommunications: CogInfoCom 2019, pp. 41–46. IEEE (2019)
14. Kiss, D., Baranyi, P.: 3D webspace VS 2D website. In: 2020 11th IEEE International Conference on Cognitive Infocommunications (CogInfoCom), pp. 517–518. IEEE (2020)
15. Szabó, P., Baranyi, P., Sik-Lanyi, C.: Dos-windows-virtual desktop. In: 2023 IEEE 2nd International Conference on Cognitive Aspects of Virtual Reality (CVR). IEEE (2023)
16. Lanier, J., Biocca, F.: An insider's view of the future of virtual reality. J. Commun. **42**(4), 150–172 (1992)
17. Dalgarno, B., Hedberg, J., Harper, B.: The contribution of 3D environments to conceptual understanding. In: 19th Annual Conference of the Australasian Society for Computers in Learning in Tertiary Education, Winds of Change in the Sea of Learning, pp. 1–10. UNITEC Institute of Technology (2002)
18. Radianti, J., Majchrzak, T.A., Fromm, J., Wohlgenannt, I.: A systematic review of immersive virtual reality applications for higher education: design elements, lessons learned, and research agenda. Comput. & Educ. **147**, 103778 (2020)
19. Horváth, I.: Experiences and guidelines for developing 3D VR curricula. In: Accentuated Innovations in Cognitive Info-Communication, pp. 99–123. Springer International Publishing (2022)
20. Villena-Taranilla, R., Tirado-Olivares, S., Cózar-Gutiérrez, R., González-Calero, J.A.: Effects of virtual reality on learning outcomes in k-6 education: a meta-analysis. Educ. Res. Rev. **35**, 100434 (2022)
21. Horváth, I., Vér, C., Balogh, B.: Empowering circular economy goals through cognitive infocommunications in dual training programs. In: 2023 14th IEEE International Conference on Cognitive Infocommunications (CogInfoCom), pp. 000 051–000 056 (2023)
22. van der Kruk, S.R., Zielinski, R., MacDougall, H., Hughes-Barton, D., Gunn, K.M.: Virtual reality as a patient education tool in healthcare: a scoping review. Pat. Educ. Couns. (2022)
23. Venkatesan, M., Mohan, H., Ryan, J.R., Schürch, C.M., Nolan, G.P., Frakes, D.H., Coskun, A.F.: Virtual and augmented reality for biomedical applications. Cell Rep. Med. **2**(7), 100348 (2021)
24. Geraets, C.N., van der Stouwe, E.C., Pot-Kolder, R., Veling, W.: Advances in immersive virtual reality interventions for mental disorders: a new reality? Curr. Opin. Psychol. **41**, 40–45 (2021)
25. Vayssiere, P., Constanthin, P.E., Herbelin, B., Blanke, O., Schaller, K., Bijlenga, P.: Application of virtual reality in neurosurgery: patient missing. A systematic review. J. Clin. Neurosci. **95**, 55–62 (2022)
26. Idi, D.B., Khaidzir, K.A.M.: Critical perspective of design collaboration: a review. Front. Archit. Res. **7**(4), 544–560 (2018)
27. Dianatfar, M., Latokartano, J., Lanz, M.: Review on existing VR/AR solutions in human–robot collaboration. Procedia CIRP **97**, 407–411 (2021)
28. Delgado, J.M.D., Oyedele, L., Demian, P., Beach, T.: A research agenda for augmented and virtual reality in architecture, engineering and construction. Adv. Eng. Inform. **45**, 101122 (2020)
29. Guo, Z., Zhou, D., Zhou, Q., Zhang, X., Geng, J., Zeng, S., Lv, C., Hao, A.: Applications of virtual reality in maintenance during the industrial product lifecycle: a systematic review. J. Manuf. Syst. **56**, 525–538 (2020)
30. Firu, A.C., Tapîrdea, A.I., Feier, A.I., Drăghici, G.: Virtual reality in the automotive field in industry 4.0. Mater. Today: Proc. **45**, 4177–4182 (2021)
31. Baranyi, P., Persa, G., Csapo, A.: Definition of cognitive infocommunications and an architectural implementation of cognitive infocommunications systems. Int. J. Electr. Commun. Eng. (2011)
32. Baranyi, P., Csapo, A.B.: Revisiting the concept of generation CE - generation of cognitive entities. In: 2015 6th IEEE International Conference on Cognitive Infocommunications (CogInfoCom), pp. 583–586. IEEE (2015)

33. Berki, B.: 2D advertising in 3D virtual spaces. Acta Polytech. Hungarica **15**(3), 175–190 (2018)
34. Berki, B.: Better memory performance for images in MaxWhere 3D VR space than in website. In: 2018 9th IEEE International Conference on Cognitive Infocommunications (CogInfoCom), pp. 000 281–000 284. IEEE (2018)
35. Horvath, I., Sudar, A.: Factors contributing to the enhanced performance of the maxwhere 3d vr platform in the distribution of digital information. Acta Polytechn. Hungarica **15**(3), 149–173 (2018)
36. Horváth, I., Berki, B.: Investigating the operational complexity of digital workflows based on human cognitive aspects. Electronics **12**(3) (2023). https://www.mdpi.com/2079-9292/12/3/528
37. Horváth, I., Berki, B.: Unlocking the potential of immersive VR systems: The role of gesture and natural controls. In: 2023 IEEE 2nd International Conference on Cognitive Aspects of Virtual Reality (CVR). IEEE (2023)
38. Koles, M., Hercegfi, K., Logo, E., Tovolgyi, S., Szabo, B., Hamornik, B.P., Komlodi, A., Baranyi, P.Z., Galambos, P., Persa, G.: Collaboration experience in immersive vr environment in the frame of the neurocogspace project. In: 2014 5th IEEE Conference on Cognitive Infocommunications (CogInfoCom). IEEE (2014)
39. Vámos, Á.N., Fülöp, I.M., Reskó, B., Baranyi, P.Z.: Collaboration in virtual reality of intelligent agents. Acta Electrotechnica et Informatica **10**(2), 21–27 (2010). https://eprints.sztaki.hu/6255/
40. Torok, A.: From human-computer interaction to cognitive infocommunications: a cognitive science perspective. In: 2016 7th IEEE International Conference on Cognitive Infocommunications (CogInfoCom), pp. 433–438. IEEE (2016)
41. Blohm, I., Leimeister, J.M.: Gamification. Bus. & Inf. Syst. Eng. **5**(4), 275–278 (2013)
42. Deterding, S., Dixon, D., Khaled, R., Nacke, L.: "From game design elements to gamefulness: defining gamification. In: Proceedings of the 15th International Academic MindTrek Conference: Envisioning Future Media Environments, pp. 9–15 (2011)
43. Brockman, J.: Possible Minds: Twenty-Five Ways of Looking at AI. Penguin (2020)
44. Harris, T.: How technology hijacks people's minds—from a magician and google's design ethicist. Med. Mag. (2016)
45. Mclellan, H.: Cognitive issues in virtual reality. J. Vis. Lit. **18**(2), 175–199 (1998)
46. Silver, D., Schrittwieser, J., Simonyan, K., Antonoglou, I., Huang, A., Guez, A., Hubert, T., Baker, L., Lai, M., Bolton, A., et al.: Mastering the game of go without human knowledge. Nature **550**(7676), 354–359 (2017)
47. Graves, A., Wayne, G., Reynolds, M., Harley, T., Danihelka, I., Grabska-Barwińska, A., Colmenarejo, S.G., Grefenstette, E., Ramalho, T., Agapiou, J., et al.: Hybrid computing using a neural network with dynamic external memory. Nature **538**(7626), 471–476 (2016)
48. Horváth, I., Csapó, Á.B.: Structured template language and generative AI driven content management for personalized workspace reconfiguration. In: 2023 IEEE 2nd International Conference on Cognitive Aspects of Virtual Reality (CVR). IEEE (2023)
49. Gonzalez-Franco, M., Lanier, J.: Model of illusions and virtual reality. Front. Psychol. **8**, 1125 (2017)
50. Won, A.S., Bailenson, J., Lee, J., Lanier, J.: Homuncular flexibility in virtual reality. J. Comput.-Mediat. Commun. **20**(3), 241–259 (2015)
51. Tversky, B.: Mind in Motion, 1st edn. Basic Books, New York (2019). literaturverzeichnis: Seite 295–358
52. Kóczy, J.B.: Reference point constructions in the meaning construal of hungarian folksongs. Cognit. Linguistic Stud. **3**(1), 113–133 (2016)
53. Mathewson, J.H.: Visual-spatial thinking: an aspect of science overlooked by educators. Sci. Educ. **83**(1), 33–54 (1999)
54. Cisco: The Internet of Everything - Cisco IoE Value Index Study (2013).https://www.cisco.com/c/dam/en_us/about/business-insights/docs/ioe-value-index-faq.pdf
55. Srinivasan, C., Rajesh, B., Saikalyan, P., Premsagar, K., Yadav, E.S.: A review on the different types of Internet of Things (IoT). J. Adv. Res. Dyn. Control Syst. **11**(1), 154–158 (2019)
56. Baranyi, P., Csapo, A., Sallai, G.: Cognitive Capabilities in the Future Internet, pp. 173–185. Springer International Publishing (2015)

Part II
Human Cognitive Abilities in Desktop VR

In the current part, we explore the relationship between cognitive abilities and nonimmersive virtual reality. Each chapter presents a distinct viewpoint, examining different aspects of how cognitive processes interact within the context of desktop VR technology.

Chapter 5 studies how task performance and navigational experience are influenced by individual spatial abilities in desktop VR. We investigate conflicting hypotheses about VR's impact on people with different spatial abilities through an experiment and theoretical discourse.

Chapter 6 examines presence in virtual reality, its link to navigation, and how it varies across VR systems. It covers theoretical foundations, measurement tools, and experimental results, offering insights into presence factors in desktop VR.

Chapter 7 focuses on the potential of desktop VR to improve memory and learning in advertising and education. Empirical studies showVR's unique impact on memory retention compared to 2D settings, with implications for VR-based learning practices.

Chapter 8 examines the cognitive advantages of non-immersive desktop VR technology. It compares memory recall between desktop VR and traditional 2D environments, highlighting the benefits of virtual reality.

Chapter 5
Relationship Between Navigational Experience and Spatial Abilities

Abstract This chapter investigates the interplay between task performance, navigational experience, and individual spatial abilities within a desktop virtual reality (VR) environment. Two competing hypotheses, the Ability-As-Compensator and the Ability-As-Enhancer, are discussed concerning how VR impacts individuals with varying spatial abilities, alongside a proposed unifying theory. The analysis includes an exploration of navigation modes in desktop VR and an experiment assessing spatial abilities, subjective navigation experience, and task performance. However, the findings suggest that, in this specific VR context, no significant correlation was observed among these variables, indicating that this type of VR technology may not distinctly influence individual spatial memory and mental rotation abilities. The chapter concludes by discussing these outcomes, addressing limitations, and highlighting potential areas for future research.

Keywords Virtual reality · Desktop VR · Navigation · Spatial memory · Mental rotation

5.1 Introduction

Human spatial abilities play a fundamental role in our interaction with the world around us, influencing everything from navigation to problem-solving. In the context of virtual reality (VR), understanding the relationship between individual spatial abilities, navigational experience, and task performance is crucial for optimizing user experiences and designing effective VR environments [1–3], and promote the coevolution of human cognitive characteristics and infocommunication devices [4–8].

The present chapter investigates the potential relationship between task performance, navigational experience, and individual spatial abilities in a desktop virtual reality environment, and it is based on the following papers: [9–11]. The chapter is structured as follows: in Sect. 5.2, the relationship between human spatial abilities and the use of virtual reality (VR) technology, specifically in educational contexts is

I. Horváth et al., *Cognitive Aspects of Virtual Reality*, Studies in Big Data 156,
https://doi.org/10.1007/978-3-031-68130-1_5

discussed. The Ability-As-Compensator Hypothesis (Sect. 5.2.1) suggests that individuals with lower spatial abilities may benefit more from VR technology as it can compensate for their deficiencies and enhance their spatial perception. In contrast, the Ability-As-Enhancer Hypothesis (presented in Sect. 5.2.2) proposes that individuals with higher spatial abilities may benefit more from VR technology as they have sufficient cognitive capacity for constructing mental models. The last subsection (Sect. 5.2.3) presents a recent theory that can unify these hypotheses about the relationship between human spatial abilities and spatial visualization. Section 5.3 provides a brief analysis of navigation modes in desktop virtual realities. Section 5.4 presents the detailed experimental design and procedure. The study analyzes individual spatial abilities, subjective navigational experience, and task performance in a desktop virtual reality platform. Spatial ability was assessed with an online version of the Corsi-block tapping test and mental rotation task. The results, presented in Sect. 5.5, did not reveal any significant relationship between these variables, suggesting that the use of this type of VR is neutral concerning individual spatial memory and mental rotation ability. This is explained in more detail in Sect. 5.6. The limitations and further possible research directions are also presented here.

5.2 Theoretical Background

Human spatial abilities play a crucial role in our daily lives, enabling us to navigate through complex environments, understand spatial relationships, and interact with objects in our surroundings. With the advent of virtual reality (VR), these abilities take on new significance as users are immersed in simulated 3D environments. However, the extent to which an individual's spatial abilities affect their experience in VR remains an open question.

Spatial ability refers to the skill required to orient oneself in space, perceive one's body's position, and detect or reason about relationships within or between objects in a spatial context. For instance, spatial ability comes into play when reading maps, navigating, and assembling jigsaw puzzles [12]. Two main hypotheses exist concerning spatial ability and learning: the ability-as-compensator hypothesis and the ability-as-enhancer hypothesis.

5.2.1 Ability-As-Compensator Hyphotesis

The Ability-As-Compensator Hypothesis proposes that individuals with lower spatial abilities in the real world may benefit more from 3D and VR technology. This technology can compensate for their deficiencies and enhance their spatial perception and navigation skills.

According to this hypothesis, students with lower spatial abilities gain an advantage from the additional visual information provided, which aids them in mentally constructing visualizations. Without this supplementary information, these students would experience difficulty, which would place a greater demand on their working memory and increase their cognitive load [13–15]. In a study conducted by Hays [13], three different types of presentation—textual, static images, and animated graphics—were compared to evaluate their impact on learning performance. The results indicated that low spatial ability learners who were presented with animated graphics performed significantly better than those who were not. Moreover, spatial ability was found to be a significant factor in short-term comprehension, whereas the type of presentation had a statistically significant effect on long-term understanding. A study [16] with high school students aimed to compare the effectiveness of different learning environments based on spatial abilities. The study utilized a pretest/posttest experimental design to compare a conventional classroom learning method using PowerPoint slides with a desktop virtual reality-based learning environment. The findings, which were based on a comparative analysis of the two environments, revealed that low spatial ability learners benefited more significantly from the unique learning environment than their high spatial ability counterparts.

The use of virtual realities presents a unique opportunity for interaction with 3D models that extends beyond passive viewing. This direct manipulation of virtual environments has been found to provide significant benefits, particularly for students with lower spatial abilities. Through direct manipulation, the embodiment of anatomical structures becomes more tangible, enabling a clearer frame of reference while interacting with the virtual environment [17]. Such benefits have important implications for educational outcomes, as they can enhance students' ability to comprehend complex anatomical structures and facilitate deeper learning experiences. Therefore, the incorporation of direct manipulation in virtual reality-based educational materials may prove to be a valuable tool for improving learning outcomes for students of varying spatial abilities.

These findings have important implications for instructional design and education. They suggest that the use of animated graphics and other visual aids can enhance comprehension, reduce cognitive load, and facilitate more effective learning experiences for learners of varying spatial abilities.

5.2.2 Ability-As-Enhancer Hyphotesis

On the other hand, the Ability-As-Enhancer Hypothesis proposes that individuals with higher spatial abilities in the real world may benefit more from VR technology, as it can provide a more challenging and engaging spatial environment that allows them to further enhance their skills. Individuals who are already skilled may find 3D presentations beneficial, as they have sufficient cognitive capacity available for constructing mental models [18].

The use of 3D models in educational materials is advantageous for helping students to comprehend complex spatial arrangements. However, students with lower spatial abilities may experience cognitive overload when exposed to such materials. Conversely, students with higher spatial abilities are likely to benefit from the use of 3D models, as their cognitive load is less likely to exceed working memory limits. Thus, while the use of 3D models in educational materials can be beneficial for enhancing spatial understanding, the potential impact on students' cognitive load must be carefully considered, particularly for those with lower spatial abilities [15].

Anatomy is a prime example of a subject that requires the understanding of complex structures with critical spatial relations. The traditional approach to learning anatomical structures involves 2D figures and sections obtained from predetermined planes, requiring students to mentally reconstruct a 3D model from various sections. The number of views presented is also a crucial factor affecting learning outcomes. A study conducted by Levinson and colleagues [19] examined the effects of learner control and the number of views on the study of brain anatomy, where the best learning performance was observed when students were presented with key views and had low control over the learning environment. Interestingly, low-skilled students performed better when presented with only key views instead of multiple views, as the higher cognitive load required to match unfamiliar views with familiar representations was reduced. The worst performance was observed when students had no control over the presentation and were exposed to multiple views. In contrast, virtual realities offer a more interactive learning experience beyond simply viewing an object from different angles. Users are also able to manipulate objects within a complex virtual environment, facilitating a deeper understanding of spatial relations.

In virtual reality studies, navigational tasks are frequently used to examine the relationship between visualization abilities and spatial skills. The ability-as-enhancer hypothesis posits that individuals with better spatial skills can construct more effective mental models of virtual spaces, which can translate into superior performance in related tasks such as navigation and orientation. Modjeska and Chignell [20] investigated this hypothesis by comparing a 2.5D map view and a 3D fly-through view in a search task. The study found no significant difference between the two conditions, but spatial ability measures were consistent with the ability-as-enhancer hypothesis. Participants with low spatial ability performed worse in the 3D condition, indicating that a minimum level of spatial ability is necessary for effective navigation. However, once this threshold is met, other factors likely account for performance differences [20].

5.2.3 The Unifying Hypothesis: The Role of Optimal and Suboptimal Visualizations

Overall, both of these hypotheses suggest that 3D technology has the potential to enhance spatial abilities, but the extent and nature of this enhancement may depend on individual differences in spatial abilities and other factors, such as the type of

VR environment and the tasks involved. A recent study [21] presented a unifying conceptualization for these hypotheses. They distinguished optimal and suboptimal visualizations. In their regard, the ability-as-enhancer hypothesis means that to profit from an optimized visualization, a certain amount of spatial ability is required. Learners with medium ability have the resources to profit from the optimized design. Learners with low ability, lack these resources, and they cannot have this benefit. The high-ability learners can compensate for the suboptimal design, thus they can be successful, regardless of the visualization. This conceptualization relates to the results of [20] who also declared, that a minimum level of spatial knowledge is required to complete their experimental task.

5.3 Navigation in Desktop VR

This section is based on the analysis of navigation modes in desktop VRs, presented first in [11]. Desktop virtual realities are a form of virtual reality that display a high-resolution simulated 3D environment on a 2D screen, with a wide range of control devices available, including traditional navigational controls like mouse, keyboards, joysticks, and touchscreens, as well as specialized motion controls like datagloves or hand tracking systems. Headphones can also be used to provide a 3D sound experience, and gesture-based interaction can be integrated to provide a more natural control in the simulated environment.

Desktop VRs offer several advantages, including their technical and financial accessibility, the absence of simulator sickness, and the fact that they do not isolate the user from the outside world, unlike head-mounted display-based VRs. Consequently, these VRs are gaining popularity in education. Furthermore, they support collaborative work. The selection of VR technology for educational purposes is highly dependent on the type of educational material being taught. For instance, motion tracking and full-immersive features are more crucial for motion and practice-based content compared to general or theoretical educational materials. Therefore, careful consideration is necessary when selecting the appropriate VR technology for educational settings.

Accurate and precise navigation is a fundamental requirement in desktop virtual realities. Navigation is a cardinal feature that allows users to access various remote locations within the virtual environment and manipulate objects with ease. Cumbersome navigation can adversely impact the overall user experience, making an effective navigation method a crucial requirement of virtual reality systems.

Locomotion is an interaction component that allows navigation in virtual environments. Various hardware and software techniques have been developed to create seamless and user-friendly navigation experiences. The interaction type of VR locomotion determines how users trigger navigation. Physical interaction involves physical motion that is translated into VR motion, while artificial interaction employs input devices. Another aspect of interaction is the nature of virtual motion, which can be continuous or non-continuous. Continuous virtual motion is smooth and uninterrupted, whereas non-continuous motion involves teleportation-like transitions between locations [22].

The following part of this section presents an analysis of MaxWhere desktop virtual reality in terms of interaction fidelity, controller mapping, and finally, the navigational operations of this VR platform.

5.3.1 Interaction Fidelity

Interaction fidelity refers to the degree of accuracy with which real-world actions are replicated in an interactive system [23]. Interaction designers aim to create authentic and seamless interactions to enhance usability and user performance. However, interaction fidelity follows a U-shaped curve with user performance, implying that even non-natural interaction techniques can yield high user performance [24]. In the context of MaxWhere VR, real-world actions are emulated using a pointing device, primarily a mouse, as advised by the VR developers. This approach constitutes a non-natural interaction technique.

The level of interaction fidelity can be evaluated using the Framework for Interaction Fidelity Analysis (FIFA) [24, 25]. Given that FIFA compares navigation in the virtual environment to real-world navigation (e.g., walking), it is unsurprising that a desktop VR system employing a mouse as the control device exhibits relatively low interaction fidelity, particularly concerning biomechanical symmetry. Biomechanical symmetry encompasses anthropometric symmetry (involvement of body segments in real-world tasks and interaction), kinematic symmetry (faithful reproduction of body motions), and kinetic symmetry (representation of involved forces). These components are markedly limited in a desktop virtual reality environment when compared to real walking.

As interaction fidelity follows a U-shaped curve with user performance, a technology characterized by lower interaction fidelity can outperform a mid-fidelity counterpart in terms of user performance [24].

5.3.2 Controller Mapping

Steuer [26] coined the term "mapping" to describe how user actions in interactive media connect with corresponding changes in the mediated environment. Mapping can vary from arbitrary to natural. In an arbitrary mapping, the executed function lacks a direct link to the user's action. For instance, using a QWERTY keyboard with randomly assigned keys to control a video game. On the other hand, natural mapping implies a strong connection between the user's action and the function performed by the interactive media. For instance, employing a steering wheel as a controller in a car racing game exemplifies a naturally mapped interface.

Within the context of natural mapping, users perceive interactions as intuitive and aligned with their prior expectations. The concept of naturalness encompasses both technological aspects and individual differences [27]. Given its dependence on both technology and individual factors, repeated use over time can lead users to find specific interfaces highly natural, such as keyboards.

The significance of naturally mapped controllers ties into the concept of mental models. Mental models serve as cognitive representations of situations [27]. Comprehension of actions and events is influenced by these mental models, which develop through direct or mediated experiences [28, 29]. Naturally mapped controllers enable users to readily access mental models of real-world behaviors, providing more accurate guidance on interacting with games or virtual reality environments [30]. This can enhance the sense of spatial presence and minimize the user's focus on controls [27].

The mouse is a widely familiar device for most computer users. Due to this familiarity, experience contributes to a strong perception of the mouse as a very natural input method [31]. The perceived naturalness of a control device can outweigh its objective naturalness or its specific type of mapping. For instance, Seibert and Shafer's study [32] indicated that spatial presence values were linked to perceived controller naturalness, regardless of the type of controller used.

Within the typology of controller mapping, both the mouse and touchpad, that are used in MaxWhere desktop VR, fall under the category of natural, directionally mapped interfaces. The term directional mapping implies a correspondence between the control device's input directions and the corresponding on-screen outcomes [27]. An advantageous aspect of this type of natural mapping is that it leverages physical analogies, leading to immediate comprehension [33]. Naturalness, being a psychological state influenced by technology and individual differences [27], encompasses factors like prior experiences and perceived naturalness [32].

5.3.3 Navigational Operations in VR

The navigational operations in the utilized VR system are characterized by their continuous nature rather than simple on/off characteristics. This cardinal feature involves the translation of mouse motion in an analog manner, which contributes to enhanced understanding and usability. Human perceptual and nervous systems are naturally attuned to interpret analog systems, making them easily comprehensible [34–36]. In the following, we present the fundamental navigational operations, each of which can be performed single-handedly. These operations are intuitive to understand and can be seamlessly combined.

- **Forward and backward movement:** To move forward or backward, users engage the scroll button of the mouse. This directionally mapped navigation is readily graspable due to its widespread usage. However, for covering long distances, the continuous scrolling might become cumbersome. To address this, users can employ a double click to activate distant smartboards, which automatically switch to full-screen mode (gesture-based activation mode).

- **Lifting:** The "lifting" navigational operation involves pressing and moving the left mouse button, resulting in vertical movement. Although this vertical movement doesn't directly mimic a natural human action, it serves a practical purpose in the virtual reality environment. Particularly useful when interacting with 2D content on smartboards situated in the same plane, this operation facilitates effortless position adjustment and refocusing. Lifting, too, follows a directionally mapped navigation approach.
- **Spherical orbit:** The spherical orbit function enables users to navigate around a chosen point of an object by holding down the right mouse button and manipulating the device. Commonly found in CAD software, this orbital movement is well-suited for observational tasks [37]. MaxWhere's spherical orbit empowers users to view objects from various angles, or, with a broader radius, obtain an overview of the virtual environment. While not directly tied to natural human motion, this navigational operation finds its roots in the everyday behavior of observation. In the real world, we manipulate smaller objects in our hands or walk around larger objects to observe them closely.

MaxWhere's virtual reality environment provides a unique navigational experience by offering an easily accessible orbit function, in addition to the conventional zooming and panning operations. All of these functions can be controlled using a common device—a mouse equipped with two buttons and a scroll wheel. While the spherical orbit is prevalent in engineering software, its application in virtual realities is less common due to its departure from specific human movements, particularly in contexts where high interaction fidelity is a primary goal.

The choice of the mouse as a control device in MaxWhere proves advantageous, especially given its use as a desktop virtual reality platform. The familiarity of the mouse to users fosters a sense of naturalness, and the analog characteristics of the navigation mechanism contribute to intuitive learning and ease of use.

5.4 Spatial Abilities and Navigation: Experimental Design and Procedure

The study utilized MaxWhere[1] desktop virtual reality to investigate navigation and task performance. Participants were required to navigate the virtual environment and complete tasks, including answering questions related to the digital content's placement. Following the task, participants rated their navigational experience. The study also measured participants' spatial ability using the Corsi-block tapping test and mental rotation task. The following sections present the details of the experiment.

[1] www.maxwhere.com.

A total of 36 participants participated in the experiment. However, the data of five individuals were excluded due to data loss or a misunderstanding of the experimental task, and the final analysis included 31 participants. The mean age of the participants was 20.5 years, with a standard deviation of 3.4. There were 25 male and 6 female participants in the study. All participants were native Hungarian speakers.

5.4.1 Navigation and Task Performance

In this study, the Glassy space (Fig. 5.1) of MaxWhere desktop virtual reality (version: 2.4.0-b16) was used. The Glassy space is a three-floor, modern glass and concrete building situated above a serene sea, which was designed for presentations and online exhibitions. It features a relatively high number of smartboards, with 40 distributed throughout the building's open walls, inside and outside. These smartboards can display text, images, and multimedia content, and users can interact with them using various input devices, such as a mouse or touch screen. In this study, participants used an external mouse with two buttons and a scroll wheel to navigate in virtual reality.

The participants had to complete various tasks within the virtual space, which necessitated navigating the entire environment. Upon initial entry, users were situated on the top floor of the virtual building where they encountered instructions and a webpage containing information regarding the spatial measures, experimental tasks, and other inquiries. Displayed on additional smartboards were images sourced from unsplash.com, featuring different objects, numbers, or titles presented against plain-colored backgrounds. In addition, two smartboards contained PDF documents. To successfully complete the tasks, participants were required to navigate through the space and answer questions related to the digital content and its placement (e.g., "How many images with yellow backgrounds are located on the ground floor?" or

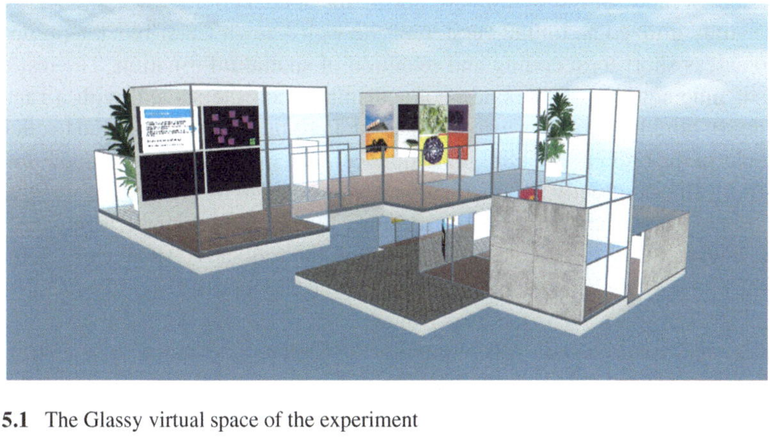

Fig. 5.1 The Glassy virtual space of the experiment

"How many pages are in the given document?"). Participants indicated their answers on the third-floor webpage, accessible by using the TAB key to return to the main area. A maximum of twelve points were possible for these tasks.

5.4.2 Subjective Navigational Experience

Following the completion of the task in the virtual reality environment, participants were asked to rate their navigational experience by indicating their level of agreement with five statements related to their virtual experience. Navigational experience means how natural and automatic the movement in virtual space is. A better navigation experience means more natural, automatic navigation. The rating was done on a ten-point Likert scale.

- "I moved confidently in the space."
- "I felt that I was controlling my movements in space."
- "I had difficulty navigating to the place where I wanted to go." (reverse-scored)
- "Navigation in virtual space was automatic for me."
- "I felt natural moving in the virtual space."

With a Cronbach's α of 0.883, the navigational experience scale used in the study is considered reliable. The scale was used in Hungarian.

5.4.3 Spatial Ability Measurement

The spatial ability tests, questionnaires, and experimental tasks were administered using version 2.5.2 of the Psytoolkit software [38]. To minimize any visual distraction from the virtual environment in the background, participants were instructed to utilize the fullscreen mode of the smartboard during the spatial ability tests.

The study utilized an online version of the well-known Corsi-block tapping test [39] to assess short-term coding and retention of spatial information. The test consisted of nine squares of the same color displayed on the screen, with a random sequence of blinking blocks in a different color. Participants were required to click on the blocks in the order of the blinking sequence, with the sequence length increasing if they performed correctly. A "done" button was provided for participants to indicate the end of the sequence. Participants received feedback on their performance, and the test ended if they repeatedly failed to repeat the sequence. Their score, referred to as the Corsi span, was calculated based on the number of correctly memorized blocks.

The mental rotation task [40] required the subjects to visualize how a stimulus would appear when rotated. In the online version, participants were presented with three figures on the screen. The top one was the target image, and they had to identify which of the two at the bottom matched the target image at the top of the screen. Matching involved mentally rotating the stimuli. The mental rotation time, measured in milliseconds, was the time taken to mentally rotate one item. The stimuli used in this implementation were in 2D. Participants completed five practice trials before proceeding to the ten experimental tasks.

5.4.4 Experimental Procedure

At the start of the experiment, the experimenter provided a brief overview of the procedure and instructions. The instructions were also visible on a smartboard in the virtual environment. Participants were instructed to use full-screen mode for the first two tasks, the Corsi test, and the mental rotation task. After completing these tasks, participants were informed that using full-screen mode was optional for the remaining tasks. Next, they completed the experimental tasks in the same VR environment. They then rated their navigational experience and indicated their familiarity with the software and how much time they had spent using it. Participants also provided information about their regular use of other 3D software (e.g., games or designer programs). Lastly, they answered basic demographic questions. The average duration of the experiment was 14 min (SD: 4.7), and the entire experiment was conducted in Hungarian.

5.5 Spatial Abilities and Navigational Experience—Experimental Task Performance

The descriptive statistics are presented in Table 5.1. A Spearman's rank-order correlation was used to assess the relationship between the Corsi span and the score of the experimental task. No significant correlation was found between the two variables, with $r_s(28) = 0.09$, $p = 0.635$, 95% CI [−0.279, 0.436]. Similarly, no significant correlation was found between the score of the experimental task and the mental rotation

Table 5.1 Descriptive statistics of the examined variables of the role of spatial abilities experiment

	Mean	Standard deviation
Corsi span	6.2	1
Mental rotation score	8.4	1.4
Reaction time of correct answers in the mental rotation task	6573 ms	2087 ms
Score of the experimental task	9.6	1.6
Navigational experience (1–10)	7.04	2.19

score ($r_s(29) = -0.174$, $p = 0.351$, 95% CI [−0.497, 0.193]) nor the reaction time of it ($r_s(29) = 0.148$, $p = 0.426$, 95% CI [−0.217, 0.478]). Overall, it was found that there was no significant correlation between the examined variables, indicating that the obtained scores for the experimental tasks are not related to either of the measured spatial abilities.

According to Spearman's rank-order correlation, there was no significant relationship between the score of the experimental task and navigational experience, with $r_s(29) = 0.068$, $p = 0.718$, 95% CI [−0.294, 0.412].

Furthermore, the correlation analysis revealed that the navigational experience was not significantly related to any of the measured spatial abilities, including Corsi span ($r_s(28) = -0.137$, $p = 0.471$, 95% CI [−0.474, 0.235]), mental rotation score ($r_s(29) = 0.09$, $p = 0.63$, 95% CI [−0.273, 0.431]), or the reaction time of the correct mental rotation answers ($r_s(29) = 0.133$, $p = 0.475$, 95% CI [−0.232, 0.466]). Thus, the present experiment did not demonstrate any statistically significant correlation between navigational experience and individual spatial abilities or the performance of the experimental task.

5.6 Limitations and Future Directions of Virtual Reality and Spatial Abilities

This chapter investigated the potential relationship between task performance, navigational experience, and individual spatial abilities in a desktop virtual reality environment. The results, presented in Sect. 5.5, did not reveal any significant relationship between these variables, suggesting that the use of this type of VR is neutral for individual spatial memory and mental rotation ability.

The lack of a statistically significant relationship between the measured task performance and spatial abilities supports neither the ability-as-compensator nor the ability-as-enhancer hypotheses [18]. The former suggests that 3D visualization can help reduce cognitive load, while the latter suggests that a minimum level of spatial ability is necessary for the effective use of virtual reality. A possible explanation could be, that all participants reached a minimum level of spatial ability, that was required to complete the task. Concerning the unifying hypothesis of Kühl and colleagues [21], the present research applied only an optimized 3D presentation of the task. Thus, if the participants reached a medium level of spatial ability, all of them had the cognitive resources to profit from this optimized design. In future research, it would be worthwhile to revisit this measurement and compare an optimal and suboptimal visualization, to prove the hypothesis.

The present study aimed to model the everyday use of a desktop virtual reality and measure its relationship with individual spatial abilities. However, it should be noted that the virtual space used in the experiment was relatively large, whereas educational settings typically utilize smaller, more concentrated spaces. Additionally, the use of CogiNav technology in MaxWhere VR, designed to facilitate navigation in 3D virtual spaces, may have played a role in the lack of a significant relationship between spatial abilities and task performance observed in this study, as the navigation could have had a moderating effect between these variables.

Overall, the results of this study suggest that students with varying levels of spatial abilities can benefit similarly from using this kind of virtual reality. Desktop VRs appear to be accessible to anyone, regardless of their spatial memory or mental rotation skills. Furthermore, this study contributes to the ongoing discourse surrounding the relationship between virtual reality and spatial abilities, highlighting the need for further research in this area.

References

1. Horváth, I., Csapó, Á.B., Berki, B., Sudár, A., Baranyi, P.: Definition, background and research perspectives behind 'Cognitive Aspects of Virtual Reality'(cVR). Infocommunications J. **15**, no. Special Issue, 9–14 (2023)
2. Baranyi, P., Csapó, Á., Budai, T., Wersényi, G.: Introducing the concept of internet of digital reality – part I. Acta Polytech. Hungarica **18**(7), 225–240 (2021)
3. Wersényi, G., Csapó, Á., Budai, T., Baranyi, P.: Internet of digital reality: Infrastructuralbackground – part ii. Acta Polytechn. Hungarica **18**(8), 91–104 (2021)
4. Baranyi, P., Csapó, Á.: Definition and synergies of cognitive infocommunications. Acta Polytech. Hungarica **9**(1), 67–83 (2012)
5. Baranyi, P., Csapo, A., Sallai, G.: Cognitive Infocommunications (coginfocom). Springer (2015)
6. Baranyi, P., Csapó, Á.B., Berki, B.: Concepts of cognitive infocommunications. Infocommunications J. (2024). Accepted, in press
7. Baranyi, P., Persa, G., Csapo, A.: Definition of cognitive infocommunications and an architectural implementation of cognitive infocommunications systems". Int. J. Electr. Commun. Eng. (2011)
8. Csapo, A., Baranyi, P.: A unified terminology for coginfocom applications. In: 2011 2nd International Conference on Cognitive Infocommunications (CogInfoCom), pp. 1–6 (2011)
9. Berki, B.: Does effective use of MaxWhere VR relate to the individual spatial memory and mental rotation skills? Acta Polytech. Hungarica **16**(6), 41–53 (2019)
10. Berki, B.: Humán kognitív jellemzők szerepe a virtuális terek használata során. In: Networkshop 2019, pp. 1–6. Hungarnet Egyesület (2019)
11. Berki, B.: Navigation power of MaxWhere: a unique solution. In: Baranyi, P. (ed.) Proceedings of the 11th IEEE International Conference on Cognitive Infocommunications (CogInfoCom), pp. 487–491. IEEE (2020)
12. American Psychological Association: Spatial ability. Accessed 08 March 2023. https://dictionary.apa.org/spatial-ability
13. Hays, T.A.: Spatial abilities and the effects of computer animation on short-term and long-term comprehension. J. Educ. Comput. Res. **14**(2), 139–155 (1996)
14. Hegarty, M., Sims, V.K.: Individual differences in mental animation during mechanical reasoning. Memory & Cognition **22**(4), 411–430 (1994)
15. Huk, T.: Who benefits from learning with 3D models? The case of spatial ability. J. Comput. Ass. Learn. **22**(6), 392–404 (2006)
16. Lee, E.A.-L., Wong, K.W.: Learning with desktop virtual reality: low spatial ability learners are more positively affected. Comput. & Educ. **79**, 49–58 (2014)

17. Jang, S., Vitale, J.M., Jyung, R.W., Black, J.B.: Direct manipulation is better than passive viewing for learning anatomy in a three-dimensional virtual reality environment. Comput. & Educ. **106**, 150–165 (2017)
18. Mayer, R.E., Sims, V.K.: For whom is a picture worth a thousand words? Extensions of a dual-coding theory of multimedia learning. J. Educ. Psychol. **86**(3), 389–401 (1994)
19. Levinson, A.J., Weaver, B., Garside, S., McGinn, H., Norman, G.R.: Virtual reality and brain anatomy: a randomised trial of e-learning instructional designs. Med. Educ. **41**(5), 495–501 (2007)
20. Modjeska, D., Chignell, M.: Individual differences in exploration using desktop VR. J. Amer. Soc. Inf. Sci. Technol. **54**(3), 216–228 (2003)
21. Kühl, T., Fehringer, B.C.O.F., Münzer, S.: Unifying the ability-as-compensator and ability-as-enhancer hypotheses. Educ. Psychol. Rev. **34**(2), 1063–1095 (2022)
22. Boletsis, C.: The new era of virtual reality locomotion: a systematic literature review of techniques and a proposed typology. Multimodal Technol. Inter. **1**(4) (2017). https://www.mdpi.com/2414-4088/1/4/24
23. McMahan, R.P., Bowman, D.A., Zielinski, D.J., Brady, R.B.: Evaluating display fidelity and interaction fidelity in a virtual reality game. IEEE Trans. Visual. Comput. Graph. **18**(4), 626–633 (2012)
24. McMahan, R.P., Lai, C., Pal, S.K.: Interaction fidelity: the uncanny valley of virtual reality interactions. In: Lecture Notes in Computer Science, pp. 59–70. Springer International Publishing (2016)
25. McMahan, R.P.: Exploring the effects of higher-fidelity display and interaction for virtual reality games. Ph.D. Dissertation, Virginia Tech (2011)
26. Steuer, J.: Defining virtual reality: dimensions determining telepresence. J. Commun. **42**(4), 73–93 (1992)
27. Skalski, P., Tamborini, R., Shelton, A., Buncher, M., Lindmark, P.: Mapping the road to fun: natural video game controllers, presence, and game enjoyment. New Media & Soc. **13**(2), 224–242 (2011)
28. Halford, G.S.: Children's Understanding: The Development of Mental Models. Psychology Press (1993)
29. Wyer, R.S., Radvansky, G.A.: The comprehension and validation of social information. Psychol. Rev. **106**(1), 89–118 (1999)
30. Tamborini, R., Skalski, P.: The role of presence in the experience of electronic games. In: Vorderer, P., Bryant, J. (eds.) Playing Video Games: Motives, Responses, and Consequences, pp. 225–240. Lawrence Erlbaum (2006)
31. Vorderer, P.: Interactive entertainment and beyond. In: Zilmann, D., Vorderer, P. (eds.) Media Entertainment: The Psychology of Its Appeal. LEA's Communication Series Lawrence Erlbaum, pp. 21–36. Associates Publishers (2000)
32. Seibert, J., Shafer, D.M.: Control mapping in virtual reality: effects on spatial presence and controller naturalness. Virtual Real. **22**(1), 79–88 (2017)
33. Norman, D.A.: Cognitive engineering. In: Norman, D.A., Draper, S.W. (eds.) User Centered System Design: New Perspectives on Human-computer Interaction, pp. 31–61 (1986)
34. Norman, D.A.: The Invisible Computer: Why Good Products Can Fail, the Personal Computer is so Complex, and Information Appliances are the Solution. MIT Press (1998)
35. Sudár, A., Csapó, Á.: Interaction patterns of spatial navigation and smartboard use in VR workspaces. In: Accentuated Innovations in Cognitive Info-Communication, pp. 149–166. Springer International Publishing (2022)
36. Horváth, I., Berki, B.: Unlocking the potential of immersive VR systems: the role of gesture and natural controls. In: 2023 IEEE 2nd International Conference on Cognitive Aspects of Virtual Reality (CVR). IEEE (2023)
37. Khan, A., Komalo, B., Stam, J., Fitzmaurice, G., Kurtenbach, G.: Hovercam: interactive 3D navigation for proximal object inspection. In: Proceedings of the 2005 Symposium on Interactive 3D Graphics and Games, pp. 73–80 (2005)

38. Stoet, G.: PsyToolkit: a novel web-based method for running online questionnaires and reaction-time experiments. Teach. Psychol. **44**(1), 24–31 (2016)
39. Corsi, P.M.: Human memory and the medial temporal region of the brain. Ph.D. dissertation, McGill University, Montreal (1972)
40. Shepard, R.N., Metzler, J.: Mental rotation of three-dimensional objects. Science **171**(3972), 701–703 (1971)

Chapter 6
Sense of Presence in Desktop VR

Abstract This chapter provides a comprehensive exploration of the sense of presence in virtual reality, emphasizing navigational experience and comparing the presence profile of MaxWhere desktop VR with other VR systems. It thoroughly investigates theoretical foundations, technological and human factors influencing presence, measurement tools, and experimental findings related to the sense of presence. The theoretical section offers foundational definitions of presence, exploring its relationship with immersion. Discussion on technological and human factors influencing presence includes recent results highlighting the correlation between the sense of presence and task performance in VR. The chapter introduces commonly used measurement tools and methods, such as the IPQ (Igroup Presence Questionnaire), while the navigational experience was assessed through a 10-point Likert scale based on five statements. The study included 31 individuals (mean age 20.5 years, SD: 3.4). The results support the hypothesis that navigational experience correlates with a stronger sense of presence. In this examination, participants were divided into two groups based on their proficiency with MaxWhere VR: distinguishing between novices and experienced users. Novices perceived the virtual environment as less realistic compared to experienced users. Lastly, a comparison of the presence profile between MaxWhere desktop VR and other VR systems extends and complements the earlier findings.

Keywords Virtual reality · Desktop VR · Presence · IPQ · Navigation · Prior experience

6.1 Introduction

Virtual reality (VR) technology has revolutionized the way we interact with digital environments, offering immersive experiences that blur the boundaries between the physical and virtual worlds [1–6]. Central to the effectiveness of VR is the concept of presence, which refers to the subjective feeling of "being there" in the virtual environment. This chapter provides a comprehensive examination of the sense of presence in virtual reality, with a particular focus on navigational experience, and a

I. Horváth et al., *Cognitive Aspects of Virtual Reality*, Studies in Big Data 156,
https://doi.org/10.1007/978-3-031-68130-1_6

comparative analysis of the presence profile of MaxWhere desktop VR with other VR systems.

The chapter is based on the following papers: [7–9]. Section 6.2 presents the theoretical foundations and definitions of the sense of presence. This is followed by the description of the technological and human factors that influence the experience of presence in Sect. 6.3. As the sense of presence is also related to the effectiveness of performing tasks in virtual reality, some recent results are also presented in Sect. 6.4. The available and most commonly used measurement tools and measurement methods are presented in Sect. 6.5.

This is followed by a thorough description of the instruments used in the experiment, the participants, and the process Sect. 6.6. Section 6.7 provides the results, that support the hypothesis that navigational experience comes along with a stronger sense of presence. In this analysis, participants were also categorized into two groups, based on their prior usage of MaxWhere VR: novices and experienced users. Novice users perceived the virtual environment to be less realistic compared to the evaluation of more experienced users (Sect. 6.8). Finally in Sect. 6.9, a comparison of the presence profile of MaxWhere desktop virtual reality and other VRs is shown. This comparison extends the results of the previous sections.

6.2 Theoretical Background of Presence

To understand the phenomenon of sense of presence, this section presents the theoretical foundations, and definitions, alongside its relation to immersion.

Presence, the subjective experience of reality, is the focus of researchers from various fields such as computer science, psychology, digital entertainment, and literature [10]. In the scientific field of psychology, the first book chapter on presence was published in 1890, in William James' Principles of Psychology [11]. The present chapter focuses on the virtual reality-induced presence, although this state of consciousness can be experienced in other situations as well (e.g.: during reading, watching movies, dream states).

A recent analysis [12] used the Interpretative Phenomenological Approach to reveal the background of users' experiences and identify the characteristics that enable the feeling of presence in virtual environments. The results showed that participants considered presence as a positive feature of VR. Three main themes were identified among users: visual satisfaction, freedom of interaction, and suspension of real life. Visual satisfaction relates to visual realism, and a realistic environment conveys more believability. When virtual environments (VE) are considered real, the users describe them as a space to act within. The interaction options allow a higher similarity to reality, with a more natural feeling [12].

6.2.1 Scientific Definitions of Presence

Several different definitions of presence are available in the current scientific literature. Felton and Jackson [10] summarized these diversified definitions in a recent paper. Furthermore, they aimed to create a new, unified definition of presence, to overcome the contradictory and sometimes arbitrary definitions used over the years by researchers from different scientific disciplines. They offer the following unified definition of presence:

> The extent to which something (environment, person, object, or any other stimulus) appears to exist in the same physical world as the observer.

This definition implicitly incorporates that it is a state of consciousness, which was the essence of Slater's previous definition: "a state of consciousness, the (psychological) sense of being in the virtual environment" [13].

Presence researchers created different typologies to categorize the different aspects of the presence experience. According to Felton and Jackson [10], these can be synthesized into two dimensions: spatial and social.

Spatial presence is a definitive characteristic of virtual reality, as it aims to be indistinguishable from reality [14]. It means the sense that one is physically located within the perceived environment. Users often report after VR sessions that they felt like they actually visited a place, rather than just viewing it on a screen [15].

Social presence refers to the users' sensation of being socially connected to the other participants of the virtual environment. This means that users apply everyday social norms, and basic communicational affordances, while users perform real-time interaction with their avatars. Social-focused VR-s allow more natural social interaction among users compared to classic virtual realities [16].

6.2.2 Relations and Differences Between Immersion and Presence

Immersion is a related term, that is sometimes used interchangeably with presence. Immersion is a more objective variable compared to presence as immersion is based on technological aspects. Thus, the assessment of immersion is also more objective, the different applied instruments and technology are more comparable than the subjective presence.

Slater [17] summarized the relationship of the two terms as presence is the subjective reaction to immersion. That means that due to individual differences, people can experience different levels of presence in environments with the same immersivness. From the users' perspective, presence is an experience. For example, after a virtual reality session, the users remember the virtual environment as a place rather than as a set of pictures or as a computer program [15].

Wilkinson et al. [18] created a thematic content analysis of presence and immersion definitions. Their results showed that presence is generally described as an experiential quality in the virtual environment. Four meaningful words were revealed for presence: environment, illusion, experience, and subjective feeling. In the reviewed definitions they highlighted the detachment from normal reality and the attachment to a different reality. While in the case of immersion, the technological aspects were more dominant. The revealed meaningful words were experience, system, and environment. These results fall in line with the theories behind the following questionnaires: Presence Questionnaire (PQ) of Witmer and Singer [19] and also the Igroup Presence Questionnaire [20].

6.2.3 Process of Presence Development

The process of formulating the sense of presence according to Schubert and colleagues [20] is based on constructing a spatial-functional mental model of the virtual environment. First, the technology immerses the user's sense and body into the VR. Then, different cognitive processes create a mental model of the body in the virtual environment. The major cognitive processes are the representation of bodily actions and possible actions in the simulated environment, and the suppression of incompatible sensory input. Then the user experiences the so-called sense of presence, the sense of being in the virtual environment. The sense of presence increases with the time spent in the virtual environment [21].

Slater [17] highlighted that there are two concurrent ways to achieve a high sense of presence in virtual reality. Either create such high fidelity, that the simulated environment is indistinguishable from reality; or apply the scientific knowledge of the human perceptual system to deliver a high sense of presence even in low immersion systems. By identifying the crucial characteristics of the human representation of reality, developers of VR can focus on enhancing the features that target the appropriate perceptions. This also includes the options for increasing the immersion, such as the multisensory feedback, audio stimuli (such as ambient sound in VR), and emotional responses [18].

6.3 Understanding Technological and Human Factors

The purpose of this section is to describe the technological and human factors that influence the experience of presence.

From a technological point of view, two major dimensions enable the development of presence: vividness and interactivity. Vividness means that the virtual environment is sensorially rich, and the human perceptual system can process these features. The other dimension is interactivity, where users can influent the content of the environment or the form of the simulation. This is closely related to user characteristics, and individual differences may occur [14].

These technological variables are also called external determinants of presence, that relate to the sensory and content variables [10]. The content of the virtual reality could invoke emotions too. The dominant emotion induced by the virtual environment is positively correlated with presence. Additionally, the agency has a significant positive effect on presence directly [22].

Internal determinants consist of demographic, psychological, and cultural variables [10]. Among demographic variables, the effect of sex on presence rating is widely discussed [23–25], although there is no conclusive evidence. These differences in the level of presence may also be sensitive to the content and depend on other factors [24].

The level of absorption, which refers to how much people get involved and engaged, influences the users' likelihood of accepting the media stimuli as their primary point of reference. Furthermore, less anxious users in new media environments can more easily create mental representations of the mediated media world and experience higher levels of presence as a result [26].

General experiences with computers and games are surprisingly not related to the sense of presence [27]. Although, the previous experience with virtual reality helps to guide the users' attention to the details of virtual reality, instead of navigational tools [26]. This result relates to the concept of perceived controller naturalness, which also predicts the spatial presence. The explanation behind this, it that a natural interface does not require the user to think about how to control the simulation, they can effortlessly feel inside the virtual environment [28].

6.4 The Impact of Presence on Human Performance in Virtual Environment

The sense of presence is also related to the effectiveness of performing tasks in virtual reality. Presence is positively correlated with human performance, and learning outcome [29–31]. The first-person view, non-symbolic interaction, and sense of presence are key features of virtual environments that can significantly impact learning outcomes [30]. The higher sense of presence helps them to complete their learning tasks more efficiently and successfully and construct knowledge [32]. A 2019 study by Selzer and colleagues showed a medium positive correlation between learning outcomes and presence. In the ten-minute session, those who experienced a higher sense of presence reached higher scores on the test [31].

6.5 Measuring Presence in Virtual Reality

Presence has been in the focus of scientific research since the early 1990s. However, as several definitions of presence exist, its measurement also has different variations. Nevertheless, this is not necessarily a bad thing, concluded van Baren and IJsselsteijn in [33]. In their collection, they provide a thorough examination of presence measurement approaches. In addition to questionnaires, they included continuous, psychophysical, and physiological measurements, as well as neural correlates [33].

Questionnaires are the most frequently used methods for the assessment of presence [34]. Usually, these are ordinal scales, such as a Likert scale ranging from 1 to 7. Questionnaires are advantageous due to their simplicity in administering, applicability to any virtual reality, and cost-effectiveness in comparison to psychophysiological measures [35]. In this review, the authors found that there is no physiological measure (EEG, EMG, fMRI, galvanic skin response, heart-rate variability, skin temperature) that collected enough evidence to be reliably used alone, without the subjective evaluation of the participants [35]. The high number of questionnaires (28) in the above-mentioned compendium [33], represents the theoretical diversity of the present concept. Questionnaires used as assessment methods assume that presence is a subjective factor, as it can be evaluated through subjective ratings that necessitate introspection from the participants. Additionally, most of these are post-test questionnaires, meaning they are given after the intervention, i.e., the participants fill them out after the virtual reality session. A limitation of these subjective post-immersion measures is that it is sensitive to recall error [33].

Continuous assessment is another form of measurement in which the participants must assess the changes in their sense of presence during the experience. This version has the benefit of providing time information, though it could potentially disrupt the user experience. Furthermore, the user can only rate one aspect of the experience, contrary to the post-test questionnaires [33].

Behavioral measures are based on recording the participants' actions and responses in the virtual environment. These assessment types are based on the concept, that users who experience the sense of presence, similarly behave in the virtual reality as they would do it in the physical world [33, 36].

Physiological measures are also studied (skin conductance, heart rate fluctuation, body temperature), and it could be a more objective assessment of presence. However, there is not yet an accepted measurement protocol for this, and construct validity is still lacking [10]. Until further proof, researchers should utilize physiological measures in combination with other corroborative measures. Neurological measures (fMRI, EEG) are also present, and this also can increase objectivity, and it can provide continuous temporal measurement. These are quite rare due to practical limitations, specific equipment requirements, high costs, and the difficulty of operating simultaneously with virtual reality.

Witmer and Singer developed the Presence Questionnaire (PQ), which is a post-test questionnaire that is frequently used [19]. PQ measures the degree to which the subject is "...experiencing the computer-generated environment rather than the actual physical locale". It has three subscales: Involved/Control (11 items), Natural (3 items), and Interface Quality (3 items). The Involved/Control subscale measures the perceived control of events in virtual reality, the responsiveness of the VE to user-initiated actions, and the involvement of the participant during the VR session. The items of the Natural scale assess the extent to which the interactions and locomotion were felt natural. The Interface Quality subscale measures whether the control and display devices interfered with or distracted the users from their tasks.

The Igroup Presence Questionnaire (IPQ) is a combination of pre-existing questionnaires and newly developed items and is a commonly employed tool for measuring presence [20, 37]. In contrast to the previously presented Presence Questionnaire [19], This questionnaire was created to analyze the components of the presence construct, with a particular focus on the separation between spatial-constructive and attention components. Furthermore, the IPQ questionnaire was statistically validated and thoroughly tested [20, 37]. The IPQ questionnaire is a fourteen-item questionnaire with three factors: spatial presence, involvement, and experienced realism. Spatial presence assesses the relationship between the virtual environment and the users' bodies, the sense of being physically present in the simulated environment. The involvement factor measures the awareness devoted to virtual reality. These two factors support the distinction between the spatial-constructive and attention component [38]. The third factor, experienced realism, assesses the subjective experience of realism in VR. There is also a general item: 'In the computer-generated world, I had a sense of "being there"' [20].

6.6 Measures of Navigational Experience and Sense of Presence

6.6.1 Subjects

A total of 36 participants participated in the experiments, but the final analysis only included data from 31 individuals. The remaining five participants were excluded due to data loss or a misunderstanding of the experimental task. The average age of the participants was 20.5 years (SD: 3.4). Among the participants, there were 25 men and 6 women who took part in the experiment.

6.6.2 Igroup Presence Questionnaire

To assess the participants' sense of presence, the Igroup Presence Questionnaire (IPQ) [20] was used, in Hungarian. This is a fourteen-item questionnaire, with a seven-point Likert scale. It has three subscales: spatial presence, involvement, and experienced realism. The spatial presence subscale assesses the feeling of being physically present within a virtual environment. The involvement subscale measures the degree of attention and engagement focused on the virtual space. On the other hand, the subscale of experienced realism is used to measure the subjective perception of realism within virtual reality [20, 37].

6.6.3 Navigational Experience

At the end of the experimental session, the participants were required to evaluate their experience of navigating the virtual space using a ten-point Likert scale. They were presented with five statements and asked to express their level of agreement with each statement using a ten-point scale. The statements were: “I moved confidently in the space”, “I felt that I was controlling my movements in space”, “I had difficulty navigating to the place where I wanted to go” (reverse-scored item), “Navigation in virtual space was automatic for me”, “I felt the natural moving in the virtual space” [7]. The Cronbach’s α of the scale was 0.883. Therefore this scale can be considered reliable.

6.6.4 Virtual Experimental Procedure

The experimenter provided a brief overview of the procedure and instructions to the participants, which were also displayed on a smartboard within the virtual reality. The participants were required to perform a series of simple tasks within the MaxWhere virtual reality environment, which involved navigating through a relatively complex virtual space (as depicted in Fig. 6.1).

Fig. 6.1 Overview of the virtual environment showcasing a 3-floor building with glass and concrete walls and multiple digital displays for content

The tasks involved questions regarding the digital content and its arrangement. Upon initial entry into the virtual environment, participants found themselves on the top floor of a virtual building where they could access instructions and the web page containing the tasks and the IPQ. This main area was easily accessible by using the TAB key. SmartBoards located throughout the virtual environment displayed images from unsplash.com, depicting various objects and numbers. Additionally, two SmartBoards contained PDF documents.

After completing all experimental tasks, participants were instructed to proceed to the top floor of the building. There, they were asked to rate their navigational experience, share their familiarity with MaxWhere software, specify the amount of time they had spent with it, and mention if they regularly used any other 3D software. They also answered basic demographic questions and filled out the IPQ questionnaire. On average, the experiment lasted for fourteen minutes (SD: 4.7 min). The experiment was conducted in the Hungarian language.

6.7 Results Regarding the Subjective Navigational Experience

Given that previous research found a positive association between the naturalness of the controller and spatial presence [28], the present section shows an investigation [7, 8] of the connection between presence and navigational experience, which is closely related to the naturalness of the controller. Navigational experience is hypothesized to come along with a stronger sense of presence. This section provides evidence for this hypothesis.

On a ten-point Likert scale, the participants reported an average score of 7.04 ($SD = 2.19$) for their navigational experience. Means and standard deviations for the IPQ subscales are presented in Table 6.1. To explore the relationship between subjective navigational experience and presence scores, Pearson's correlation was employed. The analysis revealed a significant and positive correlation between the two variables, with a correlation coefficient of $r(29) = 0.419$ and $p = 0.019$. There was also a statistically significant positive correlation between the navigational experience score and the score of the spatial presence subscale, $r(29) = 0.497$, $p = 0.004$.

Table 6.1 Means and standard deviations of the results of the Igroup Presence Questionnaire (IPQ)

	Mean IPQ score ± SD
General presence	2.71 ± 1.77
Spatial presence	3.06 ± 1.29
Involvement	2.6 ± 1.29
Experienced realism	1.81 ± 1.05

There was a positive relationship between the subjective navigational experience and the sense of presence or spatial presence in the virtual environment. Participants who rated their navigational experience as more natural and automatic also reported a higher sense of spatial presence. This finding is consistent with a previous study by Skalski and colleagues [28], which demonstrated that natural mapping of controllers and interfaces in VR requires less effort to navigate and contributes to a more immersive experience for users.

6.8 Prior Experience

In this analysis, participants were categorized into two groups, based on their prior usage of MaxWhere VR: novices and experienced users. Novices were defined as individuals with less than one hour of experience, while those who had at least one hour of prior experience with MaxWhere VR were classified as experienced users.

6.8.1 Results of the Analysis

Table 6.2 displays the means and standard deviations for the navigation experience scale and IPQ subscales in two groups, namely nineteen people in the novice group and twelve in the experienced group. To compare the variable means between these groups, an independent sample t-test or Mann-Whitney U test was employed.

The score for navigational experience was 6.64 in the novice group and 7.67 in the experienced group. However, an independent sample t-test ($t(29) = -1.26, p =$

Table 6.2 Descriptive statistics of results on the Igroup Presence Questionnaire and the subjective navigational experience scale in the novice and experienced group

	Novice[a]	Experienced[b]
Navigational experience[c]	6.64 ± 2.44	7.67 ± 1.53
General presence[d]	2.4 ± 1.5	3.2 ± 2.0
Spatial presence[d]	2.67 ± 1.06	3.67 ± 1.34
Involvement[d]	2.45 ± 1.3	2.83 ± 1.2
Realism[d]	1.5 ± 0.9	2.32 ± 1.04

[a]Novice: None, or less than one hour of experience with MaxWhere VR prior to the experiment
[b] Experienced: At least one hour of experience with the MaxWhere virtual reality
[c]Navigational experience was measured on a 10-point Likert scale
[d]The general presence, spatial presence, involvement, and realism, were rated on a 7-point Likert scale from 0 (not at all) to 6 (very much)

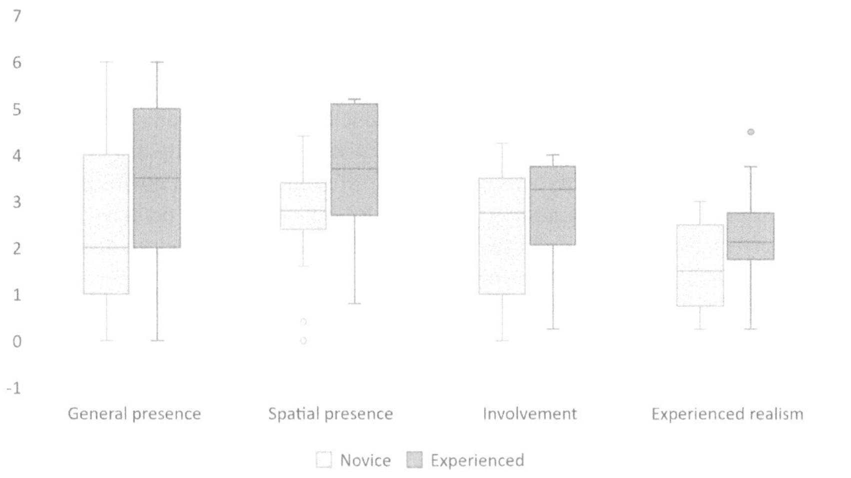

Fig. 6.2 Scores of presence factors among novice and experienced users. The difference was significant between the two groups in the case of spatial presence and experienced realism

0.217) showed no significant difference between the two groups. Thus, differences in sense of presence cannot be attributed to differences in navigational experience.

An independent sample t-test indicated that the general presence scores did not differ significantly between the novice (2.4) and the experienced (3.2) group ($t(29) = -1.37$, $p = 0.182$). The involvement scores (attention devoted to the environment.) did not differ between the novice group (2.447) and the experienced group (2.833), according to the Mann-Whitney U test ($U = 92.5$, $p = 0.392$). The level of general presence and involvement was similar in both groups.

In terms of spatial presence, the mean score was 2.67 in the novice group and 3.67 in the experienced group, and an independent sample t-test revealed a statistically significant difference ($t(29) = -2.21$, $p = 0.035$).

The experienced group had significantly higher experienced realism (2.323) than the novice group (1.5), according to an independent sample t-test ($t(29) = -2.23$, $p = 0.034$). A box plot depicting the mean scores of IPQ subscales in the novice and experienced groups can be found in Fig. 6.2. Therefore, novice users experienced less spatial presence than those with at least one hour of prior experience with the software and rated the virtual environment as less realistic than experienced participants.

6.8.2 *Role of Prior Experience*

The analysis revealed that there were no significant differences between the two groups in terms of their general sense of presence in virtual reality and their levels of involvement. The question used to measure the general presence was "In the computer-generated world, I had a sense of 'being there'." Additionally, the involvement scores were similar, indicating that all participants were equally focused and engaged in the virtual reality task.

Experienced users had significantly higher mean scores for spatial presence and experienced realism than novice users. In contrast to an earlier virtual reality study [39], where navigation ability was the primary difference between novice and experienced users, this study found no significant difference in users' perceived navigational performance between the two groups. However, since the present study did not include behavioral results or objective measures of navigational performance, it is possible that experienced users were more effective in interacting with the virtual environment but that the navigational experience scale could not detect these differences. For example, users who are new to a system may consider the navigation effective if they can reach a specific location. On the other hand, experienced users may deem it effective if they can arrive at the selected smartboard with just a double-click, without having to physically navigate through the room.

Novice users perceived the virtual environment to be less realistic compared to the evaluation of more experienced users. This could be due to their prior exposure to the same virtual reality, which sensitizes observers to the different features and characteristics of the medium, leading to different presence scores [40]. Similarly, the repeated use of the software could have sensitized and familiarized experienced users with the different features of the virtual environment, resulting in more realistic ratings. Additionally, novice users could compare the virtual experience with reality, while experienced participants may unconsciously consider their prior experiences with VR.

6.9 Sense of Presence in a Desktop VR Based on the Igroup Presence Questionnaire

This section shows the presence profile of the MaxWhere desktop virtual reality and compares it with other desktop virtual realities. This comparison extends the results of the previous sections.

Although presence is a subjective variable and not directly related to the technological aspects of virtual reality, it can provide valuable insight into the user's experience within a particular application or virtual environment. Subjective ratings can be used to plot the presence profile of an application, which highlights its strengths and provides an overview of the different presence components. In addition, the IPQ scores of MaxWhere VR were compared to two other desktop virtual reality studies with similar duration (up to 35 minutes) and presentation mode (monoscopic monitor), allowing for a more accurate understanding of the presence experience in MaxWhere VR [9].

6.9.1 Verifying the Experimental Environment

The presence results of the MaxWhere VR were compared with the online IPQ dataset [41]. Two studies meeting the criteria of nonclinical participants and applying monoscopic monitor were selected for analysis from the available online data, but their specific names were not provided in the description and were referred to as IPQ1 and IPQ2 studies in the following sections. The virtual experience duration in each study was limited to a maximum of 35 minutes, and longer sessions were excluded from the analysis. There were no statistically significant differences between MaxWhere (M = 15.7 min, SD = 6.7), IPQ1 (M = 18.6 min, SD = 8.3), and IPQ2 (M = 15.95 min, SD = 6.2) mean duration as determined by one-way ANOVA ($F(2, 59) = 0.696$, $p = 0.503$).

The presence scores in the MaxWhere virtual reality were compared to those of the IPQ1 and IPQ2 studies in the analysis. Descriptive analyses were conducted on the variables in the three groups, and the results are presented in Table 6.3. The reliability analysis of the MaxWhere data showed that all subscales of the IPQ had high reliability. All Cronbach's α coefficients for the subscales of the IPQ questionnaire were above 0.7 (spatial presence: 0.872, involvement: 0.729, experienced realism: 0.716).

A one-way analysis of variance (ANOVA) was conducted to compare the mean IPQ scores of MaxWhere with the corresponding data of the two selected studies from the IPQ dataset. The one-way ANOVA is a statistical test that compares the means of two or more independent groups to determine if there is significant evidence that the associated population means differ from each other. No statistically significant difference was found between the three studies in terms of general presence ($F(2, 59) = 2.192$, $p = 0.121$), spatial presence ($F(2, 59) = 1.557$, $p = 0.219$), involvement ($F(2, 59) = 0.125$, $p = 0.883$), or experienced realism ($F(2, 59) = 1.556$, $p = 0.219$). The means and standard deviations of each group are presented in Table 6.3.

Table 6.3 Means and standard deviations of the measured variables in the three compared studies: MaxWhere, IPQ1, and IPQ2

	MaxWhere	IPQ1	IPQ2
Age of participants (years)	20.52 ± 3.46	22.5 ± 6.03	22.05 ± 5.77
VR session length (min.)	15.73 ± 6.74	18.6 ± 8.3	15.95 ± 6.2
General presence[a]	2.71 ± 1.77	3.6 ± 2.07	2.19 ± 1.57
Spatial presence[a]	3.06 ± 1.29	2.98 ± 1.12	2.48 ± 1.07
Involvement[a]	2.6 ± 1.29	2.75 ± 0.75	2.75 ± 1.24
Realism[a]	1.81 ± 1.05	1.6 ± 0.54	2.1 ± 0.97

[a]The general presence, spatial presence, involvement, and realism, were rated on a 7-point Likert scale from 0 (not at all) to 6 (very much)

This suggests that within the given time frame of up to 35 minutes per VR session, the level of immersion provided by MaxWhere desktop VR is comparable to the other two desktop virtual realities.

6.9.2 Presence Profile

The presence profile shows the strengths and weaknesses of a VR, and it can provide an overview of the ratio of the different presence components. The presence profile of MaxWhere virtual reality (Fig. 6.3, plotted from the scores from Table 6.3) shows that the spatial presence score was the highest among the subscales. That means that the users had a relatively higher sense of being physically present in the VR. On the other hand, the experienced realism factor was the lowest (i.e., the users did not rate the VR very realistic).

Vividness and interactivity, are two essential dimensions according to Steuer [14]. MaxWhere virtual reality achieves its vividness through its visual design, which is suited for desktop VR. Its colors are vibrant, and the virtual spaces are thoughtfully designed based on their intended purposes. The learning spaces have a minimal and uncluttered design that aids students in concentrating on their learning materials. The exhibition spaces provide a muted backdrop for the displayed objects or create an ambiance that aligns with the exhibition's theme.

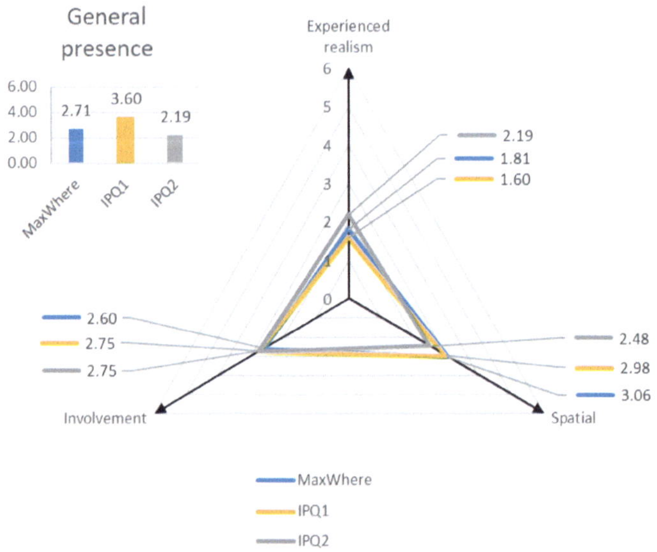

Fig. 6.3 Presence profile of MaxWhere and the two other analyzed desktop virtual reality

Interactivity has a more enhanced role in this virtual reality. The users can freely navigate in the virtual space, using the CogiNav technology [42]. Users can control the smartboards to load any type of web content, such as videos, images, or documents. Additionally, several web applications that enable interactivity are also supported. In certain virtual spaces within MaxWhere, users are provided with a control panel that enables them to modify the content displayed on the smartboards. This feature can be used to navigate through different chapters of learning materials, for instance.

6.10 Conclusions

The level of presence experienced within the MaxWhere virtual reality aligns with the average presence typically encountered in other desktop-based virtual reality environments. Analysis of the presence profile reveals relatively high general and spatial presence within this VR setting. However, the experienced realism scale demonstrates comparatively lower ratings due to the VR's focus on facilitating work and learning rather than high-fidelity immersive experiences.

The navigational experience in MaxWhere VR positively correlated with spatial presence. Participants who perceived navigation as more automatic and natural tended to experience a heightened sense of spatial presence. Furthermore, a significant outcome of this study was the distinction between novice and expert users. Experienced users reported higher levels of presence and realism. Nonetheless, both novice and expert users displayed similar levels of involvement and general presence, a factor particularly advantageous in educational contexts, as based on previous studies, presence significantly correlates with improved learning outcomes.

References

1. Baranyi, P., Csapó, Á.B., Berki, B.: Concepts of cognitive infocommunications. Infocommunications J. (2024). Accepted, in press
2. Horváth, I., Csapó, Á.B., Berki, B., Sudár, A., Baranyi, P.: Definition, background and research perspectives behind 'Cognitive Aspects of Virtual Reality' (cVR). Infocommunications J. **15**, no. Special Issue, 9–14 (2023)
3. Baranyi, P., Csapó, Á., Budai, T., Wersényi, G.: Introducing the concept of internet of digital reality – part I. Acta Polytech. Hungarica **18**(7), 225–240 (2021)
4. Wersényi, G., Csapó, Á., Budai, T., Baranyi, P.: Internet of digital reality: Infrastructuralbackground – part ii. Acta Polytech. Hungarica **18**(8), 91–104 (2021)
5. Baranyi, P., Csapo, A., Sallai, G.: Cognitive infocommunications (coginfocom). Springer (2015)
6. Baranyi, P., Csapó, Á.: Definition and synergies of cognitive infocommunications. Acta Polytechn. Hungarica **9**(1), 67–83 (2012)
7. Berki, B.: Sense of presence in maxwhere virtual reality. In: Baranyi, P. (ed.) Proceedings of the 10th IEEE International Conference on Cognitive Infocommunications: CogInfoCom 2019, pp. 91–94. IEEE (2019)

8. Berki, B.: Experiencing the sense of presence within an educational desktop virtual reality. Acta Polytech. Hungarica **17**(2), 255–265 (2020)
9. Berki, B.: Level of presence in MaxWhere virtual reality. In: Baranyi, P. (ed.) Proceedings of the 11th IEEE International Conference on Cognitive Infocommunications (CogInfoCom), pp. 511–545. IEEE (2020)
10. Felton, W.M., Jackson, R.E.: Presence: a review. Int. J. Human-Comput. Inter. **38**(1), 1–18 (2021)
11. James, W.: The Principles of Psychology, vol. I. Henry Holt and Co (1890)
12. Kelly, N.J.: Using interpretative phenomenological analysis to gain a qualitative understanding of presence in virtual reality. Virtual Real. (2022)
13. Slater, M., Wilbur, S.: A framework for immersive virtual environments (FIVE): speculations on the role of presence in virtual environments. Presence: Teleoper. & Virtual Env. **6**(6), 603–616 (1997)
14. Steuer, J.: Defining virtual reality: dimensions determining telepresence. J. Commun. **42**(4), 73–93 (1992)
15. Slater, M., Pertaub, D.-P., Steed, A.: Public speaking in virtual reality: facing an audience of avatars. IEEE Comput. Graph. Appl. **19**(2), 6–9 (1999)
16. Barreda-Ángeles, M., Hartmann, T.: Psychological benefits of using social virtual reality platforms during the covid-19 pandemic: the role of social and spatial presence. Comput. Hum. Behav. **127**, 107047 (2022)
17. Slater, M.: A note on presence terminology. Presence Connect **3**(3), 1–5 (2003)
18. Wilkinson, M., Brantley, S., Feng, J.: A mini review of presence and immersion in virtual reality. Proc. Hum. Factors Ergon. Soc. Ann. Meet. **65**(1), 1099–1103 (2021)
19. Witmer, B.G., Singer, M.J.: Measuring presence in virtual environments: A presence questionnaire. Presence: Teleoper. Virtual Env. **7**(3), 225–240 (1998). https://doi.org/10.1162/105474698565686
20. Schubert, T., Friedmann, F., Regenbrecht, H.: The experience of presence: factor analytic insights. Presence: Teleoper. & Virtual Environ. **10**(3), 266–281 (2001)
21. Thorp, S., Ree, A.S., Grassini, S.: Temporal development of sense of presence and cybersickness during an immersive VR experience. Multimodal Technol. Inter. **6**(5), 31 (2022)
22. Jicol, C., Wan, C.H., Doling, B., Illingworth, C.H., Yoon, J., Headey, C., Lutteroth, C., Proulx, M.J., Petrini, K., O'Neill, E.: Effects of emotion and agency on presence in virtual reality. In: Proceedings of the 2021 CHI Conference on Human Factors in Computing Systems. ACM (2021)
23. Felnhofer, A., Kothgassner, O., Beutl, L., Hlavacs, H., Kryspin-Exner, I.: Is virtual reality made for men only? exploring gender differences in the sense of presence. In: International Society for Presence Research Annual Conference ? ISPR 2012, Philadelphia, Pennsylvania, USA (2012), we earned the BEST PAPER AWARD! http://eprints.cs.univie.ac.at/3557/
24. Grassini, S., Laumann, K., Luzi, A.K.: Association of individual factors with simulator sickness and sense of presence in virtual reality mediated by head-mounted displays (HMDs). Multimodal Technol. Inter. **5**(3) (2021). https://www.mdpi.com/2414-4088/5/3/7
25. Slater, M., Usoh, M., Steed, A.: Depth of Presence in Virtual Environments. Presence: Teleoper. Virtual Env. **3**(2), 130–144 (1994). https://doi.org/10.1162/pres.1994.3.2.130
26. Sacau, A., Laarni, J., Hartmann, T.: Influence of individual factors on presence. Comput. Hum. Behav. **24**(5), 2255–2273 (2008)
27. Ling, Y., Nefs, H.T., Brinkman, W.-P., Qu, C., Heynderickx, I.: The relationship between individual characteristics and experienced presence. Comput. Hum. Behav. **29**(4), 1519–1530 (2013)
28. Skalski, P., Tamborini, R., Shelton, A., Buncher, M., Lindmark, P.: Mapping the road to fun: natural video game controllers, presence, and game enjoyment. New Media & Soc. **13**(2), 224–242 (2011)
29. Krassmann, A.L., Melo, M., Pinto, D., Peixoto, B., Bessa, M., Bercht, M.: What is the relationship between the sense of presence and learning in virtual reality? a 24-year systematic literature review. Presence: Virtual Augment. Real. **28**, 247–265 (2019)

30. Mikropoulos, T.A., Bellou, J.: The unique features of educational virtual environments. In: IsaíÂas, P., McPherson, M., Bannister, F. (eds.) Proceedings of the IADIS International Conference on e-Society, pp. 122–128. IADIS Press, Dublin, Ireland (2006)
31. Selzer, M.N., Gazcon, N.F., Larrea, M.L.: Effects of virtual presence and learning outcome using low-end virtual reality systems. Displays **59**, 9–15 (2019). http://www.sciencedirect.com/science/article/pii/S0141938218300398
32. Mikropoulos, T.A., Strouboulis, V.: Factors that influence presence in educational virtual environments. CyberPsychol. & Behav. **7**(5), 582–591 (2004)
33. van Baren, J., IJsselsteijn, W.: Measuring Presence: A Guide to CurrentMeasurement Approaches. OmniPres project forThe International Society for Presence Research (ISPR) and Eindhoven Universityof Technology (2004)
34. Hein, D., Mai, C., Hußmann, H.: The usage of presence measurements in research: a review. In: Proceedings of the 17th of the International Society for Presence Research (ISPR). The International Society forPresence Research, Prague (2018)
35. Grassini, S., Laumann, K.: Questionnaire measures and physiological correlates of presence: a systematic review. Front. Psychol. **11** (2020)
36. Slater, M., Linakis, V., Usoh, M., Kooper, R.: Immersion, presence and performance in virtual environments. In: Proceedings of the ACM Symposium on Virtual Reality Software and Technology - VRST '96, vol. 163, p. 72. ACM Press (1996)
37. Schubert, T., Friedmann, F., Regenbrecht, H.: Embodied presence in virtual environments. In: Visual Representations and Interpretations, pp. 269–278. Springer (1999)
38. Schuemie, M.J., van der Straaten, P., Krijn, M., van der Mast, C.A.P.G.: Research on presence in virtual reality: a survey. CyberPsychol. & Behav. **4**(2), 183–201 (2001)
39. Greenhalgh, C., Benford, S.: MASSIVE: a collaborative virtual environment for teleconferencing. ACM Trans. Comput.-Hum. Inter. (TOCHI) **2**(3), 239–261 (1995)
40. Freeman, J., Avons, S.E., Pearson, D.E., IJsselsteijn, W.A.: Effects of sensory information and prior experience on direct subjective ratings of presence. Presence: Teleoper. & Virtual Env. **8**(1), 1–13 (1999)
41. Igroup Presence Questionnaire (IPQ) database. http://www.igroup.org/pq/ipq/data.php. http://www.igroup.org/pq/ipq/data.php
42. Baranyi, P., Galambos, P., Csapó, Á., Jaloveczki, L.: Cognitive navigation and manipulation (coginav) method. US Patent App. 15/658,579. Accessed 1 Feb 2018

Chapter 7
Memory Performance in Desktop VR Versus 2D Browser

Abstract This chapter explores the potential of desktop virtual reality in enhancing memory and learning, along with an examination of advertising evolution and online formats. Two studies, with a total of fifty participants, aimed to compare memory retention of advertisements in 3D highlighted and embedded formats, in both 2D browser and desktop VR environments. The findings suggest a distinct impact of 3D highlighting in VR contexts on memory retention compared to traditional 2D settings. This aspect holds significant importance in educational contexts, particularly as traditional e-learning is gradually transitioning towards VR-based education. In virtual educational settings, instructors can continuously present essential diagrams, making necessary information more accessible and promoting improved comprehension among learners.

Keywords Virtual reality · Advertisement · Memory · Browser

7.1 Introduction

This chapter discusses desktop virtual reality's potential to improve memory and learning, and it is based on the following publications: [1–3]. The emergence of desktop virtual reality (VR) has opened up new opportunities for enhancing memory and learning experiences, especially in the field of advertising. With the continuous evolution of technology, the modalities through which we interact with information and promotional material also transform [4–11]. This chapter focuses on the intersection of virtual reality, advertising effectiveness, and memory retention.

First, Sect. 7.2 discusses historical memory aids and explores the potential of virtual environments, especially virtual reality, supported by studies on their effectiveness in enhancing memory and learning, offering advancements beyond conventional mnemonic techniques. The following Sect. 7.3 provides an in-depth exploration of the historical evolution and contemporary landscape of advertising, with a particular focus on online advertising, digital ad formats, and the influence of visual elements. It further delves into the potential of VR-based advertising, discussing its efficacy,

I. Horváth et al., *Cognitive Aspects of Virtual Reality*, Studies in Big Data 156,
https://doi.org/10.1007/978-3-031-68130-1_7

influence on emotional engagement, and the role of presence in virtual reality, highlighting the advantages of 3D visualization and comparing traditional 2D web-based advertisements with VR-based counterparts. Then the objective of the current study is formulated in Sect. 7.4, which compares the efficacy of advertising content between virtual reality (VR) and traditional web-based formats using a between-subject design to assess participant recall performance. The first study (Sect. 7.5) contrasted memory retention in two VR-based groups: one exposed to a highlighted ad within the 3D environment, and the other group encountering the ad embedded within a webpage, both situated in the VR space. It found significantly better recall in the group exposed to the highlighted ad, emphasizing the efficacy of direct VR ad integration for memory retention.

The second study, presented in Sect. 7.6, aimed to compare the impact of highlighting information between 3D VR and 2D browser environments regarding memory retention of highlighted versus embedded advertisements. The results indicated a significant difference in recall proportions between highlighted and embedded images within VR but not within the 2D browser, suggesting a potentially unique effect of 3D highlighting on memory retention.

7.2 Memory Performance

The present section discusses historical memory aids and explores the potential of virtual environments, especially virtual reality, supported by studies on their effectiveness in enhancing memory and learning, offering advancements beyond conventional mnemonic techniques.

Memory represents a fundamental cognitive function in our daily lives. It allows us to employ previously encoded information in various mental operations, forming the very foundation of our thought processes. Memory allows the recollection of diverse facts, skills, and habits acquired in the past. Moreover, it enables us to learn from our previous experiences. The three stages of human memory encompass encoding, storing, and retrieving. Encoding involves entering information into the human memory system, storing pertains to upholding the information, and retrieving signifies the ability to locate and recall [12]. These stages adhere to a precise sequence, as the successful retrieval of specific information necessitates proper encoding and effective storage of the information.

The desire to expand human memory capacity is not a recent endeavor. Even in ancient times, cave paintings served a similar purpose. These depictions encoded stories and information, effectively storing them within the environment, thereby enhancing information retrieval. Subsequently, advancements like paper, printing, and audio or video recording further extended human capabilities. The trajectory of progress may next lead to virtual environments, which can serve not only as a medium of communication but also as tools for memory management Furthermore, they might serve as a foundation for the "algorithmification" of human learning. For instance, virtual reality (VR) could provide a platform for the classical virtual memory palace method [13]. This method is a well-known mnemonic strategy, that serves to assist the recall of information in the desired order. For creating a memory palace, first, the items (the concepts that are aimed to be remembered) should be placed in the imagined palace. During the recall, the person has to mentally walk through the memory palace, and by looking at the salient cues they can recall the items. The method is based on the spatial organization of thought and concept by adding mental visualization to each of them. Displaying a memory palace in virtual reality is a great visualization that can be shared, or saved for the long term. Another advantage, that is provides constant visual feedback [13]. Several studies examined the effectiveness and optimal characteristics of virtual memory palaces [14–18], and the behaviors of the different generations [19–21]. The degree of immersion serves as another significant factor within virtual memory palaces. As an instance, a study undertook a comparison between immersive (HMD) and desktop VR employing a within-subject design. Their results demonstrated that the HMD condition yielded enhanced memory retrieval in contrast to the desktop condition. Moreover, participants also expressed an elevated level of engagement with the task in the HMD condition [17].

Human visual perception serves as the foundation for visual memory. Within a complex environment, the process of scanning a visual scene may appear seamless and continuous; however, it relies on a sequence of swift and discrete eye movements. These brief glances are integrated into cohesive representations within the visual short-term memory. This involves linking various perceptual attributes of objects and anchoring them to a spatial framework. The visual short-term memory preserves this data temporarily. In terms of semantic coding, the significance of the item is processed, establishing connections to other information within the long-term memory [12].

When an abstract concept is presented, it is verbally stored. In contrast, when a concrete concept is depicted as an image, it is encoded into both verbal and visual systems. This framework corresponds to the dual-coding theory. This theory forms the basis for the picture superiority effect, which is the phenomenon wherein images tend to be better recalled than words. The advantage of images over words applies to both the coding and retrieval stages [22]. Images are more readily encoded, capitalizing on human memory's susceptibility to symbolically presented information [23]. Another interpretation of the picture superiority effect lies in the greater familiarity and prevalence of pictorial objects in our daily lives [24].

The relationship between distinct levels of presence and memory performance appears to be complex. When the virtual experience incorporates a broader spectrum of modalities such as tactile, olfactory, auditory, and visual elements, participants report a higher sense of presence, coupled with heightened memory performance concerning the entities within the environment. However, the heightened degree of visual details does not appear to influence either the sense of presence or the efficacy of memory performance [25]. Similarly, the fidelity of the simulation did not impact the subjective sense of presence or the memory performance [26]. Conversely, in the context of tasks involving procedural memorization, an elevated level of immersion distinctly enhances performance [27, 28]. Moreover, these learned procedures seamlessly transferred to the real world [27].

7.3 Theoretical Background of Online Advertising

The current section provides an in-depth exploration of the historical evolution and contemporary landscape of advertising, with a particular focus on online advertising, digital ad formats, and the influence of visual elements. It further presents the potential of VR-based advertising, discussing its efficacy, influence on emotional engagement, and the role of presence in virtual reality, highlighting the advantages of 3D visualization and comparing traditional 2D web-based advertisements with VR-based counterparts.

7.3.1 Web-Based Advertising

Throughout history, advertisements have appeared on various platforms in order to connect with the general public and generate interest in the goods being promoted. From sales narratives written on Egyptian papyrus to the visual-slogan approach of the 19th century, advertising has undergone various transformations to identify the most effective mechanisms of impact on consumers. In the current era, the focus within the advertising sector is on television advertising and online promotional activity. With the growing attraction to virtual realities, advertisers are increasingly turning their attention to this innovative medium to exploit its potential.

The global internet advertising revenue experienced a substantial increase, nearly doubling from 2018 to 2022, and is anticipated to expand by more than 2.5 times by the year 2027 [29]. This expansion is largely attributed to the widespread adoption of digital technologies. Also, mobile devices have become more common, alongside the targeted advertising of social media platforms. Search engine advertising remains influential, securing visibility through paid search ads. Native advertising seamlessly integrates with platform content, elevating user engagement across websites and social media. Data-driven insights and personalized ads are pivotal, leveraging analytics to comprehend consumer behavior. In addition, banner ads and popup ads remain prevalent across various digital platforms.

The advertising industry widely uses the picture superiority effect, mentioned earlier (Sect. 7.2). Images are foundational components of advertisements. Visual messages were more effective for less motivated audiences and for consumers with a lower ability to semantically process information [30]. Another advantage emerges in terms of long-term memory impact, as visual messages necessitate comparatively lesser exposure than their verbal counterparts [31]. The sequence of information processing further contributes: visual elements typically take precedence over verbal content in terms of viewing. Thus, the consumers can have expectations about the verbal component primarily [32].

Web-based advertising derives from printed advertisements, thus in terms of arrangement and graphic design, several similarities can be found. Besides these similarities, the new media allows other digital elements besides text and still images, such as animation, video clips, sounds, and hyperlinks. All these enhancement has one major role, to capture the consumers' attention and direct it to the advertised product or service. A prevalent variant of digital advertisements is banner ads, in which advertisers compensate for the allocation of space across single or multiple web pages to showcase a stationary or hyperlinked banner or logo [33]. A multitude of investigations have been conducted to examine these advertisements [34–37]. Distinct consumer perspectives towards diverse ad formats have been observed; nevertheless, it appears that banner ads are endowed with the highest information assimilation propensity, coupled with an elevated comprehensive favorable disposition [34].

The effectiveness of an advertisement is also based on the credibility of the medium and the alignment of its content with the web page. With the widespread appearance of e-commerce, consumers lost the possibility to make decisions based on first-hand, physical experiences. This elevated the role of credibility in online marketing. As online advertisements are mostly not a standalone content type, they are mostly embedded in non-advertising content; thus the reliability of the advertiser and the website are linked. Advertisements placed on credible sites produce better advertising outcomes. Consumers find more valuable the product if it matches the website content [38]. An interesting remark is that the positive sentiment is not toward the advertisement, but for the brand and the intention to make a purchase [39].

Traditional web-based advertisements exhibit certain common attributes, such as embeddedness, a trait that can be construed as persistently disrupting not only the general content but also content that is thematically linked. These attributes present a potential drawback, as they can irritate users and trigger ad avoidance behaviors. Irritation constitutes a notable factor contributing to the effectiveness of advertising efforts. Advertisements can elicit substantial resistance from users, given their propensity to employ tactics of blocking or disregarding. When individuals are engrossed in accomplishing specific objectives, the emergence of advertisements is commonly construed as an intrusive disruption, particularly by those users who possess an understanding of online persuasive strategies such as larger dimensions and animations [40]. Generally occupying peripheral positions, banner ads do not directly impede the engagement of web visitors. Conversely, pop-up ads emerge as one of the most vexatious forms of web-based advertising. According to the definition by the Interactive Advertising Bureau, pop-up ads, or formally interstitials, manifest as advertisements that materialize within a distinct window automatically superimposed upon an existing content window [33]. Thus users are forced to react to them. In response to this disruption, individuals can exhibit cognitive, affective, or behavioral reactions, which can hold both positive and negative implications for advertisers. A favorable outcome might involve the unexpected interruption fostering heightened cognitive processing and enhanced memory retention for the pop-up advertisement [38]. Conversely, interstitials tend to instigate unfavorable attitude formation, consequently paving the way for ad avoidance tendencies [41].

Determining why some advertisements irritate consumers while others do not center around the concept of intrusiveness, which refers to interference with editorial content. Paradoxically, this notion seems somewhat contradictory, since the primary purpose of advertising is to attract attention and maximize exposure; therefore, disruption of editorial content should seemingly be inherent in all advertising. However, if the advertisements are closely aligned with the user's ongoing task, participants perceived the advertisements to be less intrusive [41]. The emergence of negative attitudes and reactance can consequently trigger ad avoidance behaviors and evoke sentiments of irritation. Reactance theory, based on social psychology, defines human behavior as a reaction to perceived restrictions on freedom within the environment. When an individual feels that their options or freedom are being restricted, reactance serves as a mechanism to restore the appearance of control and the perception of freedom [42]. This phenomenon of reactance extends to the domain of advertising as well. Consumers actively interpret advertisements and respond to the persuasive threats they represent. In line with this, advertisements that relate harmoniously to the web user's task are perceived as less obtrusive. In parallel, an ad's value is augmented (whether through increased informational content or entertainment), leading to a reduced sense of intrusiveness in its evaluation.

7.3.2 VR-Based Advertising

As previously mentioned, marketers often invent fresh approaches and innovative platforms to convey their commercial messages to potential customers. Virtual reality is a rapidly expanding field where advertisers are striving to discover more efficient methods of advertising. For example, a virtual experience can allow customers to get some kind of first-hand information, before the actual acquisition [43]. In a VR-based shop, compared to a traditional web shop, the user can see a 3D model of products, and can easily compare their size for example. Furthermore, in a virtual store, avatars can help the customers in the shopping experience by providing information on the products and answering the questions that may arise [44]. Also, destination marketing uses virtual reality as a marketing tour. In this case, VR experiences try to promote the real-life visits of these destinations, by impressive visual imagery [45]. In VR-based advertising, the fundamental success factors are the value for customers, community management, and highly interactive advantages [46].

Informational advertising refers to promotional efforts that primarily focus on conveying factual details, features, benefits, or informative content about a product, service, or brand to the target audience. The main objective of informational advertising is to provide consumers with valuable information that aids in making informed purchasing decisions. In the context of informational advertising, the utilization of 3D representations of products within a virtual environment holds considerable persuasive potential. Although, there are some differences in terms of product type. The spatial presentation of geometric products, e.g. products whose spatial layout is their main distinctive feature, is more beneficial than presenting material products, such as clothes [47, 48]. Naturally, individual differences also play a role in these cases. There are customers, who have a high need for touch, to decide on the purchase. For them, due to the missing tactile experience, the 3D version of a product is less impressive, than for customers with different preferences [47]. The 3D display also enhances the feeling of presence which influences consumer comprehension and decision-making. This greater sense of presence is irrespective of the type of product (geometric or material) [49].

A distinct type of advertising is transformational advertisements, which prioritize the experiential aspect of using a product or service. The focus is on the feelings of the person, who uses the product. It influences both the attitude toward the advertisement and the brand. The 3D immersive transformational advertisements are more efficient than their 2D counterparts [49]. Also in destination marketing, VR-based ads were found to evoke more emotional responses toward the featured destination and were perceived as more engaging [45].

The exposure effect is a psychological phenomenon wherein individuals develop a preference or positive attitude toward something merely due to its repeated exposure. In essence, the more people are exposed to a stimulus, whether it's an object, person, or idea, the more familiar and comfortable they become with it, leading to a perceived liking or positive evaluation of that stimulus. This effect highlights the subconscious influence of familiarity on human preferences and behaviors. Put simply, individuals tend to interpret regular exposure to an advertisement or product as indicative of their liking for that product [50]. Another advantage of product display is that it can be used to enhance the realism of virtual environments. Accordingly, in video games, the prevalence of actual product advertisements exceeds that of fictional products, as this addition contributes to an enhanced perception of presence in an alternate reality [51]. An additional merit of product placement is its unobtrusiveness, which diminishes the likelihood of media users developing the aforementioned psychological reactance in response to the advertisement [38]. In contrast, when comparing product placement and billboard placement in virtual reality, the recall of objects and brands of products presented on billboards was found to be more significant than for product placement [52].

Presence refers to the subjective feeling of the individual that they exist in the virtual environment presented. This sense of presence can potentially be a mediating factor between the format of the advertisement (2D or 3D) and its effectiveness [47, 53]. In addition, the experience of presence and enhanced immersion may also contribute to evoking the emotional engagement of users [52].

The efficacy of an advertisement is also influenced by the type of virtual reality (VR) device employed. Devices such as head-mounted displays or CAVE systems contribute to a heightened sense of presence due to their immersive nature compared to desktop monitors. However, this increased presence does not necessarily translate to greater advertising effectiveness in VR environments. Curiously, studies have demonstrated improved memory retention and more effective brand advertising outcomes when utilizing VR displayed on desktop monitors. Notably, an intriguing finding from this research indicates that, within immersive systems, the positioning of an advertisement holds greater significance than its actual size [54].

In summary, prior research has predominantly centered around investigating the effects of 3D visualization in advertisements [43, 47, 48], or the utilization of virtual prototypes [51, 52]. However, there has been limited focus on comparing the outcomes of traditional web-based 2D advertisements with those based on VR technology.

7.4 Study Objectives

The primary objective of this investigation was to assess the efficacy of advertising content presented in two distinct contexts: within a virtual reality (VR) environment and within the traditional web-based format. It was hypothesized that advertisements integrated into virtual spaces would exhibit a higher degree of effectiveness when compared to those displayed on conventional websites. The criterion for evaluating effectiveness was the recall performance of participants, specifically, the number of individuals capable of recalling the advertisements encountered.

To test this hypothesis, a between-subject design was implemented, resulting in the division of participants into two groups. This division enabled a simultaneous evaluation of advertising effectiveness in both the VR and classic web-based settings. The hypothesis posited that a larger number of individuals within the VR group would recall encountering the advertisements in comparison to the other group. This first comparison was entirely in virtual reality, the classic web-based setting was also implemented in a virtual environment to avoid the novelty effect of VR. Then these two settings were implemented in a 2D browser environment with the other two groups, to verify the result of the VR-based study.

7.5 VR-Based Study with Banner Ads and Incidental Encoding

This section provides a detailed overview of a virtual reality study that used banner ads to evaluate memory performance through incidental encoding. It includes information on the experiment's objective and design, the participants involved, and the results, which are presented in the final subsection.

The objective of this investigation was to assess human memory performance within a 3D virtual environment. To enhance ecological validity, instead of conducting a controlled laboratory memory test, a real-life scenario with incidental encoding was employed, involving the reading of online articles and subsequent inquiries regarding the advertisements encountered. It is hypothesized that participants' memory performance will exhibit improvement when exposed to target images presented directly within the 3D virtual environment. To evaluate this hypothesis, the same advertisement was integrated both directly into a 3D VR space and within articles in the form of classic banner ads. Banner ads were selected due to their favorable reception compared to various other online advertising methods [34].

In addition, to reduce the repetition effect, a between-subjects design was used, whereby participants were divided into two different groups. One group performed the task with an advertisement highlighted in the 3D space, while no advertisements appeared within the articles (hereafter "VR group"). The second group, on the other hand, encountered the memory test in the form of banner ads integrated into the web pages containing the articles (hereafter "classic web group").

Both groups executed the task within the MaxWhere VR environment.[1] Consequently, any perceived distinctions between the groups were not attributed to the novelty of the software. By eliminating the novelty effect, a direct comparison of memory performance could be conducted, contrasting a webpage banner ad with an advertisement seamlessly incorporated within the 3D virtual realm. Following the hypothesis, it is anticipated that a greater number of participants from the VR group will retain the memory of the advertisement compared to the classic web-based group. This anticipation stems from the belief that heightened attention and a stronger sense of presence contribute to enhanced memory retention [55]. The VR group stands to benefit from this phenomenon, as the advertisement is directly immersed within the virtual space, fostering a heightened sense of presence, as opposed to being embedded within another webpage contained within the VR environment.

7.5.1 Virtual Reality Applied in the Study

For the experiment, the CogInfoCom17 space within MaxWhere (version 1.4.x) was employed. This space is primarily intended for academic lectures, coursework, and presentations. The experiment took place within three key locations within this virtual environment: the central area, the original personal corner, and the information desk, which served as the interface for the presentation of the experimental content.

A table with five smartboards occupied a central position in the virtual space. These smartboards displayed the four articles. The ad on the fifth smartboard was only visible to the VR group. It is important to note that for this group, the advertisements were not integrated into the articles (see Fig. 7.1 for a visual representation). In contrast, within the classic web group, participants saw the advertisements only as banner ads for the articles and were not available to them on a separate board in the space (as depicted in Fig. 7.2).

To facilitate the distribution of these layouts to the participants, pack files, serving as the archive file format for exporting content within the MaxWhere environment, were utilized. These pack files contained duplicates or references to the pinned

[1] www.maxwhere.com.

Fig. 7.1 Experimental configuration for the VR-highlighted group: Central smartboard featuring advertisement surrounded by four articles

Fig. 7.2 Experimental setup for the VR-embedded group: Four articles with embedded advertisements, leaving the fifth smartboard blank

content, along with precise spatial coordinates within the 3D environment. The participants navigated in the virtual environment by mouse [56].

7.5.2 Participants

A total of 22 individuals, participated in this study. Their ages ranged from 19 to 28 years (mean = 23.45 years, SD: 2.26), with 18 being male and 4 being female. All participants were native Hungarian speakers, and the entire experiment was conducted in Hungarian language. Each participant was familiar with the MaxWhere software and did not encounter any issues while using it. The study employed a between-subjects design, with ten participants viewing the image as a webpage banner ad and twelve participants observing it directly within the VR environment.

7.5.3 Experimental Procedure

To compare participant responses across different conditions, a between-subjects experimental design was employed. Participants were randomly assigned to one of two groups: the classic web-based advertisement group or the VR advertisement group.

In the classic web-based advertisement condition, participants viewed four online articles within the central area of the virtual space. Each of these articles featured a banner ad positioned on the right side of the page, and the same advertisement was presented in all four articles (refer to Fig. 7.2). In both conditions, the total size of the advertisements was kept consistent.

In contrast, participants in the VR advertisement group encountered the same four articles with an identical spatial layout, but without the banner ads. Instead, the advertisements were displayed on a separate webtable alongside the four articles (see Fig. 7.1).

During the experiment, participants completed an online questionnaire integrated into the virtual reality environment. This questionnaire was uniform for all participants, irrespective of their assigned group. It began with a brief task description, omitting any reference to the study's purpose. Participants indicated their willingness to participate with a simple click. Subsequently, they answered general questions about the content of the four articles.

Following this, participants were instructed to import another content pack, replacing all previous materials except for the questionnaire with a masking image. This step ensured that participants could not view the articles or the advertisements during the subsequent part of the study. To verify compliance, participants were asked a question about the imported image. They were then queried about whether they recalled any advertisements in the virtual space. If they answered affirmatively, they were prompted to provide as many details as possible, including the brand, product, colors, and slogan of the advertisement. Finally, participants completed some basic demographic questions.

7.5.4 Outcomes of the Virtual Reality-Based Experimental Study

The primary objective of this experiment centered on evaluating the comparative effectiveness of advertisements presented within a 3D virtual environment, either through direct or indirect placement. Advertisements' efficacy was determined by assessing the participants' memory performance in a free recall examination.

The initial phase of the analysis involved examining participants' responses to the provided articles. To mitigate the influence of random completions, especially from individuals who might not have thoroughly engaged with the papers, data from participants providing incorrect responses were excluded from subsequent analysis. Consequently, the results of 22 individuals are presented from the original pool of 27 participants.

The hypothesis posited that a greater number of participants in the VR group would recollect the advertisement, given that they had encountered it within the VR space rather than in the articles. The memory performance of the two groups was compared using Fisher's Exact Test. The null hypothesis for this test posited that the proportion of participants remembering the advertisement in both groups was equivalent. With a significance threshold of 0.05, the goal was to determine whether any observed disparity in proportions was statistically significant. The findings demonstrated a noteworthy distinction in memory performance between the two groups. Specifically, 91.67% of participants (11 out of 12) in the VR group recalled the advertisement, in contrast to 40% (4 out of 10 participants) in the conventional web-based group ($p=0.02$). This outcome suggests that the null hypothesis can be rejected, indicating a difference in the proportions of individuals who recollect the advertisement in the two groups (see Fig. 7.3).

The free recall test further revealed that when an individual remembered the advertisement, they predominantly managed to recall both the promoted product and the associated brand. While attempts were made to remember the advertisement's colors, participants mainly linked the brand's color to the product. In this comparative analysis, no statistically significant difference in performance was observed between the VR-based and web-based groups (Table 7.1).

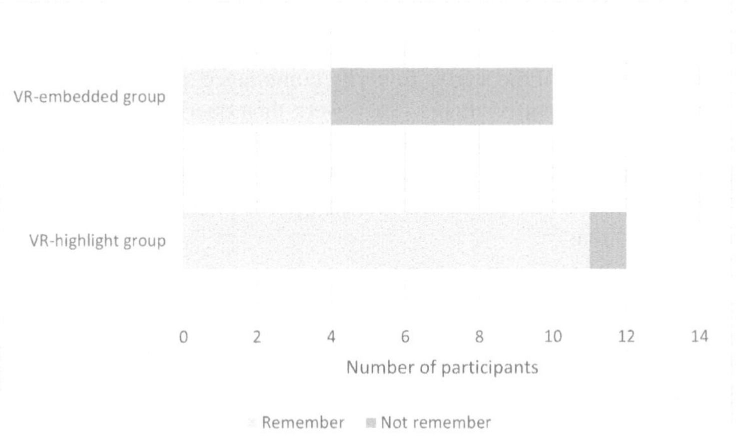

Fig. 7.3 Proportion of the remembering participants between the two virtual reality groups, error bars represent the standard error

Table 7.1 Outcome of the free recall test and the corresponding p-value from Fisher's Exact Test for participants recalling in both groups

	VR-highlight group (N = 11)	VR-embedded group (N = 4)	Fisher's Exact Test (p)
Product	10	4	1
Brand	9	3	1
Colors of the ad	4	0	0.2747
Color of product	0	0	1
Slogan	1	0	1

7.6 Comparing the Effect of Highlighting Information in 3D Virtual Reality and 2D Browser

This section presents the details of another study, which extends the results presented in the previous one. The previous study compared two groups in virtual reality, while this new study expands on the results by including two additional groups that performed the same tasks on a 2D interface using a web browser.

In this comparison, an assessment is made to determine whether the information highlighting effect observed in desktop virtual reality can be replicated in a 2D web browser. Alternatively, an investigation is conducted to ascertain whether the previously presented result, where a greater number of participants remembered highlighted images, is indicative of a unique feature of 3D virtual environments. Additionally, a comparison is made between 3D highlighting and the widely adopted 2D embedding method.

7.6.1 Procedure of the Comparative Study

For this comparative study, the same set of articles and target images used in the study presented in Sect. 7.5.3 were accessed in a traditional web browser, each in a separate tab.

Participants were randomly divided into two groups. The first group encountered the ad as a banner advertisement adjacent to all four articles, integrated within the article's content. The second group was exposed to a single large advertisement on a separate tab, with its dimensions equivalent to that of the four banner ads combined. This condition was designed to mimic the directly inserted advertisement experience in virtual reality.

Participants had to answer identical questions about the articles. As in the VR study, participants were instructed to close the article tabs and subsequently report whether they had encountered any advertisements while responding to the questions. If they recalled encountering an advertisement, they were required to provide a detailed description of the advertisement.

7.6.2 Participants

In this study, a comparison was conducted among four distinct groups, comprising two browser-based groups and two from the earlier study presented in Sect. 7.5.2. In total, fifty individuals participated in the experiments. To meet the inclusion criteria for the final analysis, participants were required to answer the article-related questions accurately. Consequently, the final analysis encompassed the involvement of 39 individuals, with an average age of 24.7 years, including 12 women.

7.6.3 Highlighting Information in a 2D Browser

To investigate the impact of information highlighting, the proportion of participants who recalled the target image (advertisement) between two groups in the browser context was compared: those for whom the target image was embedded within a browser and those for whom the target image was observed as highlighted within the browser. This comparison employed Fisher's exact test, a non-parametric statistical significance test known for its insensitivity to sample size. The null hypothesis stated that the proportion of participants remembering the target image was the same in both groups.

According to the results of the statistical test, no statistically significant difference was found in the recall proportions between the two groups ($p = 0.153$). In the highlighted image group, the image was remembered by 77.8% (7 out of 9) of participants, while in the embedded image group, it was recalled by 37.5% (3 out of 8). Consequently, although a higher proportion of participants remembered the image when it was highlighted, this difference was not statistically significant.

In contrast, when examining the highlighted and embedded groups within the virtual reality context, a statistically significant difference was observed (see Fig. 7.4). In summary, while the results obtained in virtual reality could not be fully replicated, a noticeable trend was observed in the proportion of participants recalling the highlighted image.

7.6.4 Information Highlighting in Desktop VR and 2D Browser

In desktop VR, a notable advantage is the direct insertion of digital elements, such as images, into the virtual environment, enabling users to view and walk around these elements from various angles. Conversely, in 2D browsers, the conventional approach involves embedding images or advertisements into webpages, prompting an investigation to compare these two methods. Comparing recall rates between the VR group, where images were directly inserted, and the browser group, with banner

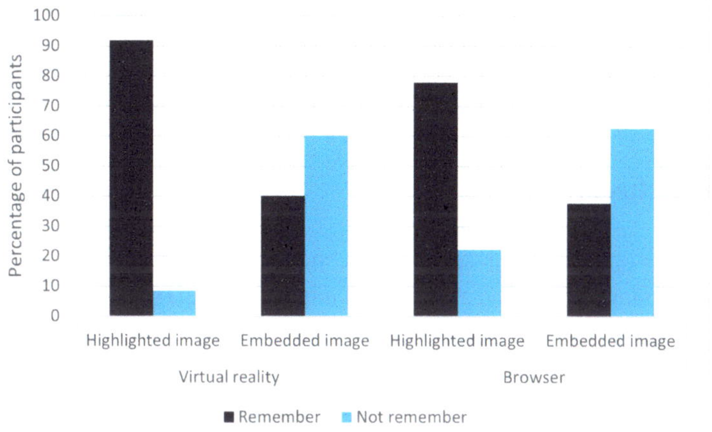

Fig. 7.4 Percentage of participants remembering the target image in the four groups

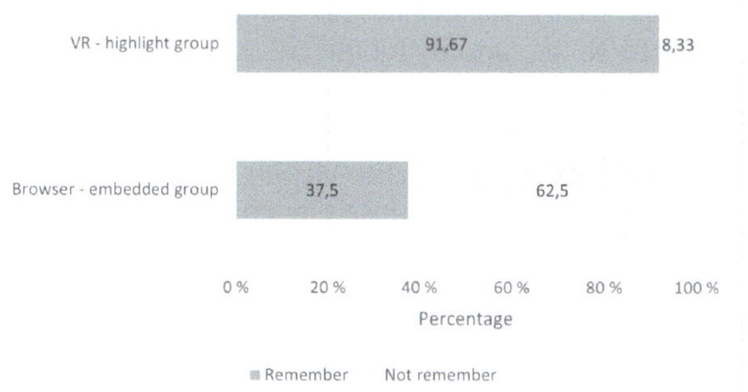

Fig. 7.5 Percentage of participants remembering the target image. The upper bar represents the group who saw the target image highlighted in virtual reality, and the second is those who saw the image embedded in a browser

ads, showed significantly higher recall (92% vs. 38%) in the VR group, substantiated by Fisher's Exact Test (Fig. 7.5). This substantial difference isn't solely due to VR novelty; it's attributed to active coding [57], allowing users in VR to actively explore the content from multiple viewpoints, fostering stronger memory retention. Images directly integrated into the virtual space were more memorable due to this active exploration and encoding. However, images inserted as classic banner ads in articles were less memorable, possibly due to users' tendency to avoid or disregard ads in everyday browsing. This discrepancy suggests that direct ads inserted directly into the virtual environment are more likely to be recalled.

Although the study has identified a distinction in memory retention, it has not directly measured the cause of this difference. Therefore, further investigations are necessary to pinpoint the exact reason behind the dissimilarity in memory retention. It could be due to the novelty of the placement, the way the area is distributed, or the virtual reality experience itself.

7.7 Conclusions

Based on the statistical analysis outcomes, the observed highlighting effect in the virtual environment exhibited a trend-level impact in the 2D browser. Notably, there was no statistically significant distinction indicating superior recall between an image displayed on a separate tab compared to an embedded one within the 2D browser context.

When comparing the most typical layouts for a given interface, there was a statistically significant difference between the virtual reality case (highlighted image in VR) and the web page with an embedded image in a 2D browser. This contrast suggests that advertisements and images presented within web pages, which are common in today's browsing experience, are less memorable than those displayed solely in a virtual space. This comparison is particularly meaningful as it juxtaposes the genuine novelty of VR, the capacity for direct 3D insertion, against conventional web environments with embedded ads.

This difference is not attributed to the novelty effect of virtual reality, as it was also observed within the VR comparison. Presumably, active coding processes underlie this discrepancy, suggesting that by inserting content into space from multiple viewpoints, users benefit from enhanced memory retention [57].

The practical importance of these findings lies in the enhanced likelihood of remembering key information through 3D highlighting. This aspect holds relevance in educational contexts, especially as traditional e-learning progressively transitions to VR-based education. In a virtual educational setting, instructors can continuously present crucial diagrams, making vital information more accessible and promoting improved comprehension among learners.

References

1. Berki, B.: 2D advertising in 3D virtual spaces. Acta Polytech. Hungarica **15**(3), 175–190 (2018)
2. Berki, B.: Better memory performance for images in MaxWhere 3D VR space than in website. In: 2018 9th IEEE International Conference on Cognitive Infocommunications (CogInfoCom), pp. 000 281–000 284. IEEE (2018)
3. Berki, B.: Információkiemelés virtuális térben és böngészőben. In: Bihari, E., Molnár, D., Szikszai-Németh, K. (eds.) Tavaszi Szél - Spring Wind,: Tanulmánykötet II. Doktoranduszok Országos Szövetsége, vol 2020, pp. 489–495 (2019)
4. Horváth, I., Csapó, Á.B., Berki, B., Sudár, A., Baranyi, P.: Definition, background and research perspectives behind 'Cognitive Aspects of Virtual Reality' (cVR). Infocommunications J. **15**, no. Special Issue, 9–14 (2023)

5. Baranyi, P., Csapó, Á.B., Berki, B.: Concepts of cognitive infocommunications. Infocommunications J. (2024). Accepted, in press
6. Baranyi, P., Csapó, Á., Budai, T., Wersényi, G.: Introducing the concept of internet of digital reality – part I. Acta Polytech. Hungarica **18**(7), 225–240 (2021)
7. Wersényi, G., Csapó, Á., Budai, T., Baranyi, P.: Internet of digital reality: Infrastructuralbackground – part ii. Acta Polytechn. Hungarica **18**(8), 91–104 (2021)
8. Baranyi, P., Gilányi, A.: Mathability: emulating and enhancing human mathematical capabilities. In: 2013 IEEE 4th International Conference on Cognitive Infocommunications (CogInfoCom), pp. 555–558. IEEE (2013)
9. Baranyi, P., Csapo, A., Sallai, G.: Cognitive Infocommunications (coginfocom). Springer (2015)
10. Baranyi, P., Csapó, Á.: Definition and synergies of cognitive infocommunications. Acta Polytech. Hungarica **9**(1), 67–83 (2012)
11. Zöldy, M., Baranyi, P.: The cognitive mobility concept. Infocommunications J. **15**, no. Special Issue, 35–40 (2023)
12. Baddeley, A., Eysenck, M.W., Anderson, M.C.: Memory, 2nd edn. Psychology Press (2009)
13. Csapó, Á.B., Horváth, I., Galambos, P., Baranyi, P.: VR as a medium of communication: from memory palaces to comprehensive memory management. In: 2018 IEEE 9th International Conference on Cognitive Infocommunications (CogInfoCom), pp. 000 389–000 394. IEEE (2018)
14. Fassbender, E., Heiden, W.: The virtual memory palace. J. Comput. Inf. Syst. **2**(1), 457–464 (2006)
15. Legge, E.L.G., Madan, C.R., Ng, E.T., Caplan, J.B.: Building a memory palace in minutes: equivalent memory performance using virtual versus conventional environments with the method of loci. Acta Physiol. (Oxf) **141**(3), 380–390 (2012)
16. Boda, I., Toth, E., Csont, I., Nagy, L.T.: Developing a knowledge base of ancient literary texts in virtual space. In: 2016 7th IEEE International Conference on Cognitive Infocommunications (CogInfoCom). IEEE (2016)
17. Krokos, E., Plaisant, C., Varshney, A.: Virtual memory palaces: immersion aids recall. Virtual Reality **23**(1), 1–15 (2018)
18. Yang, F., Qian, J., Novotny, J., Badre, D., Jackson, C.D., Laidlaw, D.H.: A virtual reality memory palace variant aids knowledge retrieval from scholarly articles. IEEE Trans. Visual Comput. Graph. **27**(12), 4359–4373 (2021)
19. Baranyi, P., Csapo, A.B.: Revisiting the concept of generation CE - generation of cognitive entities. In: 2015 6th IEEE International Conference on Cognitive Infocommunications (CogInfoCom), pp. 583–586. IEEE (2015)
20. Horváth, I.: Behaviors and capabilities of generation CE students in 3D VR. In: Baranyi, P. (ed.) Proceedings of the 10th IEEE International Conference on Cognitive Infocommunications: CogInfoCom 2019, pp. 491–494. IEEE (2019)
21. Horváth, I.: The 4th dimension of personalization in vr. In: 2022 1st IEEE International Conference on Cognitive Aspects of Virtual Reality (CVR), pp. 000 085–000 088. IEEE (2022)
22. Paivio, A., Csapo, K.: Picture superiority in free recall: imagery or dual coding? Cogn. Psychol. **5**(2), 176–206 (1973)
23. Yuille, J.C. (ed.): Imagery, memory and cognition (PLE: Memory): essays in honor of Allan Paivio, 1st edn. Psychology Press, London (2014)
24. Asch, S.E., Ebenholtz, S.M.: The principle of associative symmetry. Proc. Am. Philos. Soc. **106**(2), 135–163 (1962)
25. Dinh, H.Q., Walker, N., Hodges, L.F., Kobayashi, A.: Evaluating the importance of multisensory input on memory and the sense of presence in virtual environments. In: Proceedings IEEE Virtual Reality, pp. 222–228 (1999)
26. Mania, K., Chalmers, A.: The effects of levels of immersion on memory and presence in virtual environments: a reality centered approach. CyberPsychol. & Behav. **4**(2), 247–264 (2001)

27. Bowman, D.A., Sowndararajan, A., Ragan, E.D., Kopper, R.: Higher levels of immersion improve procedure memorization performance. In: Hirose, M., Schmalstieg, D., Wingrave, C.A., Nishimura, K. (eds.) Joint Virtual Reality Conference of EGVE - ICAT - EuroVR. The Eurographics Association (2009)
28. Ragan, E.D., Sowndararajan, A., Kopper, R., Bowman, D.A.: The effects of higher levels of immersion on procedure memorization performance and implications for educational virtual environments. Presence: Teleoper. Virtual Env. **19**(6), 527–543 (2010)
29. Statista: Global internet advertising revenue from 2017 to 2027 (2023). Accessed 29 Aug 2023. https://www.statista.com/statistics/237800/global-internet-advertising-revenue/
30. Percy, L., Rossiter, J.R.: A theory-based approach to pretesting advertising. In: Wells, W.D. (ed.) Measuring Advertising Effectiveness. Advertising and Consumer Psychology, pp. 267–281. Psychology Press (1997)
31. Childers, T.L., Houston, M.J.: Conditions for a picture-superiority effect on consumer memory. J. Consumer Res. **11**(2), 643–654 (1984)
32. Houston, M.J., Childers, T.L., Heckler, S.E.: Picture-word consistency and the elaborative processing of advertisements. J. Market. Res. **24**(4), 359 (1987)
33. Interactive Advertising Bureau: IAB internet advertising revenue report (2017). https://www.iab.com/wp-content/uploads/2017/12/IAB-Internet-Ad-Revenue-Report-Half-Year-2017-REPORT.pdf. Accessed 02 Jan 2018. https://www.iab.com/wp-content/uploads/2017/12/IAB-Internet-Ad-Revenue-Report-Half-Year-2017-REPORT.pdf
34. Burns, K.S., Lutz, R.J.: The Function of Format: consumer responses to six on-line advertising formats. J. Advert. **35**(1), 53–63 (2006)
35. Flores, W., Chen, J.-C.V., Ross, W.H.: The effect of variations in banner ad, type of product, website context, and language of advertising on Internet users' attitudes. Comput. Hum. Behav. **31**, 37–47 (2014)
36. Ha, L.: Online advertising research in advertising journals: a review. J. Current Issues & Res. Adver. **30**(1), 31–48 (2008)
37. Li, H., Bukovac, J.L.: Cognitive impact of banner ad characteristics: an experimental study. J. & Mass Commun. Q. **76**(2), 341–353 (1999)
38. Edwards, S.M., Li, H., Lee, J.-H.: Forced exposure and psychological reactance: Antecedents and consequences of the perceived intrusiveness of pop-up ads. J. Advert. **31**(3), 83–95 (2002)
39. Choi, S.M., Rifon, N.J.: Antecedents and consequences of web advertising credibility. J. Inter. Advert. **3**(1), 12–24 (2002)
40. Rodgers, S., Thorson, E.: The interactive advertising model. J. Inter. Advert. **1**(1), 41–60 (2000)
41. Ha, L.: Advertising clutter in consumer magazines: dimensions and effects. J. Advert. Res. **36**, 76–85 (1996)
42. Brehm, J.W.: A Theory of Psychological Reactance. Academic, New York (1966)
43. Choi, Y.K., Taylor, C.R.: How do 3-dimensional images promote products on the internet? J. Bus. Res. **67**(10), 2164–2170 (2014)
44. Jin, S.-A.A., Bolebruch, J.: Avatar-based advertising in second life. J. Inter. Advert. **10**(1), 51–60 (2009)
45. Griffin, T., Giberson, J., Lee, S.H.M., Guttentag, D., Kandaurova, M., Sergueeva, K., Dimanche, F.: Virtual reality and implications for destination marketing. In: Tourism Travel and Research Association: AdvancingTourism Research Globally (2017)
46. Tikkanen, H., Hietanen, J., Henttonen, T., Rokka, J.: Exploring virtual worlds: success factors in virtual world marketing. Manag. Decis. **47**(8), 1357–1381 (2009)
47. Debbabi, S., Daassi, M., Baile, S.: Effect of online 3D advertising on consumer responses: the mediating role of telepresence. J. Mark. Manag. **26**(9–10), 967–992 (2010)
48. Li, H., Daugherty, T., Biocca, F.: Impact of 3-D advertising on product knowledge, brand attitude, and purchase intention: the mediating role of presence. J. Advert. **31**(3), 43–57 (2002)
49. Kerrebroeck, H.V., Brengman, M., Willems, K.: When brands come to life: experimental research on the vividness effect of virtual reality in transformational marketing communications. Virtual Real. **21**(4), 177–191 (2017)

50. Zajonc, R.B.: Attitudinal effects of mere exposure. J. Personal. Soc. Psychol. **9**(2), Pt. 2, 1–27 (1968)
51. Lindmark, P.G.: A content analysis of advertising in popular videogames. Master's thesis, Cleveland State University (2011)
52. Grigorovici, D.M., Constantin, C.D.: Experiencing interactive advertising beyond rich media: Impacts of ad type and presence on brand effectiveness in 3D gaming immersive virtual environments. J. Interact. Advert. **5**(1), 22–36 (2004)
53. Keng, C.-J., Lin, H.-Y.: Impact of telepresence levels on internet advertising effects. CyberPsychol. & Behav. **9**(1), 82–94 (2006)
54. Borba, E.Z., Zuffo, M.K.: Advertising perception with immersive virtual reality devices. In: IEEE Virtual Reality (VR), pp. 371–372. IEEE (2017)
55. Lin, J.J.-W., Duh, H.B.L., Parker, D.E., Abi-Rached, H., Furness, T.A.: Effects of field of view on presence, enjoyment, memory, and simulator sickness in a virtual environment. In: Proceedings IEEE Virtual Reality, pp. 164–171. IEEE Computer Society (2002)
56. Baranyi, P., Galambos, P., Csapó, Á., Jaloveczki, L.: Cognitive navigation and manipulation (coginav) method. US Patent App. 15/658,579. Accessed 1 Feb 2018
57. Plancher, G., Tirard, A., Gyselinck, V., Nicolas, S., Piolino, P.: Using virtual reality to characterize episodic memory profiles in amnestic mild cognitive impairment and Alzheimer's disease: influence of active and passive encoding. Neuropsychologia **50**(5), 592–602 (2012)

Chapter 8
Assessing Memory Performance for Supplementary Information in VR

Abstract This chapter focuses on the benefits of a virtual reality working environment from a cognitive perspective. It focuses on non-immersive, or desktop virtual reality technology that uses general 2D display devices, allowing the avoidance of inconveniences associated with head-mounted displays. The study aimed to investigate whether the use of desktop virtual reality, as compared to a basic web browser, could enhance users' ability to interact with and remember information in a virtual workspace. After applying stricter inclusion criteria and disregarding participants who based their estimations on factors other than the graph, the initial hypothesis was confirmed. The study found that in a virtual reality environment, participants were able to estimate their performance in the experimental task significantly more accurately compared to those who worked in a traditional 2D digital environment.

Keywords Virtual reality · Virtual workspace · Desktop VR · Memory · Visual information

8.1 Introduction

The advancements in virtual reality (VR) technology, fueled by artificial intelligence (AI) and sensor innovations, are expanding its range of applications to various industries. This development signifies a future where humans, AI, and digital replicas collaborate within VR settings [1–8].

Integrating virtual reality (VR) into daily office work is becoming increasingly popular as organizations strive to find creative ways to improve productivity and foster collaboration among their employees. This chapter focuses on the cognitive benefits of using non-immersive desktop virtual reality technology, especially the presentation and retention of additional information within virtual workspaces. The study found that users in a virtual reality environment were able to estimate their performance in the experimental task significantly more accurately compared to those in a traditional 2D digital environment.

I. Horváth et al., *Cognitive Aspects of Virtual Reality*, Studies in Big Data 156,
https://doi.org/10.1007/978-3-031-68130-1_8

The key concepts presented in this chapter are based on publications [9, 10]. The chapter is structured as follows: Sect. 8.2 discusses the potential of using virtual reality (VR) in workspaces, presenting its advantages such as enhanced collaboration, increased productivity, and customizable environments. It also acknowledges challenges, particularly ergonomic issues associated with current VR systems, while highlighting the potential benefits of desktop VR in overcoming these drawbacks. The following paragraphs (Sect. 8.3) formulate the objective of the current study. Section 8.4 presents the experimental design that was applied to assess the memory performance for supplementary information in desktop virtual reality. This part is separated into four subsections that provide an organized overview of the experimental design, participants, experimental apparatus and software, experimental task, and the procedure of the experiment. In the following, Sect. 8.5 presents the research results that are discussed in Sect. 8.6. The chapter ends with the brief conclusion of the presented results in Sect. 8.7.

8.2 Enabling Effective Workflows: Exploring Virtual Workspaces and Collaboration Tools

This section discusses the rise of VR-based offices for remote work, citing benefits like improved user presence and productivity.

Using virtual reality as a workspace is a widely discussed topic. As remote work evolves, a paradigm shift toward VR-based offices is predicted, as this can overcome several disadvantages of current home office settings [11]. Besides the pragmatic reason, namely that less equipment is necessary to construct the setting for efficient work, there are also economic benefits for the organization in the long term [12]. On the individual level, VR allows the user to feel present in the workspace instead of merely accessing it. By providing a shared and coherent context for the essential working documents, the collaborative processes are enhanced, alongside the sense of presence and perception of togetherness [11]. Additionally, the elevated sense of presence could also enhance performance [13–15].

Over the last few decades, advancements in digital and communication technologies have allowed for a shift from traditional to remote work, where coworkers are physically separated and communicate via various devices. One main drawback of the current trend towards remote work is the various challenges that employees encounter when they have to work in a space that is not specifically designated as an office. It is easy to imagine the difficulties that come with setting up a proper workspace at home. According to a digital ethnography study [16], which analyzed 430 photos of developers' workstations, 84% of them had at least one challenge, such as limited desk space, lighting issues, or clutter. However, these issues can be resolved in a virtual reality-based workstation. For example, a well-designed virtual workspace should have optimal lighting, no clutter, unlimited desk space, and an unlimited number of conveniently placed monitors that can be easily rearranged anytime when needed.

An immersive VR blocks the distraction from the real world, which allows the user to focus on the task at hand [16, 17]. Thus, they are more engaged in their work [17]. The customizable virtual working environment allows users to choose their preferred surroundings while working. This means they can replace the dull office building scene with a view of the seaside or place their desk in a serene forest to feel closer to nature. Users also appreciate the virtual sunlight [16]. Additionally, having virtual plants in a VR workspace resulted in lower physiological stress and higher creativity and short-term memory performance scores [18, 19].

The (current) disadvantages of these systems cannot be ignored either, which are mostly ergonomic aspects. First of all, bulky, heavy wired headsets cause discomfort during elongated VR sessions [16, 17]. Then, remote jobs often require the users to work with documents, where reading and typing are elementary tasks. Although, these seem to be not optimal in virtual reality settings. According to a recent study, participants who used VR headsets during a text-intensive task tended to rotate their heads 2.7 times more frequently. Additionally, they reported higher levels of discomfort in their neck and shoulders. The study also found that 60% of the VR headset group experienced discomfort in their necks, while 17.5% reported discomfort in their shoulders. On the other hand, participants who completed the same task on a desktop computer did not report these issues. [20]. Wider use of virtual reality is limited by cybersickness, visual fatigue, muscle fatigue, and mental overload [21]. Advances in technology are likely to address these difficulties. However, these problems can mostly be avoided by using desktop VR.

According to Orel [11], the increased user productivity in a VR environment is conveyed through the sense of presence, the collaborative potential, and the knowledge transfer capacity. The following factors contribute to productivity: spatial and social design, accessibility, privacy control, customization, and interactivity. While most of these factors are easy to grasp, privacy control in the case of VR may need some explanation. When using avatars, the face of the user is covered, but their facial expressions are still visible. When working from home, one doesn't have to share their private space with colleagues since the background is not visible in VR meetings unlike in video calls. In addition to a higher level of privacy, these features put users in an equal position, so that racial or social profiling is less prevalent [11].

In virtual reality, customization allows the use of unlimited virtual monitors, placed ergonomically according to users' needs. The majority of workflows mostly require the use of separate programs, as the different types of information are fragmented. Having the required digital content visually available in the field of view instead of switching between multiple windows can improve workflow efficiency. Additionally, it's not necessary to assign formal codes or names to grouped documents. Instead, the user can simply remember their spatial location, just like in real life, and work with them accordingly [22, 23].

It has been observed that users prefer to have an overview of the whole screen. Users have stood or sat back at a distance when they used large, high-resolution displays, that allowed them to view the entire screen at once [24]. It is easy to fulfill this need by applying virtual reality as a working environment. In a VR environment, zooming into preferred details can be done effortlessly without any significant motion. The spatial arrangement of documents and other digital contents allows easier access to the information. In a previous work [25], a new methodology was proposed to measure and quantify the complexity of interactions in digital interfaces. The study found that immersive VR systems exhibit a 54% reduction in information access cost compared to classical 2D Windows systems, with information access cost defined as the weighted sum of time spent accessing information for each type of operation, considering the number of elementary operations for each type.

Digital teamwork consists of intertwining sets of real-time and asynchronous collaborations. Shared and co-editable documents or even 3D objects allow asynchronous team collaboration. Real-time collaboration relies on video/audio conferencing and instant messaging. Integrating these digital tools into the virtual workspaces was an early objective of VR developers [26]. VR-based teamwork allows members to share context, not just a document. This is known as situational awareness, which refers to the ability to comprehend the current state of the team members who are working together, which plays a crucial role in planning the subsequent task. Virtual co-presence can enhance situational awareness, creating a sense of shared space among individuals. This shared context also promotes knowledge exchange amongst the team members [27].

Individual characteristics also play a role in digital teamwork [28–31]. Younger generations who have grown up in a world where infocommunication is an integral part of their daily life possess different capabilities as they are in a synergistic relationship with infocommunication devices [7, 32]. Therefore, it is not surprising that different generations use virtual reality differently, and their preferred digital content and time management strategies also differ [33, 34].

It is essential for a computer-based workstation to help the user maintain focus and remain in context with their reasoning process. This can be achieved through direct interaction and manipulation, which keeps the user in the cognitive zone and prevents workflow interruptions. To avoid distracting the user, it is also important to refrain from using menus, particularly traditional pull-downs that require the user to sort through and consider each item. This way, the user can remain focused on the task at hand [35].

Most current virtual reality-based productivity systems use a head-mounted display (HMD) as a display device. Additionally, some systems offer a computer display and can also be accessed through a tablet or phone. These systems come with a wide range of control devices, such as controllers, gesture controls, motion capture, mouse, and keyboard. However, using a keyboard can be challenging when wearing an HMD. To solve this issue, mixed reality solutions are becoming increasingly popular in many systems. These solutions display real-time physical keyboards to assist with typing while wearing an HMD.

The non-immersive, or desktop virtual reality technology uses general 2D display devices, thus the inconveniences that derive from the use of head-mounted displays can be avoided, and still several of the previously mentioned benefits remain. For example, the users can experience the sense of presence [36]. In the following part, desktop VR-based results will be presented, that deal with the workflows and workflow communication in virtual environments. Digital workflows are defined as the sequence in which individual digital elements are accessed or processed during a digital project.

Researchers compared three types of workflow sharing: linguistic description and attachments, online interface (Moodle[1]), and 3D interface (MaxWhere desktop VR[2]). The study found that the virtual reality 3D interface allowed users to complete the same digital workflow 50% faster than the other 2D interfaces. In addition, the VR technology effectively communicated the necessary sequence more naturally [37].

A previous study [38] suggested a framework to assess the efficiency of digital workflows. The study compared the number of user operations and machine operations required for the same digital workflows in both a 3D environment and a traditional 2D interface. The results showed that the digital workflows completed in a 3D environment required 30% fewer user operations and up to 80% fewer machine operations than those in a traditional 2D interface. The lower number of required elementary operations was also observed in the case of immersive VR [25].

Both studies [37, 38] highlight the importance of digital guidance in facilitating the completion of workflows. Digital guidance refers to a logically ordered presentation of digital content that users can access automatically. This spatial organization eliminates the need for additional operations and optimizes the view for the user. Additionally, the 3D platform can enhance the effectiveness of other web applications by providing a spatial layout [39].

The research carried out by Sudár and Csapó [40] aimed to investigate user interaction in a desktop virtual reality system. Results showed that users favored the gesture-based activation technique on smartboards due to its speed and ability to transition to full-screen mode from any location, which is in line with the result of [41]. Frequency analysis also revealed that users had specific points of interest in the virtual space that they visited more frequently, and they tended to minimize movement and navigate naturally while completing tasks [40]. In a later study, they showed, that users prefer to arrange their digital content element based on their subjects while taking into consideration the 3D spatial metaphors in physical reality [42].

8.3 Aim of Study

As mentioned earlier, a desktop virtual reality can fulfill the requirements of an effective virtual workspace. It allows the user to interact with various types of information directly, which helps them maintain their workflow. By using a 3D layout, users can gain insights into documents that are not currently in focus but are visible in their visual field due to the 3D perspective. The question is whether this additional information provides benefits to the user and if they can remember it.

---

[1] https://moodle.org/.

[2] www.maxwhere.com.

The study used a between-subjects design to investigate a question. One group used a desktop VR, while the other used a general web browser. The MaxWhere[3] virtual environment was used as a desktop VR because it can load webpages on smartboards located inside of a 3D environment. These so-called smartboards have a predefined location within a space, and users can load desired web pages, documents, or web applications on them.

The experiment required participants to use three web pages, each with different content. Although this may seem like a limited number of documents, the aim was to measure differences in a simpler task with only a few documents. One of the web pages contained numerical tasks, while the second was a spreadsheet where participants had to write down the results. The third was an interactive figure that showed the percentage of completed and remaining tasks.

During the experiment, the VR group saw three web pages displayed next to each other on a virtual board, while the web browser group had three different tabs. In this case, they could arrange them. Both groups had five minutes to solve numerical tasks, after which they had to fill out a questionnaire. In this survey, they had to estimate their performance as a percentage of the completed and uncompleted tasks. For this, they had to rely on the provided figure, since the number of tasks was not mentioned anywhere. This experimental design allowed researchers to measure the use of extra information in addition to the participants' actual performance.

Participants in the VR group were able to view the performance graph continuously throughout the experiment. It was hypothesized that these participants would have a better estimation of their progress than participants in the control group. Additionally, it was hypothesized that this difference in performance estimation would still be present even if we only considered the participants who indicated the performance graph as their main source of estimation.

8.4 Experimental Design and Setup

8.4.1 Participants and Experimental Design

The study involved 58 participants, of which the data of 53 individuals (32 male, 21 female; mean age 25.2 years, SD: 5.0) were analyzed, as five subjects were excluded due to technical problems or misapprehensions related to the experimental task. All participants volunteered for the study and were native Hungarian speakers. A between-subjects design compared user performance in MaxWhere 3D VR ($N = 26$) and Google Chrome ($N = 27$), with random assignment to experimental conditions.

[3] www.maxwhere.com.

8.4.2 Apparatus and Software

The same laptop was utilized across all participants (full HD (14") display, 1920×1080 px, Windows 10 operating system, 8 GB, Nvidia GeForce 940 MX), aligning with MaxWhere engine system requirements. A computer mouse was used as a pointing device.

MaxWhere VR is a desktop VR framework that enables the display of traditional web content in a 3D virtual world. Inside the 3D space, the so-called smartboards serve to display 2D content such as webpages, PDF documents, images, and video files. The CogiNav Method [43] is used for intuitive navigation and manipulation in the 3D VR environment by a simple external mouse. In this study, the InfoSky space was used, with the three experiment webpages displayed in the same order as browser tabs.

The experiment in the VR group occurred in the InfoSky space, with smartboards arranged in a 2×3 matrix (Fig. 8.1). The upper three landscape format smartboards contained the primary experimental task. The informed consent was on another A4 format smartboard. The Tutorial—HUN virtual space served as a training VR space for those unfamiliar with the software, guiding the use of smartboards and Cognitive Navigation Technology. The VR environment only required an external mouse with two buttons and a scroll wheel.

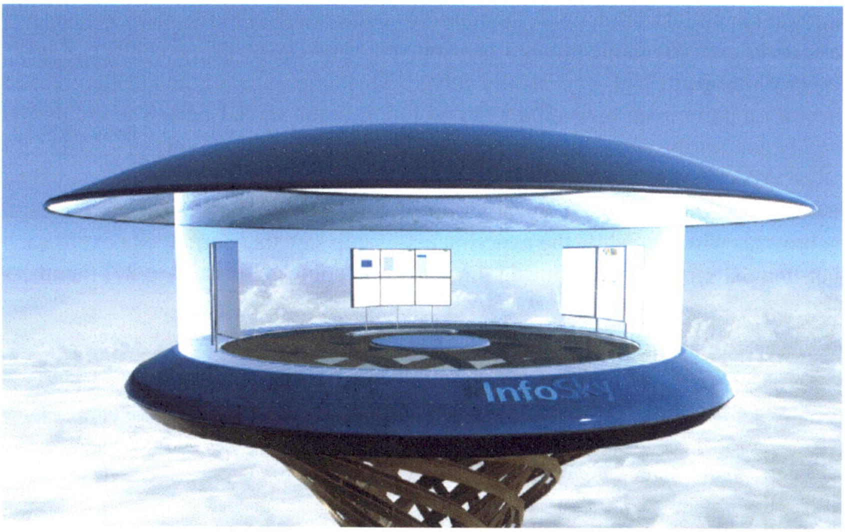

Fig. 8.1 The virtual environment of the study, the so-called InfoSky space. Participants in the VR group completed the experimental task in a spacious, modern, and clean virtual environment

For the control group and questionnaire completion, Google Chrome (version: 67.0.3396.99) was utilized. Google Chrome was chosen as it was the most widely used desktop browser in Hungary [44]. Three web pages were presented as tabs in the order of tasks, the graph, and the spreadsheet.

8.4.3 Experimental Task

The participants were required to complete mathematical tasks, such as adding two numbers, and record their results for five minutes on a displayed spreadsheet. Their progress was then shown on another display in the virtual environment. The task was presented on the first webpage. On the second webpage, a graph displayed the actual progress in percentages. The third webpage contained an online spreadsheet where the users had to write the answers. The arrangement of these three displays in the virtual space is depicted in Fig. 8.2.

The numerical tasks, such as 13 + 12 and 91 + 7, were presented one by one. Users could display the next task by clicking a button. The tasks were not numbered, and on the spreadsheet, users had to write the answers in the same column, one below the other. During the experiment, the participants had to complete several numerical tasks which were listed on a webpage. However, the webpage did not contain any numbering, which meant that the participants had no idea about the total number of tasks they were supposed to complete. On the second webpage of the experiment, a dynamic graph was displayed, showcasing the percentage of tasks that the participants had successfully solved. This graph was automatically updated whenever a participant registered a new solution to the spreadsheet. The graph was the only cue for the subsequent unheralded performance estimation.

The experimental group differed from the control group in terms of the way they viewed the web pages. The control group saw the three web pages as three tabs in one browser window, while the experimental group saw them arranged next to each other in a 3D virtual space. During the study, the control group (N = 27) members

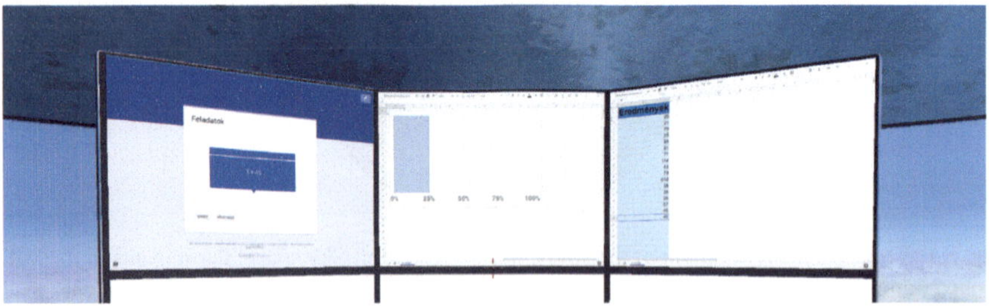

Fig. 8.2 The arrangement of the three web pages for the VR group. On the left, the tasks were presented, on the right, there is a spreadsheet for the solutions. The central display showed a graph of their progress in percentages

were given the freedom to change the order and layout of the three tabs. Out of the 27 participants, 23 preferred to change the tabs with a click, two participants alternated the windows with the alt + tab keys, and two individuals used a shared window to perform both the task and the results tab. The main objective behind giving them the freedom to choose their preferences was to ensure they worked comfortably and to map their preferences. In contrast, the experimental group saw the three tabs next to each other, with the task on the left side, the results spreadsheet on the right side, and their progress on the graph displayed in the middle. This view was saved, and with the hit of the backspace key, they could retrieve it anytime.

The final questionnaire was presented to all participants through a browser. In addition to answering sociodemographic questions, they had to estimate their performance and rank five factors in the order of their influence on their estimation. Participants were not required to rank all factors, but they had to mention at least one of them.

8.4.4 Procedure

Under the supervision of the same experimenter, all participants underwent individual testing sessions. Following a concise introduction to the experiment, participants provided informed consent. The browser group commenced the experimental task immediately, allocating five minutes for task completion, and ending with the final questionnaire. Conversely, participants in the VR group started with a tutorial, "Tutorial—HUN Where", designed to familiarize them with the MaxWhere VR environment and navigation. Participants had the flexibility to invest as much time as needed in this tutorial, with none exceeding a five-minute duration. Subsequently, participants transitioned to the InfoSky virtual space to initiate the experiment, where they had a designated five-minute period to address the experimental task before completing the final questionnaire, all conducted within the Google Chrome browser.

8.5 Results

This research aimed to examine how well participants could recall additional information using desktop virtual reality (VR) and web browsers. Throughout the study, we recorded the number of tasks each participant completed and their estimated completion percentage. After the experiment, we assessed each respondent's solutions and calculated their percentage of correctly solved tasks.

The statistical analysis did not include the accuracy of problem-solving as it varies individually and cannot draw any conclusions about the differences in workflow between the two groups. It was not expected to find a perfect solution to the numerical tasks. However, the problem-solving accuracy was found to be 96.93% (SD: 4.36). Pearson's correlation analysis indicated that there was no significant relationship between the accuracy of solving mathematical tasks and the accuracy of performance estimation ($r(51) = 0.084$, $p = 0.548$). Therefore, all data were included in further analysis regardless of the accuracy of mathematical problem-solving.

In the control group, participants were given the option to adjust the display of three web pages. Most of them (85.2%) used the mouse to switch between tabs, while the remaining 14.8% opened the task and the results spreadsheet in separate windows. A small portion of them (7.4%) switched between windows using the Alt + Tab keys, and only 7.4% used a shared display. However, the results of the Kruskal-Wallis test ($\chi^2(2, N = 27) = 0.362$, $p = 0.834$) showed no significant differences in the accuracy of performance estimation between these methods.

8.5.1 Performance Estimations in VR and Browser

The dependent variable was performance estimation accuracy, calculated as the absolute difference between actual and estimated performance, with smaller values indicating more accurate estimation and better memory of supplementary information. The purpose of the study was to compare the accuracy of performance estimation between two groups. The Shapiro-Wilk normality test indicated a significant deviation from the normal distribution in both groups (browser group: $W = 0.871$, $p = 0.003$; VR group: $W = 0.676$, $p < 0.001$). Therefore, the Mann-Whitney test was applied for the comparison. Equal variances were assumed based on Levene's test ($W = 4.02e\text{-}05$, $p = 0.995$). The Mann-Whitney rank test was used to compare the accuracy of estimation between the two groups. The results showed that there was no significant difference between the means of estimation accuracy of the two groups ($U = 300$, $p = 0.364$). This means that a desktop VR ($M = 12.37$, $SD = 16.02$) does not provide better accuracy estimation than a 2D browser ($M = 13.54$, $SD = 12.42$). Interestingly, even though all participants in the VR group could see the graph constantly, it did not improve their accuracy estimation.

8.5.2 Performance Estimations Based on the Graph

Participants in the experiment who did not refer to the performance graph were not estimating, but simply guessing. The graph was the only indicator of individual performance, and since the number of tasks was not disclosed, the tasks were not numbered. Furthermore, at the beginning of the experiment, all participants were informed about the significance of the progress bar.

Hence, we compare the estimation accuracy of those members of the two groups who indicated the graph as their primary aid to estimate their progress. The analysis included 25 participants, with 7 in the control group and 18 in the experimental group. We used a two-independent sample t-test to analyze the results. The Shapiro-Wilk normality test showed that the accuracy of performance estimation was normally distributed for both groups, both in the browser ($W = 0.958$, $p = 0.798$) and in the VR ($W = 0.929$, $p = 0.187$). Additionally, Levene's test confirmed that the variance was homogeneous ($W = 1.93$, $p = 0.178$).

The results of the two independent samples t-test indicated that those who completed the task in MaxWhere VR ($M = 5.17$, $SD = 3.47$) estimated their performance significantly more accurately ($t(23) = 2.73$, $p = 0.012$) than those who completed the task in a 2D browser ($M = 10.36$, $SD = 5.39$). In other words, the participants in the VR group were better at estimating their performance.

8.6 Memory Performance of Supplementary Information in Desktop VR and Browser

The main aim of this study was to determine whether the additional information that is placed in the visual field of the participant can provide any benefits for the users and if they can remember it. The results of the between-subject experiment showed, that although in the VR environment, the participants could see the graph constantly, surprisingly, it did not improve the estimation accuracy generally. By applying more rigorous inclusion criteria, and excluding the results of those participant, who declared that their estimation was based on anything else than the graph, the initial hypothesis was confirmed. In the virtual reality environment, the participants estimated significantly more accurately their performance in the experimental task, than the control group, who worked in a traditional 2D digital environment.

These results shed light also on the limits of this experimental design, as the users were freely navigating in the virtual environment, and the time spent on each view was not recorded. An analysis of screen recording alongside eye-tracking could help to define more precisely, the underlying factors and behaviors of this enhanced memory performance in the virtual environment. The other limitation of this study is that the participants in the browser group, could also freely organize the three windows of the task. Thus, they could easily hide the graph, as seemingly it was not the focus of the main task. This feature resulted in a lower number of participants in the second analysis, as there were several people in the browser group, who had not seen the graph and therefore they were not included in the following comparison.

Despite these limitations, the results showed, that the three-dimensional view, even in desktop virtual reality-based working environments, can provide benefits for their users in terms of memory performance. Furthermore, this experimental design provides an initial method for comparative studies in desktop VR and general 2D environments.

The better memory performance for supplementary information in virtual reality relates to the previously presented benefits of desktop VRs. Sharing a complete workflow, instead of providing several attachments to an email, can not only allow faster completion of tasks [37] but also make it easier to share extra information. For example, a supplementary graph, which relates to the shared workflow, but is not essential, though other platforms you may not share, to avoid overcrowded document sharing, but in a virtual space, it can be placed, according to its importance, without overwhelming the users.

The findings are consistent with previous research about overview tendency. According to previous studies, people tend to prefer an overview of all information when working on a screen [24]. This is commonly known as the "overview tendency" which is in line with the current results. A virtual 3D environment can fulfill this expectation, allowing users to view all documents at once. Additionally, this environment provides supplementary information that can be useful for ongoing projects.

8.7 Conclusions

This chapter presented how VR in workspaces can enhance collaboration and productivity. Desktop VR may overcome the ergonomic challenges that immersive HMD-based VR causes. The study aimed to investigate whether the use of desktop virtual reality, as compared to a basic web browser, could enhance users' ability to interact with and remember information in a virtual workspace. The study found that in a virtual reality environment, participants were able to estimate their performance in the experimental task significantly more accurately compared to those who worked in a traditional 2D digital environment.

References

1. Baranyi, P., Csapó, Á.B., Berki, B.: Concepts of cognitive infocommunications. Infocommunications J. (2024). Accepted, in press
2. Horváth, I., Csapó, Á.B., Berki, B., Sudár, A., Baranyi, P.: Definition, background and research perspectives behind 'Cognitive Aspects of Virtual Reality' (cVR). Infocommunications J. **15**, no. Special Issue, 9–14 (2023)
3. Kő, A., Szabó, I., Csapó, Á.B., Kovács, T., Lőrincz, L., Baranyi, P.: Digital & cognitive corporate reality. Infocommunications J. **15**, no. Special Issue, 2–10 (2023)
4. Baranyi, P., Csapó, Á., Budai, T., Wersényi, G.: Introducing the concept of internet of digital reality – part I. Acta Polytech. Hungarica **18**(7), 225–240 (2021)
5. Wersényi, G., Csapó, Á., Budai, T., Baranyi, P.: Internet of digital reality: Infrastructuralbackground – part ii. Acta Polytech. Hungarica **18**(8), 91–104 (2021)
6. Baranyi, P., Gilányi, A.: Mathability: emulating and enhancing human mathematical capabilities. In: 2013 IEEE 4th International Conference on Cognitive Infocommunications (CogInfoCom), pp. 555–558. IEEE (2013)
7. Baranyi, P., Csapo, A., Sallai, G.: Cognitive Infocommunications (coginfocom). Springer (2015)
8. Baranyi, P., Csapó, Á.: Definition and synergies of cognitive infocommunications. Acta Polytech. Hungarica **9**(1), 67–83 (2012)
9. Berki, B.: Desktop VR and the use of supplementary visual information. In: 2018 9th IEEE International Conference on Cognitive Infocommunications (CogInfoCom), pp. 000 333–000 336. IEEE (2018)

10. Berki, B.: Desktop VR as a virtual workspace: a cognitive aspect. Acta Polytech. Hungarica **16**(2), 219–231 (2019)
11. Orel, M.: Collaboration Potential in Virtual Reality (VR) Office Space: Transforming the Workplace of Tomorrow. Springer International Publishing (2022)
12. Poehler, L., Teuteberg, F.: Unfolding benefits of virtual reality for workplace and process design based on utility effect chains. In: Pacific Asia Conference on Information Systems (2022)
13. Krassmann, A.L., Melo, M., Pinto, D., Peixoto, B., Bessa, M., Bercht, M.: What is the relationship between the sense of presence and learning in virtual reality? a 24-year systematic literature review. Presence: Virtual Augment. Real. **28**, 247–265 (2019)
14. Mikropoulos, T.A., Bellou, J.: The unique features of educational virtual environments. In: Isaías, P., McPherson, M., Bannister, F. (eds.) Proceedings of the IADIS International Conference on e-Society, pp. 122–128. IADIS Press, Dublin, Ireland (2006)
15. Selzer, M.N., Gazcon, N.F., Larrea, M.L.: Effects of virtual presence and learning outcome using low-end virtual reality systems. Displays **59**, 9–15 (2019). http://www.sciencedirect.com/science/article/pii/S0141938218300398
16. Ruvimova, A., Fronchetti, F., Kahn, B.A., Susin, L.H., Hurley, Z., Fritz, T., Hancock, M., Shepherd, D.: Ready worker one? high-res VR for the home office. In: 29th ACM Symposium on Virtual Reality Software and Technology. VRST 2023. ACM (2023)
17. Aufegger, L., Elliott-Deflo, N.: Virtual reality and productivity in knowledge workers. Front. Virtual Real. **3** (2022)
18. Yin, J., Arfaei, N., MacNaughton, P., Catalano, P.J., Allen, J.G., Spengler, J.D.: Effects of biophilic interventions in office on stress reaction and cognitive function: a randomized crossover study in virtual reality. Indoor Air **29**(6), 1028–1039 (2019)
19. Mostajeran, F., Steinicke, F., Reinhart, S., Stuerzlinger, W., Riecke, B.E., Kühn, S.: Adding virtual plants leads to higher cognitive performance and psychological well-being in virtual reality. Sci. Rep. **13**(1) (2023)
20. Kim, E., Shin, G.: User discomfort while using a virtual reality headset as a personal viewing system for text-intensive office tasks. Ergonomics **64**(7), 891–899 (2021)
21. Souchet, A.D., Lourdeaux, D., Pagani, A., Rebenitsch, L.: A narrative review of immersive virtual reality's ergonomics and risks at the workplace: cybersickness, visual fatigue, muscular fatigue, acute stress, and mental overload. Virtual Real. **27**(1), 19–50 (2022)
22. Henderson, D.A., Jr., Card, S.: Rooms: the use of multiple virtual workspaces to reduce space contention in a window-based graphical user interface. ACM Trans. Graph. **5**(3), 211–243 (1986)
23. Card, S., Henderson Jr, D.A.: A multiple, virtual-workspace interface to support user task switching. In: Proceedings of the CHI'87 Human Factors in Computing Systems Conference. ACM Press (1987)
24. Endert, A., Bradel, L., Zeitz, J., Andrews, C., North, C.: Designing large high-resolution display workspaces. In: Proceedings of the International Working Conference on Advanced Visual Interfaces - AVI'12, pp. 58–65. ACM Press (2012)
25. Horváth, I., Berki, B.: Investigating the operational complexity of digital workflows based on human cognitive aspects. Electronics **12**(3) (2023). https://www.mdpi.com/2079-9292/12/3/528
26. Spellman, P.J., Mosier, J.N., Deus, L.M., Carlson, J.A.: Collaborative virtual workspace. In: Proceedings of the International ACM SIGGROUP Conference on Supporting Group Work: The Integration Challenge the Integration Challenge - GROUP'97, pp. 197–203. ACM Press (1997)
27. Malhotra, A., Majchrzak, A.: How virtual teams use their virtual workspace to coordinate knowledge. ACM Trans. Manag. Inf. Syst. **3**(1), 1–14 (2012)
28. Ling, Y., Nefs, H.T., Brinkman, W.-P., Qu, C., Heynderickx, I.: The relationship between individual characteristics and experienced presence. Comput. Hum. Behav. **29**(4), 1519–1530 (2013)

29. Pleva, M., Ondas, S., Juhar, J., Hudson, C.R., Carruth, D.W., Bethel, C.L.: Non-native vs native english speakers user experience in HMD vs desktop immersive training. In: 2020 11th IEEE International Conference on Cognitive Infocommunications (CogInfoCom), pp. 51–54. IEEE (2020)
30. Grassini, S., Laumann, K., Luzi, A.K.: Association of individual factors with simulator sickness and sense of presence in virtual reality mediated by head-mounted displays (HMDs). Multimodal Technol. Inter. **5**(3) (2021). https://www.mdpi.com/2414-4088/5/3/7
31. Horváth, I., Csapó, Á.B.: Motivations and tools relevant to personalized workspaces in VR environments. Electronics **12**(9), 2059 (2023). https://www.mdpi.com/2079-9292/12/9/2059
32. Baranyi, P., Csapo, A.B.: Revisiting the concept of generation CE - generation of cognitive entities. In: 2015 6th IEEE International Conference on Cognitive Infocommunications (CogInfoCom), pp. 583–586. IEEE (2015)
33. Horváth, I.: Behaviors and capabilities of generation CE students in 3D VR. In: Baranyi, P. (ed.) Proceedings of the 10th IEEE International Conference on Cognitive Infocommunications: CogInfoCom 2019, pp. 491–494. IEEE (2019)
34. Horváth, I.: The 4th dimension of personalization in vr. In: 2022 1st IEEE International Conference on Cognitive Aspects of Virtual Reality (CVR), pp. 000 085–000 088. IEEE (2022)
35. Green, T.M., Ribarsky, W., Fisher, B.: Building and applying a human cognition model for visual analytics. Inf. Vis. **8**(1), 1–13 (2009)
36. Berki, B.: Level of presence in MaxWhere virtual reality. In: Baranyi, P. (ed.) Proceedings of the 11th IEEE International Conference on Cognitive Infocommunications (CogInfoCom), pp. 511–545. IEEE (2020)
37. Lampert, B., Pongracz, A., Sipos, J., Vehrer, A., Horvath, I.: MaxWhere VR-learning improves effectiveness over classical tools of e-learning. Acta Polytech. Hungarica **15**(3), 125–147 (2018)
38. Horváth, I., Sudár, A.: Factors contributing to the enhanced performance of the MaxWhere 3D VR platform in the distribution of digital information. Acta Polytech. Hungarica **15**(3), 149–173 (2018)
39. Horváth, I.: MaxWhere 3D capabilities contributing to the enhanced efficiency of the Trello 2D management software. Acta Polytech. Hungarica **16**(6), 55–71 (2019)
40. Sudár, A., Csapó, Á.B.: Interaction patterns of spatial navigation in VR workspaces. In: Baranyi, P. (ed.) Proceedings of the 10th IEEE International Conference on Cognitive Infocommunications: CogInfoCom 2019, pp. 611–614. IEEE (2019)
41. Horváth, I., Berki, B.: Unlocking the potential of immersive VR systems: the role of gesture and natural controls. In: 2023 IEEE 2nd International Conference on Cognitive Aspects of Virtual Reality (CVR). IEEE (2023)
42. Sudár, A., Csapó, Á.B.: Elicitation of content layout preferences in virtual 3D spaces based on a free layout creation task. Electronics **12**(9), 2078 (2023). https://www.mdpi.com/2079-9292/12/9/2078
43. Baranyi, P., Galambos, P., Csapó, Á., Jaloveczki, L.: Cognitive navigation and manipulation (coginav) method. US Patent App. 15/658,579. Accessed 1 Feb 2018
44. StatCounter: Desktop browser market share Hungary (2018). http://gs.statcounter.com/browser-marketshare/desktop/hungary. Accessed 26 July 2018

Part III
Mapping Natural Cognitive Spatial Capabilities Into VR

In the past few years, Virtual Reality (VR) and its related technologies such as Augmented Reality (AR), Mixed Reality (MR), and Extended Reality (XR) have emerged as transformative forces in computing. These technologies introduce spatial content and relationships, fundamentally changing how humans perceive and interact with digital content. This part investigates the effects of this new digital environment on human cognitive abilities, examining how it alters the way individuals conceptualize and engage with digital interactions.

Exploring new environments, whether real or simulated, involves an initial phase of investigation to discern advantageous perspectives. However, this task becomes particularly challenging in virtual reality (VR) due to the intricate spatial layouts. Nevertheless, manipulating spatial arrangements and visual cues can simplify navigation. In Chap. 9, an experiment conducted in a desktop VR setting highlights the influence of spatial layout and key discovery points on task completion time. Our findings underscore the importance of strategic node identification and data-driven approaches in navigating virtual spaces effectively.

Chapter 10 introduces the potential of VR as a platform for communication and content management. Here, our focus shifts to the challenge of automating navigation path generation within VR spaces. By breaking down this issue into two components—clustering 2D displays and leveraging these clusters to identify points of interest (POIs)—we propose a novel Markov Chain Monte Carlo-based method. Through an initial implementation evaluated within the MaxWhere platform's 3D spaces, we aim to simplify navigation in complex VR environments.

As three-dimensional VR environments continue to reshape digital content interaction, Chap. 11 investigates the arrangement of digital content within these spaces. While the advantages of 3D interfaces over their 2D counterparts are evident, standardized guidelines for content arrangement remain scarce. This chapter presents findings from an experiment aimed at understanding user preferences in arranging digital content within 3D environments, offering insights for future design methodologies and potential automation solutions.

Chapter 12 presents the conversion of traditional 2D digital layouts into threedimensional (3D) dashboards to alleviate cognitive load. By comparing user performance metrics across 2D and 3D scenarios, we explore the feasibility of reducing cognitive load in VR environments. While our findings suggest promising outcomes, we emphasize the importance of optimizing 3D workflows to achieve cognitive load reduction effectively. Strategic design considerations emerge as crucial elements in enhancing user experience within VR environments.

Chapter 9
Preference Points and Interaction Patterns in Virtual Reality

Abstract During the exploration of a new environment, whether it is tangible or simulated, the initial step typically involves the act of investigation. The objective of this investigative process within this particular framework is to locate a position that offers the highest amount of informative or otherwise advantageous perspectives within the space. Depending on the intricacy of the space and the specific task at hand, this endeavour can prove to be quite challenging. However, in the realm of virtual reality, the identification of valuable viewpoints can be made more feasible by altering the arrangement of the space, as well as modifying other visual characteristics in order to reduce the cognitive burden of navigation. Within the confines of this scholarly article, we present a series of experiments carried out in a desktop virtual environment, which suggest that in addition to the spatial layout, the key points of discovery within a space can also impact the time required to successfully complete a task. This assertion is corroborated by data pertaining to the key points of navigation, as well as the activation mode employed on the various 2D displays (smartboards) situated in a three-dimensional space. In the course of this research, we demonstrated how specific nodes can be identified using a data-driven approach, thereby enabling users to concentrate on the most crucial information and gain a comprehensive overview of any given virtual space.

Keywords Virtual reality · Spatial abilities · Spatial perception · Preference points

9.1 Introduction

Since the inception of scientific inquiry, there has been an enduring question regarding the influence of the environment on human behavior. This inquiry remains pertinent in the realm of Cognitive Infocommunications (CogInfoCom), particularly when examining the utility of virtual reality (VR) environments as infocommunication systems [1, 2]. The field of cognitive science, particularly in recent years, has given rise to a new scientific discipline known as information modeling, cognitive informatics. This discipline focuses on the engineering informatics modeling of brain processes involved in perception, cognition, and comprehension. CogInfoCom, in turn, aims

I. Horváth et al., *Cognitive Aspects of Virtual Reality*, Studies in Big Data 156,
https://doi.org/10.1007/978-3-031-68130-1_9

to facilitate effective communication between engineering systems and humans by leveraging these cognitive information technology processes and established models. The ultimate objective of this field is to develop sophisticated sensor-based IT systems that enhance human-machine communication [3–15]. This is supported by novel methodologies, mathematical modeling [16, 17], learning techniques [18, 19] and associated behavioral research within the field, which contribute to a better understanding of perceptual and brain processes [20–23].

CogInfoCom solutions that rely on virtual reality have become increasingly prevalent [24–29]. These solutions involve the use of three-dimensional virtual spaces that users can explore in a manner akin to real-life environments. Navigating one's surroundings, whether familiar or unfamiliar, is a ubiquitous experience for most individuals. Therefore, VR spaces utilized as infocommunication tools can both facilitate and benefit from research on human spatial skills, including spatial cognition, orientation, and navigation.

Spatial orientation and perception have been long-standing focal points in both the cognitive sciences and CogInfoCom. Considerable progress has been made in elucidating the neuroanatomical underpinnings of spatial representations. Notably, the discovery of place cells [30] and grid cells [31] has provided valuable insights. Furthermore, researchers have made strides in unraveling the specific roles played by different brain area [32] in spatial navigation.

9.2 Classical Theories of Visual-Spatial Cognition

9.2.1 Processing, Grouping and Recall of Visual Objects

In this section, a comprehensive examination is provided of cognitive theories that pertain to the processing, organization, and recollection of information, particularly within the realm of visual perception. The primary insight derived from this examination is that the manner in which individuals process spatial information, navigate within physical space, and conceptually represent spatial relationships is contingent upon numerous factors, some of which encompass perceptual, higher-level cognitive, and even emotional elements. Consequently, the task of predicting user preferences in three-dimensional virtual environments is by no means a simplistic endeavour, and these predictions must be augmented through the inclusion of usability experiments.

9.2.1.1 Perception of Objects, Shapes and Forms

The Gestalt theory, which historically served as the initial psychophysical model to elucidate the arrangement of perceived images into objects, was formulated through the combined efforts of Wertheimer, Koffka, and Köhler. An essential tenet of the Gestalt approach is the assertion that the entirety of an object surpasses the mere summation of its constituent parts (i.e., the individual segments that comprise the objects). The work of Gestalt psychologists predominantly revolved around three primary domains of investigation [33–37]: (1) The relationships between shape and background; (2) Grouping rules; and (3) The "goodness" of shapes.

Within these domains, six principles of Gestalt were identified by them: the principle of Similarity, the principle of Good Figure, the principle of Proximity, the principle of Continuity, the principle of Closure, and the principle of Common Region.

The principles known as the Law of Common Region and the Law of Proximity hold considerable significance when it comes to comprehending the grouping of information. These principles are especially pertinent in situations where the aim is to actively enhance the speed and quality of understanding and recollection of an image. Such outcomes are crucial in the domain of virtual reality (VR) for effectively conveying information at a sophisticated level. Specifically, in accordance with the Law of the Common Region, the human brain is anticipated to unite elements that reside in a shared, enclosed area. Even in present times, designers of user interfaces and user experiences continue to rely on this principle when organizing information within a confined space and employing the same color scheme, whether it be in gaming or social interfaces, for instance. On the other hand, the Law of Proximity dictates that objects in close proximity to one another are more likely to be perceived as a group of interconnected elements. By adhering to this principle, it is feasible to clearly communicate the association of information, even in scenarios where creating enclosed regions is not feasible [38, 39].

9.2.1.2 Human Spatial Behavior

Despite the inherent constraints posed by the human sensory system and the limited capabilities of locomotion, individuals possess the dexterity to skillfully maneuver through their surroundings. This observation implies that proficient spatial orientation is not solely contingent upon the perceptual and/or motor aptitude exhibited. The capacity of individuals to adeptly and purposefully interact with their immediate habitat is frequently denoted as human environmental proficiency [40].

Human environmental competence consists of several components [40, 41]:

- The perceptual component refers to the process of recognizing, emphasizing, and assigning importance to the fundamental qualities of the surrounding surroundings.
- The cognitive component carries out the analysis of characteristics emphasized during the process of perception, in addition to the retention, arrangement, and retrieval of the characteristics that obtain significance in this manner. Consequently, this leads to the acquisition of knowledge pertaining to the environment.
- The constituents of the affective component are affirmative and adverse emotional and motivational reactions to the attributes of the surroundings, behavioral responses, and personal inclinations that act as intermediaries between the aforementioned constituents.

9.2.1.3 Categorization and Conceptual Representation

Tversky's research, which was published in 2003, examines various types of cognitive spatial reference frames employed by individuals to situate themselves in physical space. Each cognitive subsystem within the intricate and multifaceted mental space fulfills a distinct purpose and encompasses diverse spatial components and frames of reference. Although it is plausible to assume that each of these subsystems possesses a distinct cognitive structure, they are all characterized by the process of schematization. This cognitive process serves to reduce the cognitive load on memory, facilitate information processing, and enable the assimilation of a wide array of information, albeit with the trade-off of some degree of error and distortion. There are two overarching forms of schematization: categorization and chunking [42, 43].

In the world in general, as well as within one's limited surroundings, an abundance of information is encountered, which shall be required to be recollected at some point. This is substantiated by numerous cognitive processes, inclusive of categorization. Said processes primarily occur spontaneously and possess the capacity to impact all modalities. The cognitive processes that facilitate categorization render it feasible for individuals to distinguish between objects arranged within distinct categories based on specific attributes. However, in the event of objects pertaining to the same category, minor disparities are disregarded, and attention is directed towards shared qualities.

Additionally, the organization of data into segments can further enhance the efficiency of both the cognitive processes of information processing and information recall. Within the confines of this particular framework, one of the most noteworthy findings in the field of cognitive science has been the revelation of chunking, a cognitive process that entails the decomposition of information into smaller, more easily manageable units, thereby rendering the information more comprehensible [44, 45]. It is a cognitive approach for the retention of information that can be intentionally implemented, yet simultaneously manifest subconsciously as a result of exposure, thereby enabling individuals to retrieve substantial quantities of knowledge [46–48]. For an extended period, it has been understood that the ability of working memory is constrained in regards to the manipulation and retention of data. However, utilizing this cognitive approach can lead to a decrease in the quantity of information that needs to be processed simultaneously [49]. Furthermore, the effectiveness of this procedure is additionally substantiated by the reality that it frequently relies on a systematizing principle that not only employs more compact constituent units but also possesses semantic coherence, consequently leading to memories that are more readily retrievable [46, 50].

In numerous instances, nevertheless, models for acquiring information have been shown to be insufficient in the previous years due to their failure to consider the cognitive condition of individuals, as well as the cognitive approaches they might employ. The information search model proposed by Pirolli and Card in 1999 acknowledges this limitation and asserts that individuals seeking information tend to search and categorize information into clusters in order to reduce the cost associated with searching across clusters [51]. Clusters are observed in both physical and digital manifestations. Additionally, a crucial facet entails the user's cognitive framework, which elucidates the dissimilarities in information retrieval behaviors. Disparities in the cognitive framework are frequently influenced by the cognitive, demographic, and social attributes of the particular user [52].

9.2.2 *Spatial Cognition and Spatial Navigation*

Spatial ability can be understood in some sense as the ability to generate, retain, retrieve, and transform well-structured visual images. However, the concept of spatial ability is not a unitary construct. There are particular spatial abilities and each highlights different aspects of the process of image generation, storage, retrieval, and transformation [53]. Spatial ability is defined by two significant psychometric constructs: spatial orientation and spatial visualization [54]. Within the inherent limitations of the human sensory and motor system, individuals possess adeptness in navigating and orienting themselves in their surroundings. Consequently, spatial orientation is not solely a task related to perception and/or motion, but it is also closely associated with visualization and imagination [55]. Among all cognitive abilities, gender differences in spatial abilities are believed to be the most pronounced [56]. Spatial abilities and orientation skills exhibit distinct characteristics. For instance, orientation skills invariably involve an environment and entail movement (actual navigation or imagined map scanning), as well as the acquisition of information about the surroundings [57].

Even today, there remains a lack of consensus among researchers regarding the definition of spatial visualization. According to a definition from the 1990s, spatial visualization refers to the mental manipulation of spatial information in order to determine the appearance of a given spatial configuration after certain portions of that configuration have been rotated, folded, repositioned, or otherwise transformed [58]. Another definition asserts that spatial visualization involves complex, multi-step manipulations of spatially presented information [56]. A third definition states that spatial visualization is the ability to mentally manipulate an object or pattern [59]. In an attempt to reconcile some of these discrepancies, researchers have started categorizing spatial skills to account for the absence of a single, all-encompassing definition of spatial visualization skills [60].

9.2.2.1 Spatial Visualization and Orientation

By utilizing spatial visualization skills, individuals are capable of mentally manipulating an object, whereas spatial orientation skills pertain more to the mental movement of the individual's own viewpoint while the object itself remains fixed in space [60].

Spatial orientation is a complex process that relies on numerous fundamental cognitive functions [57] and it refers to one's ability to imagine the appearance of an object from different perspectives [61].

Spatial orientation and perception have been extensively studied in both the cognitive sciences and CogInfoCom, as significant progress has been made in understanding the neural basis of spatial representations. In addition to the identification of place cells [30] and grid cells [31], scientists have also made strides in comprehending the role of each brain region in spatial navigation [32].

Numerous studies have been conducted to measure spatial orientation skills in individuals, both in real and virtual environments. Real environments include a university campus [62], the interior of buildings for indoor wayfinding tasks [63], and even a woodland environment [64]. Spatial orientation tasks have experienced a growing trend towards simulation in virtual 3D environments since the early 2000s [65–67].

9.2.2.2 Spatial Navigation

Spatial navigation is also a fundamental cognitive capability observed in human beings. Navigation is defined as the ability to move oneself within the physical world by utilizing spatial information. In everyday life, the need for spatial navigation arises frequently, regardless of whether one is in a familiar environment or not [68]. The classification of human spatial navigation typically encompasses three main types of knowledge: landmark, route, and survey knowledge [69].

Spatial navigation has been a prominent focus of research for several decades due to its significant role in integrating and connecting various cognitive functions, such as cognition, memory, learning, and neuropsychology [70, 71]. Moreover, spatial navigation is closely intertwined with the previously mentioned capability of spatial orientation. Within this context, there are two distinct types of navigation that can be employed: active navigation, which enables users to freely explore the environment, and passive navigation, which restricts users to an observational position [72]. Studies have illuminated the cognitive components of active navigation that are beneficial for memory [73].

Visual-spatial cognition, a fundamental aspect of cognitive processing, plays a crucial role in how individuals perceive and navigate their surroundings. This cognitive function is intricately linked with Cognitive Load Theory, as the efficient utilization of visual-spatial processing can significantly impact the allocation of cognitive resources, thereby influencing the overall cognitive load experienced during tasks. Recognizing the interplay between visual-spatial cognition and cognitive load theory provides valuable insights into optimizing information presentation and design to enhance cognitive efficiency [74, 75].

In this section, we will describe the results of an experiment that we conducted in order to gain a better understanding of the presence and characteristics of prominent spatial viewpoints in three-dimensional virtual reality (VR) spaces, as well as common methods of interacting with two-dimensional content within the VR environment.

9.3 Aim of the Study

The aim of the study was to extend our former research presented at the 10th IEEE International Conference on Cognitive Infocommunications [76]. In our earlier study, I successfully identified specific areas within the virtual space (Fig. 9.1) that users were more inclined to visit in order to effectively oversee the space and accomplish their tasks. Additionally, I conducted an analysis on the most commonly utilized modes of activation when accessing the smartboards situated within the three-dimensional environment.

The current study complements the aforementioned research by conducting an experiment in a separate virtual space, where the layout of the two-dimensional display boards, known as "smartboards" in the MaxWhere system, differed significantly. In this new environment, the arrangement of the smartboards consisted of a greater number of distinct groups, with each group containing multiple smartboards. The primary challenge associated with passive navigation is that the creator or designer of the virtual space must determine a set of viewpoints, referred to as the story or guided tour in MaxWhere, that will be most beneficial for the majority of users. Ideally, it would be advantageous to have the capability to generate these stories solely based on patterns of past usage, without relying on human intuition.

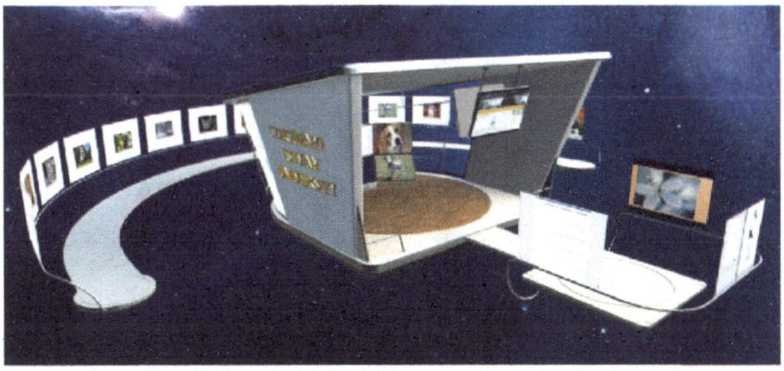

Fig. 9.1 The MaxWhere 3D VR space in which the former study was conducted

9.3.1 Modalities of Interaction at the Center of Analysis

In MaxWhere, users can interact with smartboards in two different ways as outlined below.

- Capsule-based activation On one hand, the activation and deactivation of smartboards can be accomplished by utilizing the "capsules" positioned at the top and center of each smartboard window. If the user clicks once on the smartboard, it will become activated while the user remains in the same location within the virtual space. However, in this scenario, the smartboard will be activated but not set to fullscreen. To obtain a close-up view of the smartboard's content, the user is required to click on a blue icon, or "capsule," located at the top center of the screen. To exit the smartboard and return to three-dimensional spatial navigation, users can click on a red icon, or "red capsule," adjacent to the blue one (Fig. 9.2).
- Gesture-based activation On the other hand, there exists a more natural approach to activate and deactivate smartboards. By double-clicking, the smartboard will become active and the camera will automatically zoom in towards it. Subsequently, by moving the cursor to the left or right edge of the screen, the smartboard can be automatically deactivated without the need to click a button.

We hypothesized that it would be possible to identify preference points within the three-dimensional virtual space, where users tend to spend a greater amount of time in order to accomplish a given task. Furthermore, we hypothesized that there would be variations in the modalities of activation and deactivation employed when interacting with the smartboards.

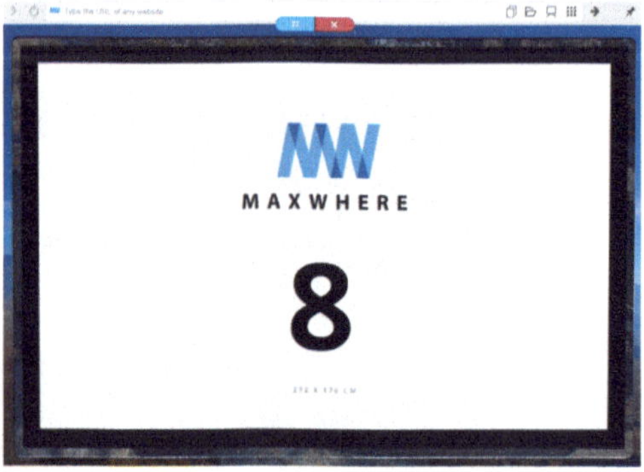

Fig. 9.2 The "Blue" and "Red" capsule used for the activation and deactivation of the smartboards

9.4 Methods

9.4.1 Subjects

Twenty-nine individuals with typical development participants (28 male and 1 female) took part in the experiment.All participants were university students from Hungary, aged between 20 and 26 years. The mean age was 22.68 (SD:1.63). Each participant voluntarily volunteered to partake in the experiment, possessing a rudimentary level of proficiency in utilizing the MaxWhere software, owing to their prior exposure to it. It is worth noting that the participants were native speakers of the Hungarian language.

9.4.2 Maxwhere Virtual Reality Environment

To facilitate the sharing and organization of documents and webpages, the experiment was conducted within a replication of the Fly Podium space. The space includes 22 Smartboards (Fig. 9.3).

9.4.3 Procedure

Initially, in early 2020, the measurements were undertaken at Széchenyi István University. However, due to the disruptions wrought by the COVID-19 pandemic, all further communication and data collection were facilitated through online means, which included remote logging incorporated into the space. In this particular scenario,

Fig. 9.3 The MaxWhere 3D VR space, where the measurement was conducted

the assigned task remained unaltered, albeit the students received the necessary information via email. Once the participants entered the virtual space, it took no longer than 10 min for them to successfully complete the assigned tasks and subsequently fill out the pertinent questionnaires.

Four questionnaires, pertaining to the content positioned within the virtual space, were administered to the participants. Additionally, a general form was given to them, wherein they provided their consent for the storage of their data, solely for the purpose of this research. Furthermore, participants were requested to disclose their video game-related habits, thereby enabling the assessment of their familiarity with Desktop VR.

From a content perspective, the tasks were deemed facile; however, they necessitated intricate navigational movements from the users. The questions pertaining to the digital content merely required a counting of diverse animal types-an approach similar to the experiment elucidated in [76, 77]. The questionnaire encompassed indispensable information concerning the spatial arrangement of the smartboards, such as the statement, "This answer sheet corresponds to the PDF document situated beneath this smartboard." These details were essential for providing answers to specific questions, such as "How many cats are depicted in the images?" It is important to highlight that no specific sequential order was prescribed for the completion of the tasks.

9.4.4 *Measurements Recorded*

9.4.4.1 Spatial Preference Points

Throughout the course of the experiment, the positions and orientations of the participants' cameras were logged at intervals of one second. Subsequently, a frequency analysis was conducted based on these records, amounting to a total of 22, 221 camera poses. Results are visualized in Figs. 9.4 and 9.5.

9.4.4.2 Activation and Deactivation of Smartboards

Recording the differences between the activation and deactivation modalities employed during interactions with the smartboards was an additional aspect of the study. Specifically, the number of clicks on the blue and red capsules, as well as the instances where participants opted to zoom in automatically through double-clicking, as opposed to single-clicking the smartboards, were documented. Results are visualized in Fig. 9.6.

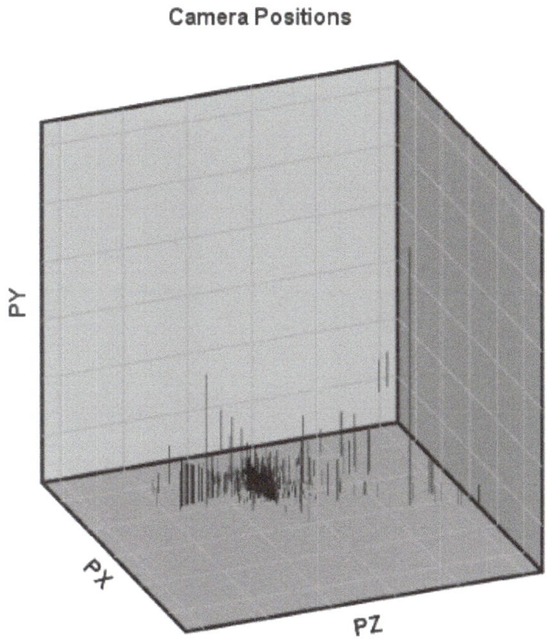

Fig. 9.4 Frequency histogram of the nodes visited in the virtual space

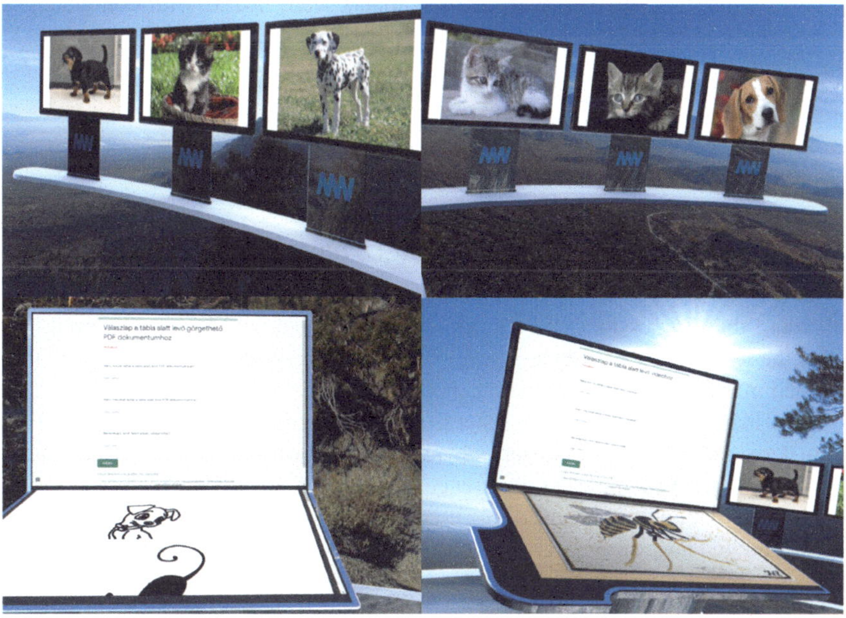

Fig. 9.5 Some of the most frequently visited camera poses within the virtual space

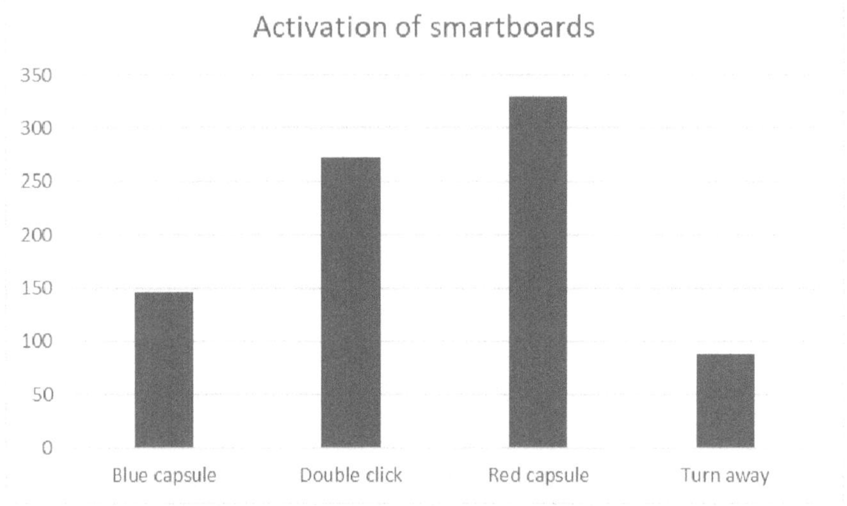

Fig. 9.6 Activation of smartboards

9.4.4.3 Video Game Usage Habits

At the beginning of the experiment, participants were tasked with completing a general questionnaire encompassing elements that surveyed their computer and video game usage habits.

9.5 Results

9.5.1 Spatial Preference Points and Orientations

Following the test sessions, the log files were analyzed, and it was observed that out of the total of 22, 221 camera poses, there were 34 nodes and orientations that occurred more frequently than the others (Figs. 9.4, and 9.5).

The visualization in Fig. 9.4 presents all the points visited in the virtual space. This figure depicts the nodes and clusters where the navigation activity was concentrated. In Fig. 9.5, several examples of preference points are displayed, indicating the position and orientation preferences of the users. These preference points were found to be concentrated near the answer sheets and in front of information groups, where the necessary information could be easily accessed in order to solve the questionnaires quickly and efficiently.

Based on an analysis of these 34 nodes (Fig. 9.5), it was possible to identify 12 areas (clusters) that were situated close to the questionnaires and different content types.

9.5.2 *Activation and Deactivation of Smartboards*

The results of a Friedman's ANOVA test showed a statistically significant difference in the deactivation (but not activation) of smartboards ($X_F^2(2) = 20.155$, $p < 0.001$) (Fig. 9.6). Post-hoc pairwise comparisons indicated that there was no significant difference between the blue capsule and the double click. However, the users were found to leave the smartboards with the red capsule click significantly more often ($Z = 3.28$, $p < 0.001$) than by turning away from them.

9.5.3 *Video Game Usage Habits*

In the general questionnaire, participants were given the opportunity to select multiple platforms that they use for playing video games. The most preferred platform was found to be the PC, with 86% of the students using a personal computer or laptop for gaming. There was no significant difference between the two consoles (PlayStation, Xbox), and out of the 29 subjects, only 4 individuals (1 female, 3 male) reported that they do not play video games at all. Additionally, only one participant mentioned that they use Nintendo consoles.

Overall, the questionnaire results support the notion that Desktop VR is a familiar environment for many users and should be further investigated for its effectiveness in everyday computing tasks.

9.6 Discussion

In the study, a total of 29 university students (28 male, 1 female), all of whom had prior experience with the MaxWhere Virtual Reality Platform, participated. These participants were instructed to complete five questionnaires (one general and four specific) while aiming for both speed and accuracy. Based on the responses to the general questionnaire, it was found that the students spend an average of 5.69 h per day using a computer. Only four participants reported not playing video games, and among those who do, 86.2% play them on a PC. The use of other gaming consoles is less common. When it comes to video game genres, the participants had varying preferences. The most popular genre was Action games, played by 88.5% of the participants, but Strategy (65.4%) and Role-Playing games (57.7%) were also quite popular. Overall, these findings reinforce the idea that Desktop VR is a widely embraced media format that merits further exploration in terms of its effectiveness in everyday computing tasks.

In this study, it was found that a total of more than 34 nodes were discovered among the camera poses. This finding supports the hypothesis that not all nodes, out of over 22 thousand candidates, would have equal measurements. The task at hand was fixed, meaning that all subjects followed the same workflow. On average, approximately 760 points were measured, which equates to 760 s of recorded interactions, given a sampling rate of 1 pose per second and a total of $22,221/29$ measurements. By considering the ratio between the 34 identified nodes and the 760 measurements, it can be concluded that the presence of these nodes can result in a compression rate of over 95% in terms of navigation complexity, specifically in relation to the camera poses visited.

In addition to the identified nodes, an additional 12 regions were clearly distinguishable where the camera positions were mainly concentrated. These preference points were situated in the vicinity of the questionnaires and the various types of content. Participants were able to locate specific areas that provided an overview of the documents, such as the middle of the space, a plan view, and two positions in the semicircles at the back. These locations allowed participants to gather the necessary information to complete their tasks and review the associated content. During the task-solving process, users minimized their movements and navigated naturally to ensure precision in completing their tasks and finding a viewpoint that enabled them to access the required content. By actively exploring the virtual space, a map can be constructed to outline a storyline consisting of preference points. This framework guides users, reduces navigational movements and operations, facilitates faster task completion, and helps users experience less cognitive load simultaneously [78]. Furthermore, some of the identified usage patterns indicate the potential emergence of a new cognitive capability in virtual reality (VR) spaces, namely the ability to position oneself in a way that allows for a holistic overview of multiple content elements simultaneously, which is not feasible in 2D interfaces.

Certain nodes that were identified were situated in front of the smartboards displaying the questionnaires. However, an interesting observation was that users chose to navigate to and position themselves in front of each distinct group of tables, even if they could have simply turned left and right to access all the necessary information for a specific questionnaire. Another counter-intuitive observation was that users did not seem to follow the configuration of the floor-like elements in the space during navigation. No clear pattern was identified in this regard, suggesting that the design of the space could have been improved to better guide users' active navigation.

Besides these findings, there were significant disparities observed between the two methods of deactivating the smartboard. Participants demonstrated a greater inclination towards utilizing the option to deactivate the smartboard through the use of a capsule, as opposed to simply turning away from the smartboard. Opting to turn away would have necessitated dragging the cursor to the outermost edge of the screen. Although the use of this alternative approach was not infrequent (and also indicated the emergence of a novel cognitive aptitude not previously encountered in two-dimensional interfaces), it transpired significantly less often. In a previous study [76], the findings revealed notable discrepancies in terms of smartboard activation, with users frequently opting for the double-clicking option (the deactivation aspect was not examined in that particular study). However, the findings of the present study did not corroborate this; nevertheless, the double-clicking option remained more prevalent.

The fact that users exhibited a preference for double-clicking in one instance, and the exit capsule in the other, appears to be contradictory as the "gesture-based" equivalent of double-clicking to activate a smartboard would entail turning away from the device. Nonetheless, taking into account users' inherent inclinations based on years of familiarity with conventional two-dimensional graphical user interface (GUI) metaphors, both the double-click and the X (exit) button for closing a window appear to be iconic metaphors that have become deeply ingrained in users' intuition. Thus, despite the potential for turning away to be a more instinctual movement from a kinesthetic perspective, the shared experience of clicking a red X button may have been the prevailing factor. Consequently, these findings may indicate the persistence of ingrained habits, thereby warranting further investigation to disentangle the influences of habit and convenience in shaping user interactions.

References

1. Baranyi, P., Csapó, Á.: Definition and synergies of cognitive infocommunications. Acta Polytechnica Hungarica **9**(1), 67–83 (2012)
2. Baranyi, P., Csapo, A., Sallai, G.: Cognitive Infocommunications (CogInfoCom). Springer (2015)
3. Poggi, I., D'Errico, F., Vincze, L.: Uncertain words uncertain texts. perception and effects of uncertainty in biomedical communication. Acta Polytechnica Hungarica **16**(2), 13–34 (2019)
4. Tariq, M., Trivailo, P.M., Shoji, Y., Simic, M.: Comparison of event-related changes in oscillatory activity during different cognitive imaginary movements within same lower-limb. Acta Polytechnica Hungarica **16**(2), 77–92 (2019)
5. Szaszák, G.: An audio-based sequential punctuation model for asr and its effect on human readability. Acta Polytechnica Hungarica **16**(2), (2019)
6. Garai, Á., Péntek, I., Adamkó, A.: Revolutionizing healthcare with IoT and cognitive, cloud-based telemedicine. Acta Polytechnica Hungarica **16**(2), 163–181 (2019)
7. Vogel, C., Esposito, A.: Interaction analysis and cognitive infocommunications. Infocommunications J. **12**(1), 2–9 (2020)
8. Kovács, Á.: Evaluation of the aggregation capability of the mpt network layer multipath communication library and multipath tcp. Acta Polytechnica Hungarica **16**(6), 129–147 (2019)
9. Suzić, S., Delić, T., Pekar, D., Delić, V., Sečujski, M.: Style transplantation in neural network based speech synthesis. Acta Polytechnica Hungarica **16**(6), 171–189 (2019)
10. Kovari, A., Katona, J., Costescu, C.: Evaluation of eye-movement metrics in a software debbuging task using gp3 eye tracker. Acta Polytechnica Hungarica **17**(2), 57–76 (2020)
11. Szücs, V., Guzsvinecz, T., Magyar, A.: Movement pattern recognition in physical rehabilitation-cognitive motivation-based it method and algorithms. Acta Polytechnica Hungarica **17**(2), 211–235 (2020)
12. Monsone, C., Csapó, Á.B.: Charting the state-of-the-art in the application of convolutional neural networks to quality control in industry 4.0 and smart manufacturing. In: 2019 10th

IEEE International Conference on Cognitive Infocommunications (CogInfoCom), pp. 463–468. IEEE (2019)
13. Kan, M., Ito, A.: Language cognition and pronunciation training using applications. Futur. Internet **12**(3), 42 (2020)
14. Ujbányi, T., Kővári, A., Sziládi, G., Katona, J.: Examination of the eye-hand coordination related to computer mouse movement. Infocommunications J. **12**(1), 26–31 (2020)
15. Erdős, F., Németh, R.: Amt-based real-time, inter-cognitive communication model. Acta Polytechnica Hungarica **16**(6), 115–127 (2019)
16. Bubnó, K., Takács, V.L.: Cognitive aspects of mathematics-aided computer science teaching. Acta Polytechnica Hungarica **16**(6), 73–93 (2019)
17. Kovari, A., Rajcsanyi-Molnar, M.: Mathability and creative problem solving in the MaTech math competition. Acta Polytechnica Hungarica **17**(2), (2020)
18. Pinter, R., Maravić, Č, Balogh, Z., Manojlović, H.: Enhancing higher education student class attendance through gamification. Acta Polytechnica Hungarica **17**(2), 13–33 (2020)
19. Nagy, B., Váraljai, M., Kollár, A.M.: E-learning spaces to empower students collaborative work serving individual goals. Acta Polytechnica Hungarica **17**(2), 97–114 (2020)
20. Hercegfi, K., Köles, M.: Temporal resolution capabilities of the mid-frequency heart rate variability-based human-computer interaction evaluation method. Acta Polytechnica Hungarica **16**(6), 95–114 (2019)
21. Kosztolányi-Iván, G., Koren, C., Borsos, A.: Can people recognize more than six road categories? Acta Polytechnica Hungarica **16**(6), (2019)
22. Costescu, C., Rosan, A., Hathazi, A., Pădure, M., Brigitta, N., Kovari, A., Katona, J., Thill, S., Heldal, I.: Educational tool for testing emotion recognition abilities in adolescents. Acta Polytechnica Hungarica **17**(2), (2020)
23. D'Errico, F., Paciello, M., Fida, R., Tramontano, C.: Effect of affective priming on prosocial orientation through mobile application: differences between digital immigrants and natives. Acta Polytechnica Hungarica **16**(2), 109–128 (2019)
24. Korečko, Š, Hudák, M., Sobota, B.: Lirkis cave: architecture, performance and applications. Acta Polytechnica Hungarica **16**(2), 199–218 (2019)
25. Hercegfi, K., Komlódi, A., Köles, M., Tóvölgyi, S.: Eye-tracking-based wizard-of-oz usability evaluation of an emotional display agent integrated to a virtual environment. Acta Polytechnica Hungarica **16**(2), 145–162 (2019)
26. Kuczmann, M., Budai, T.: Linear state space modeling and control teaching in MaxWhere virtual laboratory. Acta Polytechnica Hungarica **16**(6), 27–39 (2019)
27. Guzsvinecz, T., Orbán-Mihálykó, É., Perge, E., Sik-Lányi, C.: Analyzing the spatial skills of university students with a virtual reality application using a desktop display and the gear VR. Acta Polytechnica Hungarica **17**(2), 35–56 (2020)
28. Markopoulos, E., Luimula, M.: Immersive safe oceans technology: Developing virtual onboard training episodes for maritime safety. Futur. Internet **12**(5), 80 (2020)
29. Horváth, I.: Maxwhere 3d capabilities contributing to the enhanced efficiency of the trello 2d management software. Acta Polytechnica Hungarica **16**(6), 55–71 (2019)
30. O'Keefe, J., Dostrovsky, J.: The hippocampus as a spatial map: preliminary evidence from unit activity in the freely-moving rat. Brain Res. (1971)
31. Hafting, T., Fyhn, M., Molden, S., Moser, M.-B., Moser, E.I.: Microstructure of a spatial map in the entorhinal cortex. Nature **436**(7052), 801–806 (2005)
32. Nadasdy, Z., Nguyen, T.P., Török, Á., Shen, J.Y., Briggs, D.E., Modur, P.N., Buchanan, R.J.: Context-dependent spatially periodic activity in the human entorhinal cortex. Proc. Natl. Acad. Sci. **114**(17), E3516–E3525 (2017)
33. Wertheimer, M.: Gestalt Theory, pp. 1–11. Kegan Paul, Trench, Trubner & Company (1938)
34. Koffka, K.: Principles of Gestalt psychology, vol. 44. Routledge (2013)
35. Kohler, W.: The Mentality of Apes. Routledge (2018)
36. Arnheim, R.: Visual Thinking. University of California Press (1997)
37. Rock, I.: The Logic of Perception. MIT Press, Cambridge (1983)

38. Palmer, S.E.: Common region: a new principle of perceptual grouping. Cogn. Psychol. **24**(3), 436–447 (1992)
39. Michotte, A.: The Perception of Causality, vol. 21. Routledge (2017)
40. Pedersen, D.M.: Dimensions of environmental competence. J. Environ. Psychol. **19**(3), 303–308 (1999)
41. Steele, F.: Defining and developing environmental competence. Adv. Exp. Soc. Process. **2**(1), 225–244 (1980)
42. Tversky, B.: Structures of mental spaces: How people think about space. Environ. Behav. **35**(1), 66–80 (2003)
43. Tversky, B.: Cognitive maps, cognitive collages, and spatial mental models. In: Spatial Information Theory A Theoretical Basis for GIS: European Conference, COSIT'93 Marciana Marina, Elba Island, Italy September 19–22: Proceedings, vol. 2005, pp. 14–24. Springer (1993)
44. Gobet, F., Lane, P.C., Croker, S., Cheng, P.C., Jones, G., Oliver, I., Pine, J.M.: Chunking mechanisms in human learning. Trends Cogn. Sci. **5**(6), 236–243 (2001)
45. Alamia, A., Solopchuk, O., d'Ausilio, A., Van Bever, V., Fadiga, L., Olivier, E., Zénon, A.: Disruption of Broca's area alters higher-order chunking processing during perceptual sequence learning. J. Cogn. Neurosci. **28**(3), 402–417 (2016)
46. Bor, D., Duncan, J., Wiseman, R.J., Owen, A.M.: Encoding strategies dissociate prefrontal activity from working memory demand. Neuron **37**(2), 361–367 (2003)
47. Sakai, K., Kitaguchi, K., Hikosaka, O.: Chunking during human visuomotor sequence learning. Exp. Brain Res. **152**, 229–242 (2003)
48. Rosenbaum, D.A., Kenny, S.B., Derr, M.A.: Hierarchical control of rapid movement sequences. J. Exp. Psychol. Hum. Percept. Perform. **9**(1), 86 (1983)
49. Baddeley, A.: Working memory. Science **255**(5044), 556–559 (1992)
50. Koch, I., Hoffmann, J.: Patterns, chunks, and hierarchies in serial reaction-time tasks. Psychological research **63**, 22–35 (2000)
51. Pirolli, P., Card, S.: Information foraging. Psychol. Rev. **106**(4), 643 (1999)
52. Pan, B., Hembrooke, H., Gay, G., Gonsalves, G.C.: Bridging the gap: a conceptual model of the access of digital libraries. J. Digit. Inf. **7**(2), (2006)
53. Lohman, D.F.: Spatial ability and g. In: Human Abilities: Their Nature and Measurement, vol. 97, issue (116), p. 1 (1996)
54. Michael, W.B., Guilford, J.P., Fruchter, B., Zimmerman, W.S.: The description of spatial-visualization abilities. Educ. Psychol. Meas. **17**(2), 185–199 (1957)
55. Csépe, V., Győri, M., Ragó, A.: Általános pszichológia 1–3.–2. Tanulás–emlékezés–tudás. Osiris Kiadó, Budapest (2007)
56. Linn, M.C., Petersen, A.C.: Emergence and characterization of sex differences in spatial ability: a meta-analysis. Child Dev. 1479–1498 (1985)
57. Coluccia, E., Louse, G.: Gender differences in spatial orientation: a review. J. Environ. Psychol. **24**(3), 329–340 (2004)
58. Salthouse, T.A., Babcock, R.L., Skovronek, E., Mitchell, D.R., Palmon, R.: Age and experience effects in spatial visualization. Dev. Psychol. **26**(1), 128 (1990)
59. Kahle, J.B.: The disadvantaged majority: science education for women. AETS outstanding paper for 1983 (1983)
60. Sorby, S.A.: Developing 3-d spatial visualization skills. Eng. Des. Graph. J. **63**(2), (2009)
61. Yılmaz, H.B.: On the development and measurement of spatial ability. Int. Electron. J. Elem.Tary Educ. **1**(2), 83–96 (2017)
62. Saucier, D.M., Green, S.M., Leason, J., MacFadden, A., Bell, S., Elias, L.J.: Are sex differences in navigation caused by sexually dimorphic strategies or by differences in the ability to use the strategies? Behav. Neurosci. **116**(3), 403 (2002)
63. Lawton, C.A., Charleston, S.I., Zieles, A.S.: Individual-and gender-related differences in indoor wayfinding. Environ. Behav. **28**(2), 204–219 (1996)
64. Malinowski, J.C., Gillespie, W.T.: Individual differences in performance on a large-scale, real-world wayfinding task. J. Environ. Psychol. **21**(1), 73–82 (2001)

65. Török, Á., Nguyen, T.P., Kolozsvári, O., Buchanan, R.J., Nadasdy, Z.: Reference frames in virtual spatial navigation are viewpoint dependent. Front. Hum. Neurosci. **8**, 646 (2014)
66. Persa, G., Török, Á., Galambos, P., Sulykos, I., Kecskés-Kovács, K., Czigler, I., Honbolygó, F., Baranyi, P., Csépe, V.: Experimental framework for spatial cognition research in immersive virtual space. In: 2014 5th IEEE Conference on Cognitive Infocommunications (CogInfoCom), IEEE, pp. 587–593. IEEE (2014)
67. Ragan, E.D., Scerbo, S., Bacim, F., Bowman, D.A.: Amplified head rotation in virtual reality and the effects on 3d search, training transfer, and spatial orientation. IEEE Trans. Vis. Comput. Graph. **23**(8), 1880–1895 (2016)
68. Segond, H., Weiss, D., Sampaio, E.: Human spatial navigation via a visuo-tactile sensory substitution system. Perception **34**(10), 1231–1249 (2005)
69. Siegel, A.W., White, S.H.: The development of spatial representations of large-scale environments. In: Advances in Child Development and Behavior, vol. 10, pp. 9–55. Elsevier (1975)
70. Hartley, T., Maguire, E.A., Spiers, H.J., Burgess, N.: The well-worn route and the path less traveled: distinct neural bases of route following and wayfinding in humans. Neuron **37**(5), 877–888 (2003)
71. Liang, H.-N., Lu, F., Shi, Y., Nanjappan, V., Papangelis, K.: Evaluating the effects of collaboration and competition in navigation tasks and spatial knowledge acquisition within virtual reality environments. Futur. Gener. Comput. Syst. **95**, 855–866 (2019)
72. Burigat, S., Chittaro, L.: Passive and active navigation of virtual environments vs. traditional printed evacuation maps: a comparative evaluation in the aviation domain. Int. J. Hum. Comput. Stud. **87**, 92–105 (2016)
73. Meade, M.E., Meade, J.G., Sauzeon, H., Fernandes, M.A.: Active navigation in virtual environments benefits spatial memory in older adults. Brain Sci. **9**(3), 47 (2019)
74. Knight, M.J., Tlauka, M.: Interactivity in map learning: the effect of cognitive load. Spat. Cogn. & Comput. **17**(3), 185–198 (2017)
75. Kim, M.J., Maher, M.L.: The impact of tangible user interfaces on spatial cognition during collaborative design. Des. Stud. **29**(3), 222–253 (2008)
76. Csapó, Á.B.: Interaction patterns of spatial navigation in VR workspaces. In: 10th IEEE International Conference on Cognitive Infocommunications (CogInfoCom), vol. 2019, pp. 615–618. IEEE (2019)
77. Horvath, I., Sudar, A.: Factors contributing to the enhanced performance of the maxwhere 3d vr platform in the distribution of digital information. Acta Polytechnica Hungarica **15**(3), 149–173 (2018)
78. Horváth, I., Sudár, A.: Factors contributing to the enhanced performance of the MaxWhere 3D VR platform in the distribution of digital information. Acta Polytechnica Hungarica **15**(3), 149–173 (2018)

Chapter 10
Automatic Spatial Exploration Schemes

Abstract Virtual reality applications have a huge potential to serve as a universal platform for communication and digital content management. In this chapter, we address the problem of generating navigation paths within VR spaces in an automated way. As shown by several authors in the past, such navigation paths can help guide users' attention in high information density 3D environments. We break down the problem of generating such paths into the problem of clustering 2D displays within the spaces, and argue that such cluster information can serve as useful cues for point of interest (POI) generation. To solve the problem of clustering displays, we propose a Markov Chain Monte Carlo based method, and evaluate a first implementation of the approach in the context of 3D spaces available on the MaxWhere platform.

Keywords Virtual reality · 3D navigation · 3D points of interest (POIs) · Markov Chain Monte Carlo

10.1 Introduction

Virtual reality (VR) is a broad set of technologies capable simulating artificial 3D worlds and possibly mediating to its a sense of presence and immersion. Although the key driving force behind VR technologies have been the gaming and entertainment industries, with recent advances in dedicated graphics hardware and display technologies, the perspectives of VR in its potential application domains have broadened significantly [1]. As shown in several recent papers, VR can now be seen as a universal communication and memory management platform that can be useful in all walks of life, including education, healthcare, business management, industrial applications, the retail sector and more [2–4].

Given the spatial nature of 3D environments, a key part of user interactions in VR is spatial navigation. However, as highlighted in several recent works, users' level of comprehension and effectiveness while consuming content in VR can be greatly facilitated if they are able to jump between certain locations, or points of interest (POIs)—a capability that has been referred to as "*digital guidance* [2, 5–7]. Horváth et al. define digital guidance as any *"process that unambiguously drives the user's*

I. Horváth et al., *Cognitive Aspects of Virtual Reality*, Studies in Big Data 156,
https://doi.org/10.1007/978-3-031-68130-1_10

attention during the digital workflow and thus reduces (partially, or to 0) the time required for searching for and finding the relevant digital content".

In this chapter, we consider the problem of automatically generating navigation paths—i.e. sequences of POIs—that can be used as a form of digital guidance. Based on the assumption that the most relevant POIs would be linked to the displays showing 2D content in a given space, conceptually we break this problem into two parts: (1) Clustering the displays such that each cluster represents a set of displays that are relatively close to each other and have a similar orientation; and (2) Generating a POI for each cluster, from which the displays of that cluster are visible. Especially with a view towards solving the first problem, we propose a Markov Chain Monte Carlo (MCMC) based method for clustering 2D displays in 3D spaces based on location and orientation. Although the generation of POIs is left as future work, a brief discussion is nevertheless provided on the outlines of a solution to that problem as well.

The chapter is structured as follows. Section 10.2 introduces the theoretical background of MCMC methods. Section 10.3 describes the proposed method for clustering 2D displays based on 3D location and orientation. Section 10.4 gives an outline of what it would take to generate POIs based on the clusters of displays. Solving the latter problem in detail is left as future work.

10.2 Markov Chain Monte Carlo—Theoretical Background

In many problems that arise in inferential statistics, it is often the case that the prior distribution of a set of parameters, as well as the likelihood of specific data points (assuming certain parameter configurations) can be easily determined; however, the posterior (i.e. the dependence of the parameter values based on the measured data points) is difficult to express directly. In such cases, Bayes' theorem can be useful:

$$\begin{aligned} P(data, \theta) &= P(data \mid \theta) * P(\theta) = \\ &= P(\theta \mid data) * P(data) \\ &\Rightarrow \\ P(\theta \mid data) &= \frac{P(data \mid \theta) * P(\theta)}{P(data)} \end{aligned} \tag{10.1}$$

Note that the numerator on the right hand side includes the likelihood probability that is usually easy to compute—i.e. $P(data \mid \theta)$ can be specified based on historical data. Also, $P(\theta)$ is just the prior distribution assigned to the parameters. In general, some values will be completely improbable, while others more likely, to begin with. If one is agnostic as to the prior distribution, it can always be modeled as a uniform distribution between two sensible limits.

Note, further, that if the goal is simply to compare the probabilities of different parameter configurations, the denominator, $P(data)$ can simply be ignored: first, because it can be regarded as a constant (which has the same value regardless of the parameter configuration), and second, because its main role is to normalize the set of values obtained as the prior times the likelihood in different cases.

10.2.1 Comparing Posteriors Using Monte Carlo Methods

The key idea behind **Monte Carlo methods** in general is that since it is usually difficult to manage these distributions in analytical terms, what can often be done is to take samples from the prior distribution (each sample will have an associated prior probability), then compute the likelihood based on the specific parameter value, and the product of the two is used as measure for determining how "believable" the given parameter value is in terms of the posterior. If this is done many times, at least the unnormalized posterior values can be compared.

At the same time, the above scheme cannot be used to sample from the posterior distribution, unless the space of possible parameter values is explored along a very dense sampling grid. This is where **Markov Chain Monte Carlo (MCMC)** methods come into the picture.

10.2.2 Sampling From Posteriors Using Markov Chain Monte Carlo

The idea behind MCMC is that instead of taking independent samples from the prior distribution, one can instead take a new sample based on the previous posterior sample (in truth, a candidate prior sample is taken, which is either accepted or rejected as being a true sample from the posterior); this candidate prior, then (if accepted) can be used to generate a new sample from the posterior. Since in each case, the new (candidate) sample will depend only on the previous posterior sample, not on earlier ones, the Markov property holds and hence the name Markov Chain Monte Carlo [8].

The core tenet of MCMC methods is that if this is done in a principled way, with a theoretically sound method for accepting or rejecting candidate priors, one can arrive at true samples from the posterior, and eventually approximate the posterior distribution. It is not even necessary to explicitly compute unnormalized posteriors, the only important thing is that the criterion, according to which candidate samples are accepted or rejected, have something (reasonable) to do with the likelihood term. Different MCMC methods differ mainly in the way in which new candidate samples are generated, and the criterion based on which they are accepted or rejected. In the following, we consider two different approaches in order to highlight the areas where the general framework can be modified.

10.2.2.1 Random Walk Metropolis Hastings

In this method, the candidate value for the next prior (θ^*) is taken from a normal—or otherwise symmetric—distribution (referred to as a *generating distribution*) centered around the previous prior, θ. The variance of this generating distribution (when a normal distribution is used) is a hyper-parameter. Based on the new sample (candidate prior), an α value is computed is follows:

$$\alpha = \frac{P(\theta^*)P(data \mid \theta^*)}{P(\theta)P(data \mid \theta)} \tag{10.2}$$

This value is the unnormalized posterior, assuming the new candidate prior, divided by the unnormalized posterior of the previously generated prior. If α is greater than 1, this means that a candidate prior has been found that is more likely (given the data) than the previous prior, so the candidate prior is kept as the new prior. If α does not exceed 1, the candidate prior is kept with some probability η and rejected with probability $1 - \eta$. If the candidate prior is rejected, another sample can be generated using the previously accepted prior, and the whole process repeated afterwards.

10.2.3 Generalized Metropolis Hastings

In this method, the candidate value for the next prior (θ^*) is taken from a generating distribution (denoted by g) that is different from the normal distribution and is not necessarily symmetric about the previous prior, θ. In this case, the so-called Hastings term is also considered when calculating α:

$$\alpha = \frac{P(\theta^*)P(data \mid \theta^*)}{P(\theta)P(data \mid \theta)} \times \frac{P(g(\theta \mid \theta^*))}{P(g(\theta^* \mid \theta))} \tag{10.3}$$

Here, the second term is the Hastings term, which penalizes the generation of candidate samples that are hard to revert, so to speak. That is, if the probability of returning to the previous sample based on the new candidate sample (if it were to be accepted) is relatively low (compared to how easy it was to generate the new candidate sample), then that candidate sample is penalized. In such cases, an α value that is greater than 1 can nevertheless be obtained only if the new candidate sample leads to a higher unnormalized posterior than the previous sample did, and to an extent that offsets the Hastings term.

Note that compared to the random walk Metropolis Hastings approach, this is a more general approach, since in the special case of the random walk Metropolis Hastings method (or whenever the generating function is symmetric about the previously used prior), the Hastings term cancels out.

10.2.4 Hamiltonian MCMC

Another MCMC approach is Hamiltonian MCMC. In this case, a physical analogy from statistical mechanics is used to model the movement of an imaginary particle, which follows the shape of the inverse of the unnormalized posterior function, guided by gravity (this can be pictured as an inverted Gaussian curve, such that the particle rolls towards its bottom, where the original posterior function has its highest value).

In the case of Hamiltonian MCMC, instead of using an analytic generating distribution—like the normal distribution centered around the previous posterior sample, as in the case of random walk Metropolis Hastings—new candidate samples are generated by solving for the new location of the imaginary particle on the inverse of the unnormalized posterior function, using a physical model that includes kinetic energy (momentum) and gravitational energy (determined by the "height" of the previous posterior sample on the inverted unnormalized posterior function).

The benefit of using Hamiltonian MCMC is that this physical analogy seems to extend quite well to high-dimensional parameter spaces. Just like stochastic gradient descent has many advantages (based on empirical evidence) in high-dimensional spaces, taking an imaginary particle that "rolls" around within this inverted probability space based on its momentum and based on gravity—understandably—should display similar characteristics.[1]

The Hamiltonian model itself involves an energy level, which depends on the kinetic and gravitational energy of the particle. The kinetic energy depends on its momentum (mass times velocity), and the gravitational energy depends on the current posterior sample at any given time. Given a frictionless environment, the sum of the kinetic and gravitational energy should be a constant, expressed as:

$$
\begin{aligned}
E(\theta, M) &= Kinetic\ E(M) + Gravitational\ E(\theta) = \\
&= constant
\end{aligned}
\tag{10.4}
$$

where M stands for momentum. The probability distribution of the total energy level, in turn, is modeled based on the Boltzmann function in statistical mechanics:

$$
\begin{aligned}
P(E(\theta, M)) &= e^{\frac{-E(\theta,M)}{T}} = \\
&= e^{\frac{-KineticE(M) - GravitationalE(\theta)}{T}} = \\
&= constant
\end{aligned}
\tag{10.5}
$$

where T is the temperature of the system, which we can assume to be 1.

[1] It has been conjectured that stochastic gradient descent can be especially effective in high-dimensional spaces as the probability that a given local minima is actually a saddle point—allowing the search process to 'escape' in at least one dimension—grows as the number of dimensions is larger (see Tiernan Ray: "AI Pioneer Sejnowski says it's all about the gradient"—https://www.zdnet.com/article/ai-pioneer-sejnowski-says-its-all-about-the-gradient/).

The kinetic energy is $(1/2) * mass * velocity^2$, which is the same as $(1/2) * M * velocity$, which in turn can be expressed as $(1/2) * M * (M/mass)$. Therefore (assuming that the mass is equal to 1), we have that:

$$Kinetic\ E(M) = \frac{M^2}{2 * mass} = \frac{M^2}{2} \tag{10.6}$$

The gravitational energy, in turn, can be taken to be $-\log[P(data \mid \theta) * P(\theta)]$, which is just the negative log of the unnormalized posterior function from above.

Therefore, returning to the probability of the total energy of the hypothetical particle, we now have that:

$$\begin{aligned} P(E(\theta, M)) &= e^{-\frac{M^2}{2}} * e^{log(P(data \mid \theta)*P(data))} = \\ &= P(data \mid \theta) * P(data) * e^{-\frac{M^2}{2}} \end{aligned} \tag{10.7}$$

assuming that the mass and the temperature are 1. Now, forgetting about the energy concept, the two sides of the equation should still be in proportion. So we have that:

$$P(\theta, M) \propto P(data \mid \theta) * P(data) * e^{-\frac{M^2}{2}} \tag{10.8}$$

Now it is clear that the last factor in the equation has the functional shape of a normal distribution. But we can also see that this third factor is independent of the previous two (i.e. the first two factors depend only on the hyperparameters, whereas the third factor depends—at least directly—only on the moment of the particle), so if we want to integrate out the momentum from the joint distribution, we have that:

$$P(\theta) = \frac{1}{Z} * P(data \mid \theta) * P(\theta) * \int PDF\_norm * dM \tag{10.9}$$

Here, the last factor in the product is 1, while all other factors are independent of the momentum M, and can thus be factored out. Thus, we can see that the parameter vector θ and the momentum M are independent of each other, and the task in each iteration is to:

1. Take a sample of the momentum M from a normal distribution: $M \sim \mathcal{N}(0, 1)$
2. Consider the previous posterior sample: θ
3. Use the two to solve for the new candidate value, θ^* (which will be equal to the original θ plus a small constant times the momentum M)
4. At the same time, calculate the new momentum M^* by considering the energy of the particle (the new momentum should be in the direction of the gradient of the the gravitational energy—i.e., of the minus log probability of the unnormalized posterior—with respect to θ).
5. Repeat steps 2–4 (with M and θ replaced by M^* and θ^*) until convergence.
6. Next, the value of M^* can be discarded and the obtained value of θ^c used in itself as the new θ.

Oftentimes, a variation of the above steps are carried out based on the leapfrog algorithm, assuming discrete timesteps and modifying the momentum halfway before and after the new candidate value θ^* is selected.

10.3 Clustering Displays in 3D Spaces Using a Generalized Metropolis Hastings Approach

The main idea proposed in this chapter is to use a surrogate representation for navigation paths (from here on referred to as 'stories' based on the MaxWhere platform's terminology) based on a partitioning, or clustering of 2D displays (from here on referred to as 'smartboards' based on the MaxWhere platform's terminology) in the 3D space, with each cluster containing a 'central node'—a smartboard that is focused on in the story viewpoint. To achieve this, based on the above, a method is required for representing the possible set of smartboard clusters and the prior distribution thereof; as well as, in addition, a method for computing the likelihood term describing the probability of some derived data (based on the spatial arrangement, and spatial extent of the smartboards) given a prior partitioning. The following subsections detail one possible way to achieve these goals.

10.3.1 Characterizing Smartboard Clusters Using Cluster Vectors

The probability distribution of possible clusters (and their centroid nodes) can be represented as a probability distribution over a set of N-dimensional vectors (provided that there are N smartboards in the space) containing a positive integer, i for each smartboard belonging to the ith cluster (if the smartboard is not the central node of the partition), and a negative integer, $-i$ if the given smartboard is the central node of the ith cluster.

For example, if a space contained 3 smartboards, such that the first two belonged to one partition (with the first smartboard as its center node) and the third belonged to a second partition, the cluster vector could be written as:

$$cluster = \begin{pmatrix} -1 \\ 1 \\ -2 \end{pmatrix} \tag{10.10}$$

10.3.2 Generating Candidate Priors Based on Existing Ones

Without knowing anything about the spatial arrangement and spatial extent of the smartboards in a given space, it can still be assumed that the prior probability of some cluster vectors will be lower than that of others. For example, cluster vectors with no negative values (no cluster centers) or with many negative values (in the extreme case, with each smartboard being the center of its own cluster) can be ruled out. Also, since it can be conjectured that the designer of a given space would rarely order the smartboards (by index number) in such a way that smartboards with very different index numbers would be close to each other, or conversely, that smartboards that are close to each other would receive very different indices, a certain 'compactness' can also be assumed in a viable candidate prior.

In the case of the Metropolis Hastings approach, when sampling from the prior distribution and determining whether or not to keep the candidate prior as a sample from the posterior, both the value of the prior probability, as well as the shape of the generating distribution (which is a conditional probability given that we already have a previous prior sample) are of interest. Based on the above intuitions, the following generating distribution can be constructed—based on some key statistics of the previous and the candidate prior, and based on their proximity:

$$P(\theta^*|\theta) = \begin{cases} (1/Z) * e^{-(2*CC+MC)}, \text{ if } \theta^* \text{ is consistent} \\ 0, \text{ if } \theta^* \text{ is inconsistent} \end{cases} \tag{10.11}$$

In the above equation, cluster vectors in which a cluster has no center or has multiple centers are referred to as inconsistent. In addition, besides the number of smartboards in the space, which in the following will be denoted by NSB, two quantities are relied on to quantify the difference between θ and θ^*: the proportion of centers losing their center status or obtaining center status (denoted by CC for 'center change') and the proportion of smartboards changing cluster (denoted by MC for 'membership change'). The normalization term can be computed as the sum of all possible numerator values, depending on possible (consistent) values for CC and MC. One alternative is to also consider inconsistent values, and just discard the resulting candidate cluster vectors later. In this case, the normalization term becomes:

$$Z = \sum_{i=1}^{NSB} \sum_{j=1}^{NSB} e^{-(2i+j)/NSB} \tag{10.12}$$

where NSB is the number of smartboards in the space.

Using this approach, it can be guaranteed that both 'directions' of prior generation (from the old prior to the new prior, and from the new prior back to the old prior) are valid (the old prior must have been consistent to begin with, and the new candidate prior is simply not chosen if it is inconsistent) are valid; and that the generating distribution is symmetric (since we are dealing only with the proportion of smartboards whose status is modified, which is the same in both directions). As a result, the proposed method is an example of Random Walk Metropolis Hastings.

10.3.3 Quantifying the Prior Probability of a Candidate Partitioning

In addition to specifying the generation function for new candidate priors, the probability of those priors also needs to be specified, as this figures in the computation of the α value determining whether or not the candidate prior should be kept as a posterior sample (see Eq. 10.2).

As a prior distribution, we propose to use the following equation, depending on the number of smartboards in the largest and smallest clusters (denoted by sz_max and sz_min, respectively):

$$P(\theta) = \begin{cases} 0.2, \text{ if } sz\_min < 2 \text{ or } sz\_max > 5 \\ 0.8 \text{ otherwise} \end{cases} \tag{10.13}$$

10.3.4 Computing the Unnormalized Likelihood Terms

To determine whether the candidate prior can be taken as a sample from the posterior distribution, the following α value needs to be computed:

$$\alpha = \frac{P(\theta^*)P(data \mid \theta^*)}{P(\theta)P(data \mid \theta)} \tag{10.14}$$

Here, the probability of the data itself is irrelevant, since we are considering the relative likelihoods multiplied by the relative priors. Hence, the final missing link is how to characterize the likelihood values, even if in some unnormalized sense.

To address this remaining problem, we propose to generate some point samples, S_{in}, in the vicinity of the cluster centers (denoted by C_i), and to see how close to the different smartboards those points 'land'. Notably, if there is a smartboard X belonging to a different cluster than the ith cluster (the center of which we are generating points from), and X is located between the center C_i and the generated point S_{in}, this can signal that the clusters are incorrect. Clearly, this can happen from time to time, but if it happens often, probabilistically speaking, the likelihood of the given prior can be expected to be reduced.

In addition, the orientations of the smartboards can and ideally should also be taken into consideration. In this case, the prior should be reduced whenever a smartboard X is found that is closer to C_i than the sample S_{in}, and whose orientation is similar to the orientation of C_i.

To formalize the above intuitions, the proposed procedure consists of sampling N points in the vicinity of each cluster center C_i, such that the probability of generating a specific sample S_{in} is:

$$P\left[S_{in} = (s_{inx}, s_{iny}, s_{inz})\right] = \frac{1}{\sigma\sqrt{2\pi}} e^{-\frac{(S_{in}-C_i)^2}{2\sigma^2}} \tag{10.15}$$

where S_{in} denotes a generic sample, with n ranging from $1..N$; C_i denotes the current cluster center being considered (which is the position of one specific smartboard); and such that the standard deviation, σ is the distance between the center of the cluster and the position of the closest smartboard to it. It can be expected that 68% of all samples will fall within that standard deviation, but given enough samples, some will also fall outside of that range.

For each sample, S_{in}, we consider the proportion of all smartboards X that belong to a different cluster (from the cluster center C_i being considered) which fall between C_i and S_{in}, or whose orientation satisfies some criterion as defined by the following indicated function:

$$\begin{aligned} ind_{in} =&(dist(X, C_i) < dist(S_{in}, C_i) \text{ , AND} \\ &dist(X, S_{in}) < dist(S_{in}, C_i)) \text{ , OR} \\ &(dist(X, C_i) < dist(S_{in}, C_i) \text{ , AND} \\ &sim(ori(C), ori(X)) < \pi/4) \end{aligned} \tag{10.16}$$

where $dist()$ is a function measuring Euclidean distance, $ori()$ returns the orientation of a given smartboard, and $sim()$ is the angle between two orientations.

The idea, then, is for the likelihood function to penalize samples that had relatively high probability of being generated and for which the above indicator function still yields a true value for a relatively high proportion of other smartboards X:

$$P(data \mid \theta) = \prod_{i=1}^{I} \prod_{n=1}^{N} \left(1 - \sqrt{P(S_{in}) * Perc(ind_{in})}\right) \tag{10.17}$$

10.3.5 Simulation Example

An implementation of the proposed method was applied to the smartboard layout of the Seascape space in MaxWhere. Figure 10.1 shows two examples of partitionings yielded by the method. As shown on the figure, the method was able to discover some of the structure within the smartboard layout, e.g. smartboards 1, 2 and 3 are clustered together, and in one case smartboards 4 and 5; and 17 and 18 are also clustered together, respectively. Further, in the second example, the smartboard pairs 11 and 14, 12 and 15, and 13 and 16 are also clustered together.

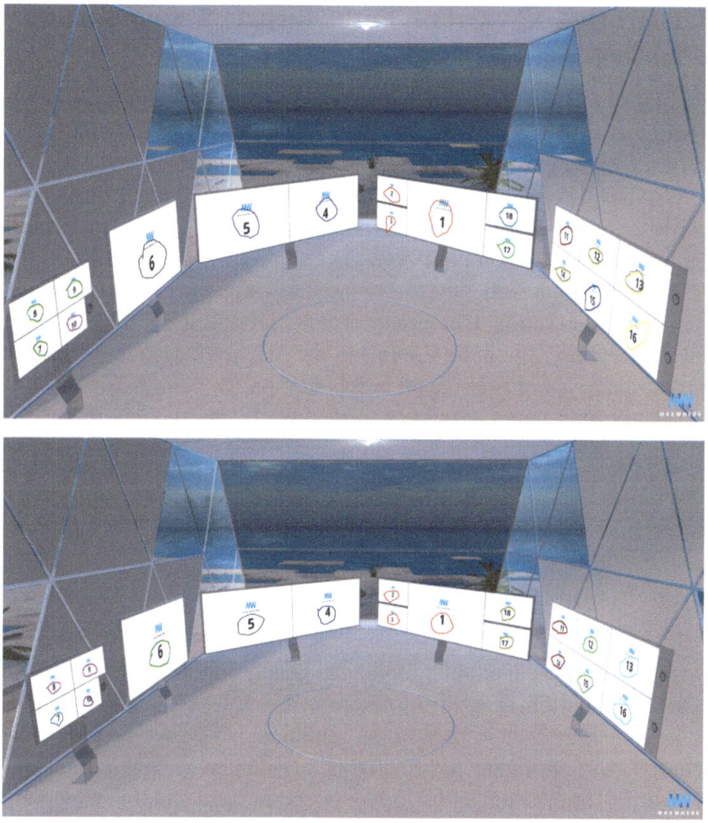

Fig. 10.1 Two different partitions yielded by the proposed method after 1,000,000 runs

At the same time, the results yielded by the method also fail to reflect some common sense relationships, such as the relationship between smartboards 7, 8, 9 and 10, which clearly belong together. This suggests that some further heuristics may be worth investigating and including in the equations for the prior and likelihood distributions. Also, the fact that the second example in Fig. 10.1 was attributed a likelihood of 1, this further suggests that while the distribution functions specified capture some useful information, the range of possible values does not optimally reflect the requirements of the use case scenario. Therefore, based on a further fine tuning of both the prior and likelihood terms, taking into consideration some geometric templates, for example, or the objects in direct proximity to the smartboards (for example, the fact that the same panel can be found behind each smartboard in a group), better results should be attainable.

Nevertheless, as the simulation examples show, the method is capable of capturing useful information, and may be useful as a first step in the automated generation of navigation paths.

10.4 Generating Navigation Paths Based on Smartboard Clusters

Smartboard clusters can serve as a good starting point for automated navigation path generation by helping to identify key points of interest (POI) and associated orientations in the 3D space. It is observable, for example, that the designers of most spaces available on the MaxWhere platform have chosen to create a default 'story' (i.e. recommended navigation path) that goes through a sequence of clusters, with the view—in the case of each cluster—pointed towards the center of the cluster.

Based on this observation, the output from the proposed MCMC-based approach can be put to direct use, by finding points in the 3D space such that the vector from a given point to a given cluster center is incident on the center. The distance of the point from the cluster center should be such that all smartboards in the cluster are visible. If this is not feasible, this should indicate that the clusters were previously incorrectly identified.

10.5 Conclusions

In this chapter, we introduced the problem of generating navigation paths as a way to guide users' attention in high information density VR spaces. To address the challenge of generating such paths in an automated way, we proposed a Markov Chain Monte Carlo approach for clustering 2D display panels within 3D spaces based on position, orientation and size. We suggested that suitable viewpoints could be generated based on such clusters. Preliminary results were promising in that useful information on the layout of 2D smartboards could be extracted, however, the solution still struggled in some cases to group together adjacent (or other formations) of displays that—based on human common sense—would appear to belong together. Following a further fine-tuning of the proposed prior and likelihood factors, the approach should nevertheless be suitable for the generation of admissible clusters for navigation path generation.

References

1. Froehlich, M.A., Azhar, S.: Investigating virtual reality headset applications in construction. In: Proceedings of the 52nd Associated Schools of Construction Annual International Conference, vol. 52, pp. 13–16 (2016)
2. Csapó, Á.B., Horvath, I., Galambos, P., Baranyi, P.: VR as a medium of communication: from memory palaces to comprehensive memory management. In: 2018 9th IEEE International Conference on Cognitive Infocommunications (CogInfoCom), pp. 000389–000394. IEEE (2018)
3. Pieskä, S., Luimula, M., Suominen, T.: Fast experimentations with virtual technologies pave the way for experience economy. Acta Polytechnica Hungarica **16**(6), 9–26 (2019)

4. Virtual reality: understanding the race for the next computing platform. Profiles in Innovation. Goldman Sachs Global Investment Research (2016). https://www.goldmansachs.com/insights/pages/technology-driving-innovation-folder/virtual-and-augmented-reality/report.pdf
5. Lampert, B., Pongracz, A., Sipos, J., Vehrer, A., Horvath, I.: MaxWhere VR-learning improves effectiveness over clasiccal tools of e-learning. Acta Polytechnica Hungarica **15**(3), 125–147 (2018)
6. Horvath, I., Sudar, A.: Factors contributing to the enhanced performance of the maxwhere 3d vr platform in the distribution of digital information. Acta Polytechnica Hungarica **15**(3), 149–173 (2018)
7. Horváth, I.: Maxwhere 3d capabilities contributing to the enhanced efficiency of the trello 2d management software. Acta Polytechnica Hungarica **16**(6), 55–71 (2019)
8. Andrieu, C., De Freitas, N., Doucet, A., Jordan, M.I.: An introduction to MCMC for machine learning. Mach. Learn. **50**(1), 5–43 (2003)

Chapter 11
Predefined Spatial Layouts—Defining Preferences

Abstract Three-dimensional virtual reality (VR) environments, whether utilized on desktop platforms or immersive screens, have been acknowledged for facilitating innovative and highly captivating methods of engaging with digital content across various domains of application. Extensive research conducted in recent years has consistently indicated that the utilization of 3D interfaces, as opposed to their 2D counterparts, can result in improvements across multiple dimensions of performance. These improvements encompass a more comprehensive comprehension and retention of information, an increased capacity for creative and efficient collaboration, and the ability to seamlessly integrate multiple sources of information into workflows in a more expedient manner. Concurrently, the integration of digital content, such as documents, audio-visual content, and web browsers, into 3D spaces is often determined by the creators of said spaces, based on aesthetic considerations or on a case-by-case basis influenced by the specific workflow. This chapter aims to present the findings of an experiment that was conducted to gain a deeper understanding of how users prefer to arrange digital content within their 3D environments, taking into account factors such as the subject matter, the format of the content (e.g., text-based, image, or audio-visual), and the presence of 3D objects within the space. The outcomes of this experiment, as outlined in this chapter, can offer valuable insights for the development of future design methodologies for 3D VR and may also offer support for the implementation of automated solutions for content arrangement.

Keywords Virtual reality · Spatial abilities · Spatial perception · Preference points

11.1 Introduction

Virtual reality (VR) is a technology that utilizes three-dimensional digital models to generate immersive experiences for users, replicating real or imagined environments. The development of VR has encountered both positive and negative phases, resembling the periods of advancement and decline in the evolution of artificial intelligence. While the exact reasons for this are not universally agreed upon, it has been

I. Horváth et al., *Cognitive Aspects of Virtual Reality*, Studies in Big Data 156,
https://doi.org/10.1007/978-3-031-68130-1_11

suggested that the main obstacles to widespread adoption of VR are the substandard performance of headsets and a limited range of applicationss [1].

Despite these challenges, the VR industry has observed significant growth in the past decade and is projected to achieve a market value of USD 50 billion by 2026 [2]. An increasing number of stakeholders now recognize the practical applications of VR beyond its entertainment potential. Initially perceived by early VR developers as a novel infocommunications platform [3], it has for a long time largely been perceived as suitable for gaming, entertainment, and professional simulations or training. However, in recent times, VR has been making inroads into a variety of sectors, including primary and higher education [4–6], healthcare [7–10], engineering, and numerous other professional industries [11–15].

This chapter focuses primarily on the utilization of VR as a digital workspace and examines the typical arrangements of two-dimensional content preferred by users within three-dimensional environments. The investigation encompasses content type, relative display size, as well as thematic cohesion, format cohesion, and the relationship with objects in the 3D environment.

The structure of the chapter is as follows. In Sect. 11.2.1, an overview of cognitive theories relevant to the processing, grouping, and recall of visual information is provided. In Sect. 11.3 offers a brief explanation of how information is generally organized and visualized in 2D and 3D environments. The main research, which entails the experiment conducted on preferred content layouts in 3D, is reported in Sect. 11.6. Finally, in Sect. 11.13 presents a concise discussion on the interpretation of the results.

11.2 Classical Theories of Visual-Spatial Cognition

11.2.1 Processing, Grouping and Recall of Visual Objects

In this section, a comprehensive examination is provided of cognitive theories that pertain to the processing, organization, and recollection of information, particularly within the realm of visual perception. The primary insight derived from this examination is that the manner in which individuals process spatial information, navigate within physical space, and conceptually represent spatial relationships is contingent upon numerous factors, some of which encompass perceptual, higher-level cognitive, and even emotional elements. Consequently, the task of predicting user preferences in three-dimensional virtual environments is by no means a simplistic endeavour, and these predictions must be augmented through the inclusion of usability experiments.

11.2.1.1 Perception of Objects, Shapes and Forms

The Gestalt theory, which historically served as the initial psychophysical model to elucidate the arrangement of perceived images into objects, was formulated through the combined efforts of Wertheimer, Koffka, and Köhler. An essential tenet of the Gestalt approach is the assertion that the entirety of an object surpasses the mere summation of its constituent parts (i.e., the individual segments that comprise the objects). The work of Gestalt psychologists predominantly revolved around three primary domains of investigation [16–20]: (1) The relationships between shape and background; (2) Grouping rules; and (3) The "goodness" of shapes.

Within these domains, six principles of Gestalt were identified by them: the principle of Similarity, the principle of Good Figure, the principle of Proximity, the principle of Continuity, the principle of Closure, and the principle of Common Region.

The principles known as the Law of Common Region and the Law of Proximity hold considerable significance when it comes to comprehending the grouping of information. These principles are especially pertinent in situations where the aim is to actively enhance the speed and quality of understanding and recollection of an image. Such outcomes are crucial in the domain of virtual reality (VR) for effectively conveying information at a sophisticated level. Specifically, in accordance with the Law of the Common Region, the human brain is anticipated to unite elements that reside in a shared, enclosed area. Even in present times, designers of user interfaces and user experiences continue to rely on this principle when organizing information within a confined space and employing the same color scheme, whether it be in gaming or social interfaces, for instance. On the other hand, the Law of Proximity dictates that objects in close proximity to one another are more likely to be perceived as a group of interconnected elements. By adhering to this principle, it is feasible to clearly communicate the association of information, even in scenarios where creating enclosed regions is not feasible [21, 22].

11.2.1.2 Human Spatial Behavior

Despite the inherent constraints posed by the human sensory system and the limited capabilities of locomotion, individuals possess the dexterity to skillfully maneuver through their surroundings. This observation implies that proficient spatial orientation is not solely contingent upon the perceptual and/or motor aptitude exhibited. The capacity of individuals to adeptly and purposefully interact with their immediate habitat is frequently denoted as human environmental proficiency [23].

Human environmental competence consists of several components [23, 24]:

- The perceptual component refers to the process of recognizing, emphasizing, and assigning importance to the fundamental qualities of the surrounding surroundings.
- The cognitive component carries out the analysis of characteristics emphasized during the process of perception, in addition to the retention, arrangement, and

retrieval of the characteristics that obtain significance in this manner. Consequently, this leads to the acquisition of knowledge pertaining to the environment.
- The constituents of the affective component are affirmative and adverse emotional and motivational reactions to the attributes of the surroundings, behavioral responses, and personal inclinations that act as intermediaries between the aforementioned constituents.

11.2.1.3 Categorization and Conceptual Representation

Tversky's research, which was published in 2003, examines various types of cognitive spatial reference frames employed by individuals to situate themselves in physical space. Each cognitive subsystem within the intricate and multifaceted mental space fulfills a distinct purpose and encompasses diverse spatial components and frames of reference. Although it is plausible to assume that each of these subsystems possesses a distinct cognitive structure, they are all characterized by the process of schematization. This cognitive process serves to reduce the cognitive load on memory, facilitate information processing, and enable the assimilation of a wide array of information, albeit with the trade-off of some degree of error and distortion. There are two overarching forms of schematization: categorization and chunking [25, 26].

In the world in general, as well as within one's limited surroundings, an abundance of information is encountered, which shall be required to be recollected at some point. This is substantiated by numerous cognitive processes, inclusive of categorization. Said processes primarily occur spontaneously and possess the capacity to impact all modalities. The cognitive processes that facilitate categorization render it feasible for individuals to distinguish between objects arranged within distinct categories based on specific attributes. However, in the event of objects pertaining to the same category, minor disparities are disregarded, and attention is directed towards shared qualities.

Additionally, the organization of data into segments can further enhance the efficiency of both the cognitive processes of information processing and information recall. Within the confines of this particular framework, one of the most noteworthy findings in the field of cognitive science has been the revelation of chunking, a cognitive process that entails the decomposition of information into smaller, more easily manageable units, thereby rendering the information more comprehensible [27, 28]. It is a cognitive approach for the retention of information that can be intentionally implemented, yet simultaneously manifest subconsciously as a result of exposure, thereby enabling individuals to retrieve substantial quantities of knowledge [29–31]. For an extended period, it has been understood that the ability of working memory is constrained in regards to the manipulation and retention of data. However, utilizing this cognitive approach can lead to a decrease in the quantity of information that needs to be processed simultaneously [32]. Furthermore, the effectiveness of this procedure is additionally substantiated by the reality that it frequently relies on a systematizing principle that not only employs more compact constituent units but also possesses semantic coherence, consequently leading to memories that are more readily retrievable [29, 33].

In numerous instances, nevertheless, models for acquiring information have been shown to be insufficient in the previous years due to their failure to consider the cognitive condition of individuals, as well as the cognitive approaches they might employ. The information search model proposed by Pirolli and Card in 1999 acknowledges this limitation and asserts that individuals seeking information tend to search and categorize information into clusters in order to reduce the cost associated with searching across clusters [34]. Clusters are observed in both physical and digital manifestations. Additionally, a crucial facet entails the user's cognitive framework, which elucidates the dissimilarities in information retrieval behaviors. Disparities in the cognitive framework are frequently influenced by the cognitive, demographic, and social attributes of the particular user [35].

11.3 Information Organization in 2D and 3D Interfaces

In this section, we present a concise overview of the general organization and visualization of information in both two-dimensional (2D) and three-dimensional (3D) environments. We emphasize the advantageous utilization of each environment to enhance comprehensibility and information throughput. Furthermore, we highlight the relative advantages of 3D environments over 2D environments in this particular context.

11.3.1 Two-Dimensional Interfaces

Due to the emergence of the digital era, the proliferation of graphical interfaces, and the exponential growth of digital information, there arose a necessity to categorize various types of digital content, stored on electronic devices such as computers, phones, and tablets, according to a specific principle. Typically, 2D interfaces adopt logically hierarchical approaches to address this need, presenting files as belonging to a designated folder within a folder hierarchy. In such cases, users are constrained to access the content of a file by considering its location in the hierarchy, as well as its extension and name [36–38].

11.3.2 Difficulties of Task-Oriented Clustering on 2D Interfaces

The primary challenge encountered when utilizing folder-file hierarchies is the difficulty in conceptually associating a file with a unique folder. Furthermore, users often struggle to recall the folder in which a particular file is stored.

Gwizdka et al. conducted a study to investigate this problem in the context of managing emails on 2D interfaces, aiming to enhance efficiency and speed. The physical environment in which individuals carry out daily activities is characterized by its spatial and flexible nature. The researchers hypothesized that integrating some of these features into the email environment would better support the various tasks performed within an email system. Their objective was to cluster messages in a task-based order, facilitating their management and response [39]. Other studies have also endeavored to address similar challenges [40].

11.3.3 Two-Dimensional Information Clustering Approaches on Computers

While folders may serve as a temporary solution for organizing documents in a structured manner, it remains challenging to provide users with universal cues that would enable strangers to locate information with the same efficiency as the system's creator.

In contemporary work settings, individuals often utilize two monitors to access information from multiple sources. In the physical realm, this would correspond to someone simultaneously opening multiple books or papers arranged side by side. However, on a computer, translating this feature can still pose difficulties.

To some extent, the split-screen feature offers a solution to this challenge and is frequently employed when specific information needs to be accessed together to solve a task. In such cases, the monitor can display multiple documents simultaneously and keep them readily accessible for the user. Nonetheless, the use of a split screen can introduce scale changes in a given document, potentially impeding its usability and interpretation. Additionally, without a specialized configuration, such layouts are typically temporary and require repeated recreation, necessitating extra effort.

If we examine the solutions offered by the two dominant operating systems for organizing applications on computers and how they are utilized, we encounter similar solutions. Both Windows and Mac OS versions provide a perspective where parallel-running applications are displayed to the user in smaller windows on one screen (e.g., Fig. 11.1). This feature is undoubtedly valuable as it allows the user to assess the number of content elements and types of documents open simultaneously. However, it does not assist the user in grouping these elements. It is conceivable that one may wish to work on multiple tasks concurrently, each involving a certain number of documents/applications. Furthermore, it is common for certain tasks to consume the

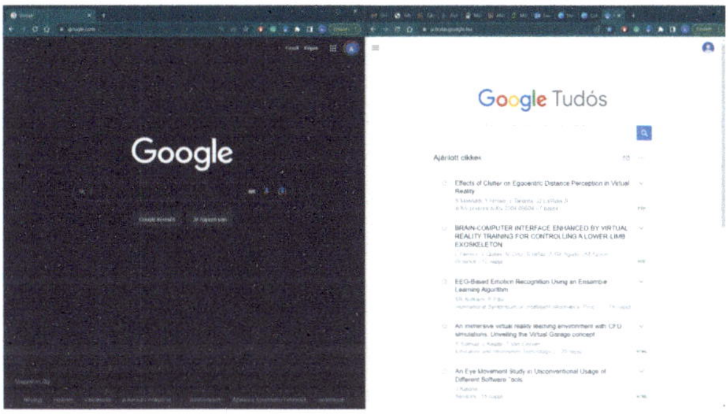

Fig. 11.1 Laptop screen with a split screen view on Windows OS

user's attention for extended periods, ranging from days to weeks or even longer. Consequently, it is crucial for the user to have continuous access to the content required to solve these tasks. Additionally, document access needs often depend on the context, such as the user's preference to consume different types of content outside of working hours.

As a resolution to this issue, the major desktop operating systems have introduced a new concept, referred to as "Virtual Desktop" by Windows and "Spaces" by Apple. The essence of these concepts is that users can have multiple desktops open simultaneously, allowing for easy switching between them. However, users are only permitted to change the order of these desktops and are unable to make any other layout modifications (Fig. 11.2).

Although this solution aligns more closely with the organization strategies observed in real-life scenarios, it remains limited to the confines of a two-dimensional environment and is unable to effectively convey relative importance or the sequential order in which tasks should be carried out. This includes the inability to differentiate content elements based on their size or their spatial arrangement.

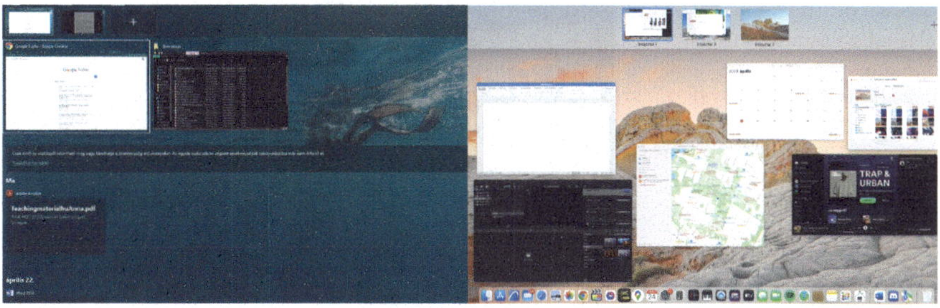

Fig. 11.2 The Windows virtual desktop and the Mac mission control and space view

11.3.4 *Three-Dimensional Interfaces*

The significance of spatial layout in processing, comprehending, and recalling information has already been demonstrated in PC and console games. However, as virtual reality becomes increasingly utilized beyond gaming, the question of its purpose in such applications becomes even more pertinent. Virtual reality closely resembles real life in many aspects, yet its major advantage lies in its ability to enhance it in various ways to assist users in managing their everyday tasks. The Virtual Desktop VR application, introduced in 2016, presented an intriguing solution to the aforementioned issue of visualizing and grouping information and documents by placing the user's desktop within a fully immersive environment (https://www.vrdesktop.net/ (accessed on 30 April 2023)). This allows users to fully utilize the available space, maximizing the sense of presence and accommodating more windows on a larger surface, similar to the configuration depicted in Fig. 11.3.

In 2021, Meta introduced the virtual office and meeting room application called Horizon Workrooms, which also operates in immersive VR and requires a headset. However, it is already suitable for collaborative work, allowing individuals to engage in shared virtual office spaces with their colleagues. This fosters cooperation and facilitates the resolution of tasks.

These solutions have the ability to harness the advantages provided by the virtual world by presenting content within a 3D environment that includes 3D objects. This feature significantly aids user recall, as demonstrated in previous research [41]. Nevertheless, the head-mounted display itself imposes notable limitations. It can induce motion sickness in a considerable number of users, has a short battery life, and remains relatively expensive.

Fig. 11.3 Virtual desktop-like control panel as viewed on the Pico 4 headset. *Source* https://en.wikipedia.org/wiki/PICO_4 (accessed on 10 February 2024)—License: Creative Commons Attribution-ShareAlike 4.0 International

Applications that enable information exchange, work, study, and event organization without the constraints of VR headsets utilize the benefits offered by virtual reality. These applications make use of traditional 2D content in desktop VR. For example, Spatial (https://www.spatial.io/ (accessed on 30 April 2023)) is a cooperative VR platform that allows the importation of 2D content, such as documents, images, and videos, into virtual workspaces. In VRChat (https://hello.vrchat.com/ (accessed on 30 April 2023)), a social VR platform, users can incorporate 2D content like images and videos into their virtual realm using the Media Player feature. Similarly, ENGAGE (https://engagevr.io/ (accessed on 30 April 2023)) permits the importation of videos, websites, and documents through the WebBrowser feature. MaxWhere VR (https://maxwhere.com (accessed on 30 April 2023)) offers smartboards that can be integrated into 3D spaces and display various content types supported by a Chrome browser, such as PDFs, images, videos, audio files, and local or remote web content.

In recent years, the effectiveness of these VR applications compared to 2D applications has been extensively studied from various perspectives. significant findings from previous research have emphasized the ability of desktop VR to alleviate cognitive load [42–45] and enhance information recall [46, 47]. Desktop virtual realities also improve digital information and content management [48–51] and can assist users with digital guidance solutions [52, 53].

11.4 The Role of VR in Research on Spatial Cognition

Over the past two decades, VR has emerged as a widely utilized technology in the study of spatial cognition and navigation. It offers researchers numerous capabilities for evaluating simulations of real-world scenarios, both in terms of environmental settings and cognitive states [54–56]. VR provides researchers with improved opportunities to monitor, control, and measure real-life situations [57] in a safer, more easily creatable, and reproducible environment.

In a previous study conducted with the participation of 431 high school students, researchers examined various factors that contribute to the effectiveness of learning in a virtual reality-based environment. They discovered significant distinctions between a research group that utilized virtual reality (VR) and a control group that utilized a 2D graphical user interface-based interaction. The findings indicated that students in the VR group performed better [58]. The utilization of VR offers numerous advantages, one of which is the facilitation of a learning mode that does not require the mental transformation of 2D objects into 3D objects, unlike more traditional e-learning environments [59]. In other words, VR promotes a spatial encoding process that comes more naturally to most users. This particular feature of VR may be crucial in reducing the cognitive load associated with various digital tasks.

11.5 Information Arrangement in 3D

Despite the advantages of 3D spaces, Setti et al. have posited a significant obstacle in terms of integrating 2D content into these spaces. The limitation lies in the inability of users to comfortably or at all modify the existing 2D layouts. Consequently, their thinking is confined within the constraints of the currently established layouts, resulting in reduced effectiveness and ease of use [48]. However, it is evident that no single layout is universally suitable for all tasks. This aligns with Krokos et al. suggestion that the ability to reconfigure spaces could have an impact on user performance [49].

Needless to say, editing and reconfiguration in a 3D environment present considerable challenges. According to Setti et al., determining how the camera (the user's viewpoint into the space) should interact with the operations used to transform the displays is not a straightforward question [48]. For instance, when a display is moved towards a wall or another object while the camera remains stationary, it becomes difficult to ascertain when the display has reached a specific distance from the wall/object. Similarly, during the rotation of the display, it is challenging to determine whether the angle between the display panel and the wall/object is as desired. On the other hand, if the camera viewpoint is automatically modified in parallel with the manipulations of the display panel, users will have limited freedom to reposition themselves relative to the objects and display panels of interest. This predicament is commonly referred to as the "*camera-object independence dilemma*".

To address this and other challenges, Setti et al. proposed a comprehensive and minimal set of operations. These operations primarily involve duplicating existing displays or snapping new displays onto existing surfaces. Subsequently, fine-tuning operations such as translation of displays on their own plane or modification of size and aspect ratio are performed. Test subjects were able to utilize these operations to accurately recreate existing layouts in empty versions of the same spaces in less than 45 seconds per display.

Based on the editing methodology presented by Setti et al. (2022), a comprehensive design paradigm known as "*doing when seeing*" has been proposed [50]. This paradigm emphasizes that performing an operation within a virtual space is most effective when it is carried out in relation to pre-existing elements within that space. For instance, if there is a wall or an existing 2D display, a new 2D display can be created by aligning it with the wall or duplicating the existing display. The "doing when seeing" approach also implies that, in general, no additional operations are required, except for specific fine-tuning actions that pertain to the object itself and are independent of its relationship to other objects

Furthermore, the "*doing when seeing*" paradigm can be extended to encompass 3D objects or higher-level configurations, such as existing content groups or layouts. By assigning metadata and content to layouts, it becomes possible to establish a '*3D file system*' that expands the "doing when seeing" paradigm to include the duplication of existing projects [50]. This concept can be further developed to address the challenge of content or project search. By connecting entered search terms and subsequent interactions, a search functionality can be created that resembles a semantic search, but primarily links together search entries with temporally close behaviors [60].

With these capabilities, 3D environments offer a flexible interface for organizing various content types within a spatial arrangement, based on semantic relationships (spatial proximity) and relative importance (spatial size).

The latter sections aim to describe experimental investigations conducted to determine whether users demonstrate preferences regarding the size and location of 2D content elements placed within 3D virtual reality (VR) spaces, and if so, how factors such as content type and the spatial environment (i.e., 3D objects within the VR space) influence these preferences.

11.6 Motivations Behind Free Spatial Arrangement Study

As discussed in the literature review, recent years have witnessed a demonstration that conventional two-dimensional (2D) systems are encountering limitations in terms of the amount of information they can handle in both human and machine interactions. In contrast, three-dimensional (3D) systems possess a much greater potential for information exchange and can effectively cater to the increasing needs of humans.

With the mounting pressure of information, users often show a preference for digital environments that enable them to enhance the clarity of relationships among digital content. These environments also facilitate the organization of content into groups, not only in a hierarchical manner but also in a workflow-based order and in ways that persist over time. Although platforms that offer these capabilities are becoming increasingly available, they generally lack sufficient guidance for users on how to effectively arrange their content in the 3D space. In particular, there is a lack of understanding regarding how real-life physical experiences and preferences influence users' choices regarding layout. Consequently, these considerations are rarely taken into account in the design of 3D environments.

This study aims to address several key questions. Firstly, do the layout conventions that users are accustomed to in real-life settings also manifest in virtual spaces? Additionally, are there differences in terms of the size at which users prefer to display different types of content in virtual reality (VR)? Are these size preferences influenced by the nature of the content (e.g., text-based, image, or audio-visual), the subject matter (with content focusing on similar topics being grouped together), or the geometry of the 3D space (such that text-based documents are placed on tables while videos are placed on vertical surfaces)?

For example, based on observations of real-world physical environments, one might expect that:

- Videos would ideally be placed on large-sized screens, TV sets, or monitors, which are predominantly vertical surfaces;
- PDF files would ideally be placed either on computer screens or on horizontal, flat surfaces such as tables. This preference may vary depending on the age group

of the user and their perception of PDF files as electronic documents or physical, printed materials
- PPT files would ideally be placed on large screens, as they are typically consumed on personal computers
- Images would ideally be placed on vertical surfaces or surfaces that are inclined towards the vertical, as pictures in real life are often hung on walls or placed on furniture.

While these arrangements may seem reasonable based on the starting point of physical reality, it is also possible that users may choose to exploit the unique qualities of 3D virtual reality, resulting in content arrangements that are less bound to physical metaphors. To investigate whether this is the case and to explore the potential novel interaction patterns that users might adopt, we conducted an experiment, the details of which are described in this section.

11.7 Experimental Design

To observe potential patterns in users' preferences for organizing their digital content in virtual environments, we constructed a three-dimensional (3D) space where participants were given the freedom to arrange a partially predefined collection of digital documents. They were also allowed to supplement this collection with additional materials of their choosing.

We conducted the experiment using a virtual reality (VR) space called "Ph.D. Student Room" (Figs. 11.4 and 11.5) in the MaxWhere VR platform. The participants were first acquainted with the platform's basic functionality and the layout of the

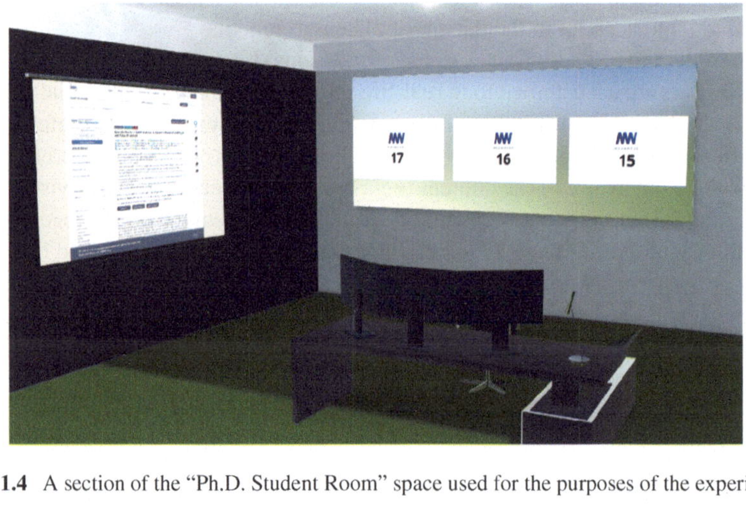

Fig. 11.4 A section of the "Ph.D. Student Room" space used for the purposes of the experiment

Fig. 11.5 A section of the "Ph.D. Student Room" space used for the purposes of the experiment

space. They were then provided with a set of "*recommended*" documents, including images, PDF files, video files, and PowerPoint files related to technology topics, particularly augmented reality and virtual reality. The task assigned to the participants was to organize these recommended documents, along with any other content they wished to include from the internet, within the 3D space.

By providing participants with a considerable degree of freedom, our aim was to create a scenario that closely resembles real-life situations where users can personalize their own 3D virtual space. The selection of recommended content on augmented reality and virtual reality was a logical choice since the experiment took place in a VR environment. The number of content elements was carefully chosen to strike a balance between being manageable for users to comprehend and providing a comprehensive overview of the subject matter. It is important to note that the participants were explicitly informed that the recommended materials were not mandatory and that they were free to choose their own content, including the topic, layout, and size.

After the experiment, we analyzed the quantity, relative size, and spatial arrangement of the different types of documents. Our focus was on examining whether the participants' organizing principle revolved around document types (e.g., separating images, PDFs, and audio-visual content), subject matter (grouping content on the same topic), or the 3D context (e.g., placing videos on monitors and PDFs on flat surfaces like tables).

11.8 Key Hypotheses

Before conducting my experiment, we formulated the following key hypotheses:

1. There is a difference in terms of the frequency with which test subjects prefer to add different content types to their spaces.
2. There is a difference in terms of the size in which subjects prefer to display various content types.
3. For at least some types of content, an interaction of the content type with certain 3D object types can be observed.

11.9 Materials and Methods

To conduct the experiment, we utilized the MaxWhere VR platform, which offers the capability to customize smartboard configurations through a dedicated layout editor interface. This feature was crucial in allowing users to explore various configurations according to their preferences.

The virtual space employed in the experiment was called the "Ph.D. Student Room" (Fig. 11.6). It is a physically open structure situated on a sand island surrounded by water. The building consists of two floors. The ground floor comprises three interconnected rooms, with either no walls or transparent glass walls separating them.

- The initial room among the three exhibits the semblance of a workspace, furnished with a desk, a chair, and three monitors placed atop the desk. Additionally, a projection screen and a sizable whiteboard are present.
- Adjacent to this, the lounge area features a small number of chairs, accompanied by a coffee table, and a prominent display adorning the wall.

Fig. 11.6 Empty version of "Ph.D. Student Room" space used for the purposes of the experiment

- The ultimate room bears the resemblance of either a meeting room or an information center, housing a large table that can be partitioned into six smaller sections. Each of these sections is furnished with a chair and a table, equipped with a document holder tilted at an angle of 45 degrees. Similar to the workroom, a whiteboard is affixed to the wall.

On the upper level, a capacious open terrace and a partially covered exhibition space occupy one side of the room, where five frames are suspended from the ceiling.

At the beginning of the experiment, the virtual space was devoid of any smartboards, which were subsequently positioned within the space by the participants themselves. The prescribed documents provided to the test subjects comprised five images, three PDF files, three videos, and two PowerPoint files, encompassing the subsequent content:

- Three PDFs:
  - Çöltekin, A. et al. (2020). "*Extended reality in spatial sciences: A review of research challenges and future directions*". ISPRS International Journal of Geo-Information, 9(7), 439;
  - Jung, T. et al. (2016). "*Effects of virtual reality and augmented reality on visitor experiences in museum*". In Information and Communication Technologies in Tourism 2016: Proceedings of the International Conference in Bilbao, Spain, 2–5, February 2016 (pp. 621–635). Springer International Publishing;
  - Morimoto, T. et al. (2022). "*XR (extended reality: virtual reality, augmented reality, mixed reality) technology in spine medicine: status quo and quo vadis*". Journal of Clinical Medicine, 11(2), 470.
- Two PPTs:
  - Future of Technology (from Canva templates);
  - Future of communications (from Canva templates);
- Three videos:
  - https://www.youtube.com/watch?v=I-EIVlHvHRM&t=1s&ab_channel=WIRED (accessed on 30 April 2023);
  - https://www.youtube.com/watch?v=WXuK6gekU1Y&ab_channel=DeepMind (accessed on 30 April 2023);
  - https://www.youtube.com/watch?v=XLP4YTpUpBI&t=9s&ab_channel=Simplilearn (accessed on 30 April 2023).
- Five images: shown in Fig. 11.7.

Throughout the task-solving process, these materials were not presented to the subjects as obligatory components, but rather as recommended resources or a baseline. They could be substituted with other supplementary web and local digital content and discarded if deemed unnecessary for the presentation of the topic, which revolved

Fig. 11.7 Five images included in the set of recommended content

around virtual reality, augmented reality, and mixed reality, while also incorporating tutorials on the utilization of artificial intelligence. All distributed materials were in English.

11.10 Subjects and Preliminaries

A total of 32 volunteers partook in the study, comprising an equal gender distribution of 16 men and 16 women. The average age was 28.594 (SD: 9.462). All participants were neurotypical individuals proficient in English and possessed at least an intermediate (B2) level of English language proficiency. Thirty-one test subjects were native Hungarian speakers, while one test subject was a native Arabic speaker. All participants possessed prior experience with 3D, acquired through their utilization of the MaxWhere software itself, as well as through engagement with 3D games in many instances. Prior to the experiment, the test subjects were acquainted with the MaxWhere platform, ensuring their familiarity and comfort with using MaxWhere spaces before undertaking the primary task. All data collected during the experiment were anonymized and solely employed as input for the subsequent statistical analyses expounded upon in later sections of this chapter.

11.11 Procedure

During the experiment, participants engaged in a task involving the creation of a personalized layout, wherein they were required to select any combination of 2D content they desired, whether from the recommended documents provided, from the internet, or from their personal devices. In all scenarios, test subjects were instructed to stay within the confines of the designated topics: virtual reality, artificial intelligence, and their collaborative potentials. Participants were unrestricted in terms of the placement of materials within the space; there were no limitations pertaining to the number, position, orientation, or size of the content. Furthermore, there was no specified time limit for the completion of the task.

To save the content, each participant generated an individual project using a random pseudonym to ensure the preservation of anonymity.

11.12 Results

In this section, we provide an analysis of the recorded data from various angles.

11.12.1 Frequency of Content Types

Figures 11.8 and 11.9 display the outcomes of a repeated measures ANOVA, indicating that the number of instances in which the content types were exhibited, in the case of different test subjects, was significantly influenced by the content types, $F(2.98, 92.89) = 11.587$, $p < 0.000$. Due to the violation of Mauchley's test of sphericity, the Greenhouse–Geisser correction was employed. The Eta squared effect size ($\eta^2 = 0.272$) also suggests that the impact of the type on the frequency of each content type used was substantial.

The pairwise comparisons reveal that PDF documents were employed significantly more frequently than PowerPoint files or any other Web-based content, and images were used significantly more often than PDF documents, PowerPoint files, or any other Web-based content. The distribution of the frequency of content types is illustrated in Fig. 11.10 using a bar chart.

Desciptive Statistics			
	Mean	Std. Deviation	N
PDF	3.0625	0.84003	32
Video	3.0625	1.77687	32
Image	4.0000	2.19971	32
PPT	1.6250	1.18458	32
Web	1.7188	2.47874	32

Fig. 11.8 Descriptive statistics of the repeated measures ANOVA—number of content types

Occurence	Occurence	Mean difference	Std. Error	Sig.	95% Confidence Interval for Difference	
					Lower Bound	Upper Bound
PDF	**Video**	**0.000**	**0.311**	1.000	**-0.634**	**0.634**
PDF	Image	-0.937	0.378	0.019	-1.709	-0.166
PDF	**PPT**	**1.438**	**0.246**	**0.000**	**0.936**	**1.939**
PDF	**Web-based content**	**1.344**	**0.432**	**0.004**	**0.462**	**2.225**
Video	PDF	0.000	0.311	1.000	-0.634	0.634
Video	Image	-0.937	0.490	0.065	-1.936	0.061
Video	**PPT**	**1.438**	**0.342**	**0.000**	**0.740**	**2.135**
Video	**Web-based content**	**1.344**	**0.437**	**0.004**	**0.453**	**2.234**
Image	**PDF**	**0.938**	**0.378**	**0.019**	**0.166**	**1.709**
Image	Video	0.938	0.490	0.065	-0.061	1.936
Image	**PPT**	**2.375**	**0.451**	**0.000**	**1.456**	**3.294**
Image	**Web-based content**	**2.281**	**0.554**	**0.000**	**1.151**	**3.411**
PPT	PDF	-1.437	0.246	0.000	-1.939	-0.936
PPT	Video	-1.437	0.342	0.000	-2.135	-0.740
PPT	Image	-2.375	0.451	0.000	-3.294	-1.456
PPT	Web-based content	-0.094	0.461	0.840	-1.034	0.846
Web-based content	PDF	-1.344	0.432	0.004	-2.225	-0.462
Web-based content	Video	-1.344	0.437	0.004	-2.234	-0.453
Web-based content	Image	-2.281	0.554	0.000	-3.411	-1.151
Web-based content	PPT	0.094	0.461	0.840	-0.846	1.034

Fig. 11.9 Pairwise comparisons of the repeated measures ANOVA in the number of content types

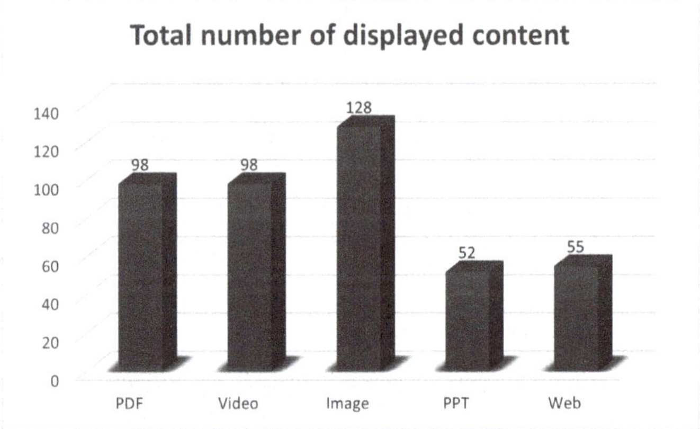

Fig. 11.10 The total number of times each given content type was used. Note that although in total, video content was added to the spaces as often as PDF files, the variance among users was such that only the distribution of the use of PDF files was statistically significantly greater than those of PowerPoint files, and Web-based content

11.12.2 Scale of Content Types

Within the same research design, the preferred size of each content type was also examined. The content types included PDF files, videos, images, PowerPoint files, and any other Web-based content. To determine the scale, the length (in centimeters) of the horizontal and vertical axes of the corresponding smartboards were multiplied together. Descriptive statistics of smartboard sizes can be observed in Fig. 11.11.

Desciptive Statistics			
	Mean	**Std. Deviation**	**N**
PDF	18,879.3349	23,497.67565	32
Video	53,336.5460	34,679.86121	32
Image	55,688.0021	52,207.77006	32
PPT	50,257.5872	36,423.23812	32
Web	21,976.3877	20,604.70064	32

Fig. 11.11 Descriptive statistics of the repeated measures ANOVA—size of the different content types

Occurence	Occurence	Mean difference	Std. Error	Sig.	95% Confidence Interval for Difference	
					Lower Bound	Upper Bound
PDF	Video	-34,457.211	8513.935	0.000	-51,821.497	-17,092.93
PDF	Image	-36,808.667	10,251.633	0.001	-57,717.011	-15,900.32
PDF	PPT	-31,378.252	7300.059	0.000	-46,266.820	-16,489.69
PDF	Web-based content	-3097.053	4862.334	0.529	-13,013.849	6819.743
Video	**PDF**	**34,457.211**	**8513.935**	**0.000**	**17,092.926**	**51,821.50**
Video	Image	-2351.456	10,365.166	0.822	-23,491.352	18,788.44
Video	PPT	3078.959	8243.670	0.711	-13,734.12	19,892.04
Video	**Web-based content**	**31,360.158**	**7389.851**	**0.000**	**16,288.46**	**46,431.86**
Image	**PDF**	**36,808.667**	**10,251.633**	**0.001**	**15,900.32**	**57,717.01**
Image	Video	2351.456	10,365.166	0.822	-18,788.44	23,491.35
Image	PPT	5430.415	12,158.689	0.658	-19,367.39	30,228.22
Image	**Web-based content**	**33,711.614**	**9562.616**	**0.001**	**14,208.53**	**53,214.70**
PPT	**PDF**	**31,378.252**	**7300.059**	**0.000**	**16,489.69**	**46,266.82**
PPT	Video	-3078.959	8243.670	0.711	-19,892.04	13,734.12
PPT	Image	-5430.415	12,158.689	0.658	-30,228.22	19,367.39
PPT	**Web-based content**	**28,281.199**	**7933.609**	**0.001**	**12,100.50**	**4461.901**
Web-based content	PDF	3097.053	4862.334	0.529	-6819.743	13,013.849
Web-based content	Video	-31,360.158	7389.851	0.000	-46431.86	-16,288.457
Web-based content	Image	-33,711.614	9562.616	0.001	-53,214.70	-14,208.531
Web-based content	PPT	-28,281.199	7933.609	0.001	-44,461.90	-12,100.498

Fig. 11.12 Pairwise comparisons of the repeated measures ANOVA size of the different content types

Results of a repeated measures ANOVA demonstrate that the size of the displayed content from one group was significantly influenced by the type of the group, $F(2.98, 92.89) = 11.587$, $p < 0.000$. Due to the violation of Mauchley's test of sphericity, the Greenhouse–Geisser correction was employed. The Eta squared effect size ($\eta^2 = 0.272$) also suggests that the effect of the type of content on smartboard size was substantial.

According to the pairwise comparisons, videos, images, and PowerPoint documents were exhibited at a significantly larger size than PDF files and Web-based content. However, there were no significant differences among the former three types (Fig. 11.12). It is worth noting that in cases where a test subject did not add a certain type of file format to their project at all, the size for that content type was set to the average of the sizes specified by all other test subjects. This adjustment was necessary to ensure that the number of samples was consistent in all cases, which is a prerequisite for the ANOVA analysis.

11.12.3 Relationships Among Dimensions of Content Type, Content Subject, and 3D Context

An examination of the projects created by the test subjects revealed a preference between given content type and spatial object in many instances. As depicted in Fig. 11.13, the projector screen inside the space was used to display video and PowerPoint content three times more frequently than any of the other predefined types. Similarly, the frames hanging from the ceiling were utilized to hold images much more often than any other type of content.

At the same time, our analysis also indicated preferences among spatial objects as receptacles for smartboards in general. Consequently, test subjects attached content (in general) more frequently to spatial objects such as tables, monitors, and projector screens than to any other type of object, including the wall itself.

Based on this observation, a distinction was made between "object-dependent" and "object-free" placement of smartboards, whereby the latter refers to walls or other surfaces that have no standalone existence. Users were then characterized based on their preference for one category or the other. Object-dependent users tended to place their content in close proximity to spatial objects inside the building, such as projector screens, monitors, and display boards. On the other hand, object-free users placed

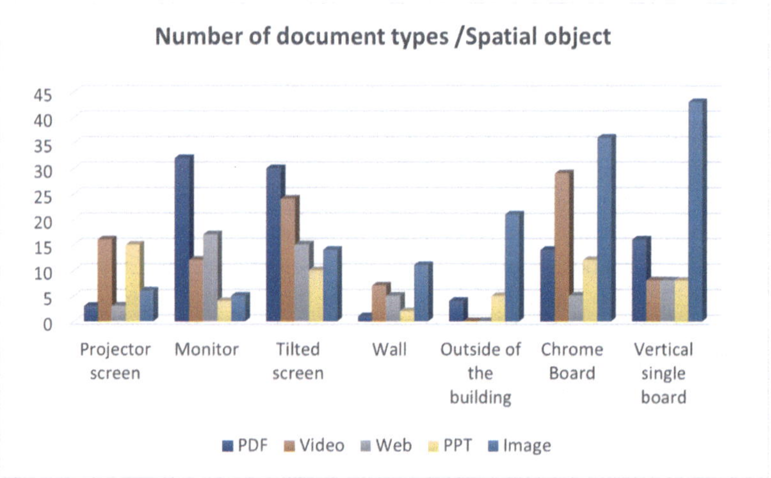

Fig. 11.13 The total number of document types used in different 3D contexts

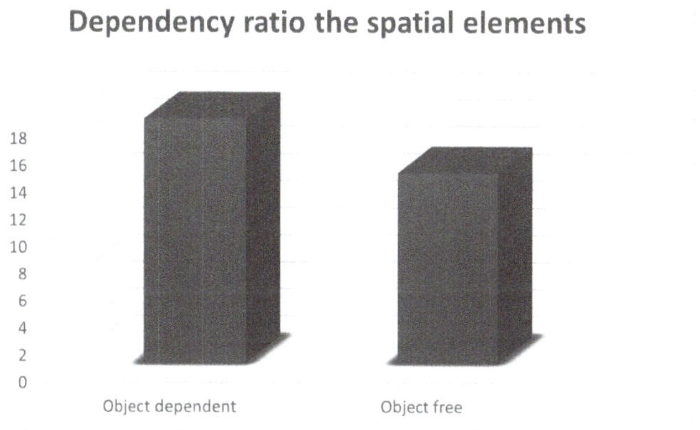

Fig. 11.14 The distribution of spatial object-dependent and spatial object-free users

Fig. 11.15 Examples of object-dependent and object-free locations within the 3D space. The first three examples in the top row and the first two in the bottom row are examples of hanging frames, monitors, and projector screens, which are 3D objects with a "standalone existence". The walls on the inside and outside of the building with no descriptive features qualify as object-free locations

most or all of their content on walls, often outside of the building, and sometimes even in mid-air.

Figure 11.14 shows that there was a relatively even split among users in terms of the object-dependent/object-free dimension. Figure 11.15 provides seven examples of object-dependent and object-independent locations in the space.

Finally, I examined the patterns with which users preferred to arrange the content, and identified three alternatives:

- In the "content case", content types that had a similar subject matter were most likely to be arranged in clusters, in close proximity to each other;

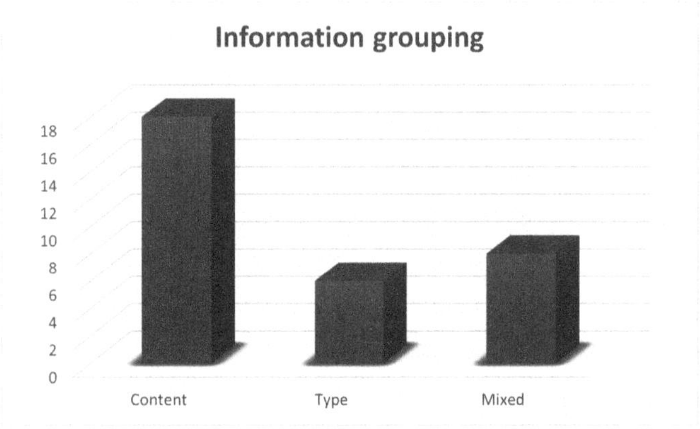

Fig. 11.16 Distribution of users among the "content", "type" and "mixed" content arrangement categories. "Content" users created smartboard clusters with content about the same or similar subject matter. "Type" users created smartboard clusters such that different file formats were clustered together. "Mixed" users employed a combination of these strategies

- In the "type case", similar content types were most likely to be arranged in clusters, close to each other;
- Finally, in the "mixed case", a combination of these two approaches was used.

Figure 11.16 shows the number of users, out of the 32, belonging to each of the three categories.

11.13 Discussion

Based on the results obtained, a reexamination of the original hypotheses was conducted to determine whether or not they were confirmed.

11.13.1 Differences in Terms of the Relative Frequency of Content Types

The experiment yielded results that supported the hypothesis, suggesting that there is a significant difference in the occurrence of certain content types within any given project.

The participants were tasked with presenting a topic in a 3D desktop virtual space, where they had the freedom to use a pre-designed virtual space and fill it with digital content. Five types of content were defined: PDF files, videos, images, PowerPoint files, and Web-based URLs. The participants were provided with sample content from each type, but they were also allowed to search for their own content and display it in the space.

The analysis of the data showed that significantly more images were laid out in the virtual space compared to PDF files, PowerPoint files, or Web content. However, there was no significant difference between the display of video content and images, as both were highly represented in the space.

PDF files were the second most common type of content, and they were added to the space significantly more often than PowerPoint files or Web content.

The prevalence of images can be explained by the use of non-figurative images that served as decorative elements in the plain concrete building. This behavior was observed in multiple cases. Additionally, audio-visual content is often perceived as more easily consumable by laypeople, which could contribute to the relative popularity of images.

To determine whether the proportion of the information grouping preferences was equal between the three groups (content, type, mixed) a chi-square goodness of fit test was performed. The proportion differed by the users' preferences ($X2 = 7750$, $df = 2$, $p = 0.021$).

11.13.2 Differences in Terms of the Relative Size of Content Types

Differences in display sizes for each content type also yielded significant results for certain types of content, thus confirming the second hypothesis.

Specifically, PDF files and Web-based content were generally smaller in size compared to smartboards containing images, videos, and PowerPoint files. However, no significant differences were found in the sizes of the latter three types of content.

The findings of this study provide support for the assumption that the preferences of users regarding the display of content in the virtual world largely reflect the patterns observed in real life. When it comes to videos, PPTs, and images, it is common for us to expect these types of content to appear in a large size, with their interpretability, level of detail, and immersion in the content becoming more prominent as their size increases.

We have observed a confirmation of this assumption. It is noteworthy that even when the space available was limited and video content was one of the most commonly added types of content, the test subjects still dedicated a significant portion of the space to smartboards displaying such content.

As for PDFs and web content, it can be assumed that users are accustomed to reading and interpreting such content on smaller devices. These types of content are typically text-based documents that are rarely seen on large surfaces in real life.

11.13.3 Interactions Between Subject Matter, Content Type and 3D Context

One particularly interesting hypothesis presented in this chapter is the third hypothesis, which suggests that certain types of content would interact with specific 3D objects. I was able to confirm this hypothesis, leading to some intriguing observations.

The 3D space used in this experiment contained several prominent spatial objects. In the majority of cases, content that was semantically related to the object type was placed on the corresponding object. This included monitors for PDF files and web-based content, projection screens for videos and PPT files, tilted screens for PDF files and videos, and vertical boards/panels for images. The most notable object in this regard was the projector projection screen, of which there were two available in the space. In many instances, a video or PowerPoint file was added to the screen.

This confirmation also aligns with the previously described theory that suggests the arrangement of the virtual space is influenced by users' experiences in real life, particularly when familiar spatial elements are present. Since this type of content is typically displayed on projectors in reality as well, my assumption is supported. On the virtual desk, the monitor predominantly displayed PDF-type textual content, appearing more than twice as often as any other type of content. On the external wall of the building, almost exclusively images were present, with minimal occurrence of other content on this particular surface.

A further highly notable aspect of the investigation relates to the positioning and alignment of material within the digital realm. Despite the option to place the material on horizontal surfaces, there was not a single instance of such placement among the various arrangements. Instead, each participant chose to position their material vertically or at an approximate 45-degree angle, in accordance with the spatial elements provided. This tendency can likely be attributed to the fact that, even in our everyday lives, we rarely place informational material on horizontal surfaces. The possibilities offered by digital environments can enhance the fulfillment of needs that are difficult to achieve in the physical world, such as a large board on which one can freely arrange material vertically to facilitate a clearer understanding of both the material itself and the relationships among its components.

Finally, our previous research was supported by the findings, as the majority of participants independently categorized the boards and their associated material into groups without any explicit guidance. This categorization was done to improve the visibility of the presented topic and to ensure that adjacent documents provided additional information to the viewer. Furthermore, in six instances, participants organized the material based on the type of documents (e.g., video, PowerPoint) rather than their subject matter. Participants in the third, most preferred group typically arranged the documents spatially according to content, but within each specific area, the primary organizing principle was the type of content.

In addition to the findings, the investigation suggests valuable insights from which a general design principle can be formulated to give users more freedom in virtual reality while addressing the limitations commonly encountered in real life. Specifically, if users are not able to freely modify the type and number of spatial elements, such as tables, monitors, or projection screens, they may face restrictions that hinder the flexible organization of material.

This was evident in the study as participants showed a preference for content-oriented arrangements, while the abundance of tables and monitor screens often led them to lean towards using specific content types. Moreover, the inherent size and alignment of such objects imposed constraints on the placement and quantity (i.e., spatial distribution) of material elements, resulting in sub-optimal solutions when there were not enough suitable objects available or when there was limited space in their surroundings, making it difficult to place other types of material nearby. Conversely, when a large billboard was affixed to a wall, participants were often able to utilize the space more freely and accommodate a wider range of content types.

Based on these observations, there is a tradeoff between using tangible, semantically significant objects and having adaptable space. When incorporating semantically significant objects, it is advisable to ensure that there is enough surrounding space to allow for the free arrangement of content types that differ from those dictated by the objects themselves.

11.14 Conclusion

In this chapter, we examined how users prefer to arrange digital content in 3D virtual environments when given a free hand in creating their own content layouts, depending on the subject matter, the format of the content (e.g., text-based, image, or audio–visual) and the 3D objects within the space. Results have demonstrated that in keeping with 3D spatial metaphors in physical reality, users most often aim to group together content elements that pertain to the same subject matter, prefer to combine written documents with audio–visual content, and have definite preferences in terms of the sizes of different content types, to the advantage of audiovisual content and presentations, which were more likely to be presented on larger displays. At the same time, the experiment also demonstrated that in some ways, users also expect to surpass the limitations of physical reality in VR. As a case in point, we have uncovered a certain tension between the availability of semantically loaded 3D objects, which entail a certain kind of functionality, and the spatial constraints or limitations any specific arrangement of such objects can engender. Thus, we have formulated a general principle, recommending that spaces be designed such that they provide ample space surrounding such "semantically charged" objects for the incorporation of a variety of content modalities. The findings of this chapter can serve as a guide for the development of future virtual spaces, by offering principles by which spatial objects can be designed and arranged to better suit subsequently created layouts of 2D content, as well as by supporting the creation of such layouts in a more principled way. The operation of future innovative solutions that leverage artificial intelligence capabilities, offering users automated content organization and document structure, can also be informed at least in part by this research. Moving forward, it would be beneficial to expand the research to include a larger number of subjects and assess the differences between the document types examined here, their size, and their relationship with spatial elements in a broader range of virtual spaces and through a broader range of (immersive, non-immersive) modalities.

References

1. Laurell, C., Sandström, C., Berthold, A., Larsson, D.: Exploring barriers to adoption of virtual reality through social media analytics and machine learning-an assessment of technology, network, price and trialability. J. Bus. Res. **100**, 469–474 (2019)
2. IDC: Worldwide Spending on Augmented and Virtual Reality Forecast to Deliver Strong Growth Through 2024, According to a New IDC Spending Guide (2020). https://www.idc.com/getdoc.jsp?containerId=prUS47012020
3. Lanier, J., Biocca, F.: An insider's view of the future of virtual reality. J. Commun. **42**(4), 150–172 (1992)
4. Dalgarno, B., Hedberg, J., Harper, B.: The contribution of 3D environments to conceptual understanding. In: 19th Annual Conference of the Australasian Society for Computers in Learning in Tertiary Education, Winds of Change in the Sea of Learning. UNITEC Institute of Technology, pp. 1–10 (2002)
5. Radianti, J., Majchrzak, T.A., Fromm, J., Wohlgenannt, I.: A systematic review of immersive virtual reality applications for higher education: design elements, lessons learned, and research agenda. Comput. & Educ. **147**, 103778 (2020)
6. Villena-Taranilla, R., Tirado-Olivares, S., Cózar-Gutiérrez, R., González-Calero, J.A.: Effects of virtual reality on learning outcomes in k-6 education: a meta-analysis. Educ. Res. Rev. **35**, 100434 (2022)
7. van der Kruk, S.R., Zielinski, R., MacDougall, H., Hughes-Barton, D., Gunn, K.M.: Virtual reality as a patient education tool in healthcare: a scoping review. Patient Educ. Couns. (2022)
8. Venkatesan, M., Mohan, H., Ryan, J.R., Schürch, C.M., Nolan, G.P., Frakes, D.H., Coskun, A.F.: Virtual and augmented reality for biomedical applications. Cell Rep. Med. **2**(7), 100348 (2021)
9. Geraets, C.N., van der Stouwe, E.C., Pot-Kolder, R., Veling, W.: Advances in immersive virtual reality interventions for mental disorders: a new reality? Curr. Opin. Psychol. **41**, 40–45 (2021)
10. Vayssiere, P., Constanthin, P.E., Herbelin, B., Blanke, O., Schaller, K., Bijlenga, P.: Application of virtual reality in neurosurgery: patient missing. A systematic review. J. Clin. Neurosci. **95**, 55–62 (2022)
11. Idi, D.B., Khaidzir, K.A.M.: Critical perspective of design collaboration: a review. Front. Arch. Res. **7**(4), 544–560 (2018)
12. Dianatfar, M., Latokartano, J., Lanz, M.: Review on existing VR/AR solutions in human–robot collaboration. Procedia CIRP **97**, 407–411 (2021)
13. Delgado, J.M.D., Oyedele, L., Demian, P., Beach, T.: A research agenda for augmented and virtual reality in architecture, engineering and construction. Adv. Eng. Inform. **45**, 101122 (2020)
14. Guo, Z., Zhou, D., Zhou, Q., Zhang, X., Geng, J., Zeng, S., Lv, C., Hao, A.: Applications of virtual reality in maintenance during the industrial product lifecycle: a systematic review. J. Manuf. Syst. **56**, 525–538 (2020)
15. Firu, A.C., Tapîrdea, A.I., Feier, A.I., Drăghici, G.: Virtual reality in the automotive field in industry 4.0. Mater. Today Proc. **45**, 4177–4182 (2021)
16. Wertheimer, M.: Gestalt Theory, pp. 1–11. Kegan Paul, Trench, Trubner & Company (1938)
17. Koffka, K.: Principles of Gestalt psychology, vol. 44. Routledge (2013)

18. Kohler, W.: The Mentality of Apes. Routledge (2018)
19. Arnheim, R.: Visual Thinking. University of California Press (1997)
20. Rock, I.: The Logic of Perception. MIT Press, Cambridge (1983)
21. Palmer, S.E.: Common region: a new principle of perceptual grouping. Cogn. Psychol. **24**(3), 436–447 (1992)
22. Michotte, A.: The Perception of Causality, vol. 21. Routledge (2017)
23. Pedersen, D.M.: Dimensions of environmental competence. J. Environ. Psychol. **19**(3), 303–308 (1999)
24. Steele, F.: Defining and developing environmental competence. Adv. Exp. Soc. Process. **2**(1), 225–244 (1980)
25. Tversky, B.: Structures of mental spaces: how people think about space. Environ. Behav. **35**(1), 66–80 (2003)
26. Tversky, B.: Cognitive maps, cognitive collages, and spatial mental models. In: Spatial Information Theory a Theoretical Basis for GIS: European Conference, COSIT'93 Marciana Marina, Elba Island, Italy September 19–22, Proceedings, vol. 2005. Springer, pp. 14–24 (1993)
27. Gobet, F., Lane, P.C., Croker, S., Cheng, P.C., Jones, G., Oliver, I., Pine, J.M.: Chunking mechanisms in human learning. Trends Cogn. Sci. **5**(6), 236–243 (2001)
28. Alamia, A., Solopchuk, O., d'Ausilio, A., Van Bever, V., Fadiga, L., Olivier, E., Zénon, A.: Disruption of Broca's area alters higher-order chunking processing during perceptual sequence learning. J. Cogn. Neurosci. **28**(3), 402–417 (2016)
29. Bor, D., Duncan, J., Wiseman, R.J., Owen, A.M.: Encoding strategies dissociate prefrontal activity from working memory demand. Neuron **37**(2), 361–367 (2003)
30. Sakai, K., Kitaguchi, K., Hikosaka, O.: Chunking during human visuomotor sequence learning. Exp. Brain Res. **152**, 229–242 (2003)
31. Rosenbaum, D.A., Kenny, S.B., Derr, M.A.: Hierarchical control of rapid movement sequences. J. Exp. Psychol. Hum. Percept. Perform. **9**(1), 86 (1983)
32. Baddeley, A.: Working memory. Science **255**(5044), 556–559 (1992)
33. Koch, I., Hoffmann, J.: Patterns, chunks, and hierarchies in serial reaction-time tasks. Psychol. Res. **63**, 22–35 (2000)
34. Pirolli, P., Card, S.: Information foraging. Psychol. Rev. **106**(4), 643 (1999)
35. Pan, B., Hembrooke, H., Gay, G., Gonsalves, G.C.: Bridging the gap: a conceptual model of the access of digital libraries. J. Digit. Inf. **7**(2), (2006)
36. Bergman, O., Gradovitch, N., Bar-Ilan, J., Beyth-Marom, R.: Folder versus tag preference in personal information management. J. Am. Soc. Inf. Sci. Technol. **64**(10), 1995–2012 (2013)
37. Dinneen, J.D., Julien, C.-A.: The ubiquitous digital file: a review of file management research. J. Assoc. Inf. Sci. Technol. **71**(1), E1–E32 (2020)
38. Whittaker, S., Massey, C.: Mood and personal information management: how we feel influences how we organize our information. Pers. Ubiquitous Comput. **24**(5), 695–707 (2020)
39. Gwizdka, J.: Taskview: design and evaluation of a task-based email interface. In: Proceedings of the 2002 Conference of the Centre for Advanced Studies on Collaborative Research, ser. CASCON '02, p. 4. IBM Press (2002)
40. Bomsdorf, B.: First steps towards task-related web user interfaces. In: Computer-Aided Design of User Interfaces III: Proceedings of the Fourth International Conference on Computer-Aided Design of User Interfaces 15–17 May 2002, Valenciennes, France, pp. 349–356. Springer (2002)
41. Csapó, Á.B., Horvath, I., Galambos, P., Baranyi, P.: VR as a medium of communication: from memory palaces to comprehensive memory management. In: 2018 9th IEEE International Conference on Cognitive Infocommunications (CogInfoCom), pp. 000389–000394. IEEE (2018)
42. Lampert, B., Pongracz, A., Sipos, J., Vehrer, A., Horvath, I.: MaxWhere VR-learning improves effectiveness over clasiccal tools of e-learning. Acta Polytechnica Hungarica **15**(3), 125–147 (2018)
43. Horváth, I.: Maxwhere 3d capabilities contributing to the enhanced efficiency of the trello 2d management software. Acta Polytechnica Hungarica **16**(6), 55–71 (2019)

44. Horvath, I., Sudar, A.: Factors contributing to the enhanced performance of the maxwhere 3d vr platform in the distribution of digital information. Acta Polytechnica Hungarica **15**(3), 149–173 (2018)
45. Sudár, A., Csapó, Á.B.: Descriptive markers for the cognitive profiling of desktop 3d spaces. Electronics **12**(2), 448 (2023)
46. Berki, B.: 2D advertising in 3D virtual spaces. Acta Polytechnica Hungarica **15**(3), 175–190 (2018)
47. Berki, B.: Desktop VR and the use of supplementary visual information. In: 2018 9th IEEE International Conference on Cognitive Infocommunications (CogInfoCom), pp. 000333–000336. IEEE (2018)
48. Setti, T., Csapo, A.B.: A canonical set of operations for editing dashboard layouts in virtual reality. Front. Comput. Sci. **3**, 659600 (2021)
49. Krokos, E., Plaisant, C., Varshney, A.: Virtual memory palaces: immersion aids recall. Virtual R. **23**(1), 1–15 (2018)
50. Setti, T., Csapó, Á.B.: Quantifying the effectiveness of project-based editing operations in virtual reality. In: 2022 1st IEEE International Conference on Cognitive Aspects of Virtual Reality (CVR), pp. 000049–000054. IEEE (2022)
51. Galambos, P., Weidig, C., Baranyi, P., Aurich, J.C., Hamann, B., Kreylos, O.: VirCA NET: a case study for collaboration in shared virtual space. In: 2012 IEEE 3rd International Conference on Cognitive Infocommunications (CogInfoCom), IEEE, pp. 273–277. IEEE (2012)
52. Sudár, A., Csapó, Á.: Interaction patterns of spatial navigation and smartboard use in VR workspaces. In: Accentuated Innovations in Cognitive Info-Communication, pp. 149–166. Springer International Publishing (2022)
53. Sudár, A., Csapó, A.: An mcmc-based method for clustering display panels with the goal of generating navigation paths in 3d. In: 2021 12th IEEE International Conference on Cognitive Infocommunications (CogInfoCom), pp. 1009–1014. IEEE (2021)
54. Cogné, M., Taillade, M., N'Kaoua, B., Tarruella, A., Klinger, E., Larrue, F., Sauzeon, H., Joseph, P.-A., Sorita, E.: The contribution of virtual reality to the diagnosis of spatial navigation disorders and to the study of the role of navigational aids: a systematic literature review. Ann. Phys. Rehabil. Med. **60**(3), 164–176 (2017)
55. Hegarty, M., Montello, D.R., Richardson, A.E., Ishikawa, T., Lovelace, K.: Spatial abilities at different scales: individual differences in aptitude-test performance and spatial-layout learning. Intelligence **34**(2), 151–176 (2006)
56. Moffat, S.D., Zonderman, A.B., Resnick, S.M.: Age differences in spatial memory in a virtual environment navigation task. Neurobiol. Aging **22**(5), 787–796 (2001)
57. Rizzo, A.A., Schultheis, M., Kerns, K.A., Mateer, C.: Analysis of assets for virtual reality applications in neuropsychology. Neuropsychol. Rehabil. **14**(1–2), 207–239 (2004)
58. Lee, E.A.-L., Wong, K.W.: Learning with desktop virtual reality: low spatial ability learners are more positively affected. Comput. & Educ. **79**, 49–58 (2014)
59. Chen, C.J.: Are spatial visualization abilities relevant to virtual reality? E-J. Instr. Sci. Technol. **9**(2), n2 (2006)
60. Setti, T., Csapo, A.B.: Outlines of a graph-tensor based adaptive associative search model for internet of digital reality applications. In: 2022 IEEE 1st International Conference on Internet of Digital Reality (IoD), pp. 000049–000054 (2022)

Chapter 12
Reducing Cognitive Load Through Content Architecture in VR

Abstract The objective of this chapter is to introduce a research where the goal was to examine the possibility of converting commonly utilized 2D digital layouts into three-dimensional dashboards in order to decrease cognitive load. In order to achieve this goal, we conducted a study in which we compared user performance metrics, pupil dilation data, and subject-reported qualitative measures in a 2D scenario and two different versions of a 3D virtual reality scenario. All three scenarios focused on a use case that involved the most prevalent digital materials and designs encountered in digital education. These scenarios made use of textual information (such as PDF files and PPT files), images, and videos. By assuming that differences in cognitive load can be verified through pupillometry measurements, we demonstrated that it is feasible to develop 3D virtual reality scenarios that result in reduced cognitive load while maintaining the same performance metrics as commonly used 2D environments. However, our experiment also revealed that these improvements are not automatic. Instead, 3D workflows that require less physical movement, even if it means increasing camera rotations, appear to be more effective in reducing cognitive load.

Keywords Virtual reality · Spatial abilities · Spatial perception · Preference points

12.1 Introduction

A substantial body of scientific inquiry, coupled with the swift evolution of practical methodologies, has revealed that virtual reality (VR) possesses a distinctive potential for transformation in the realm of education. This holds true for individuals at various stages of life, be it children and young adults in educational institutions or adults in professional settings [1–7]. The transformative power of VR lies in its ability to facilitate spatial comprehension in three dimensions, irrespective of scale, thereby surpassing the limitations of traditional text-based descriptions and two-dimensional illustrations. Furthermore, VR enables users to dynamically interact with these spatial structures, thereby promoting constructive learning approaches [8, 9], and augmenting users' motivation to learn [10, 11]. It is also worth noting that

I. Horváth et al., *Cognitive Aspects of Virtual Reality*, Studies in Big Data 156,
https://doi.org/10.1007/978-3-031-68130-1_12

a lesser-known advantage of VR is its capacity to accommodate the placement of two-dimensional documents within a three-dimensional environment. This feature allows for the simultaneous visualization of a greater amount of information, arranged spatially in a manner that effectively reflects associative and semantic relationships between different parts of the content [12–14].

Simultaneously, while some researchers have examined the ergonomics of VR in specific usage scenarios (see e.g. [15, 16], there remains a scarcity of universally applicable principles that elucidate the impact of VR design on user capabilities. It is important to underscore that this observation pertains to both objective capabilities, such as the ability to learn, as well as subjective ones, including the willingness to invest the required effort for learning. In this chapter, we direct our attention specifically to these questions by evaluating users' effectiveness and cognitive load in a learning scenario involving two versions of a desktop VR environment, as well as a two-dimensional web-based environment. Through our experiments, we demonstrate that, although users perform similarly across the three scenarios, one of the three-dimensional cases allows users to achieve comparable results with significantly lower cognitive load. Consequently, the disparities between the three-dimensional environments enable us to propose specific design recommendations for the development of educational three-dimensional virtual spaces.

The chapter is structured in the following manner. In Sect. 12.2, we provide a summary of the background information pertaining to cognitive load theory, as it relates to this particular work. In Sect. 12.6, we present the experimental framework within which the experiments discussed in this chapter were conducted, while the experiments themselves are presented in the subsequent two sections. Specifically, Sect. 12.7 provides a summary of the results from an initial experiment that has already been described in detail in [17]; additionally, the results of this experiment have now been supplemented with additional data obtained from users of the 2D interface, in order to ensure a statistically more robust basis for comparing the 3D and 2D cases. In Sect. 12.8, we then present a detailed account of the results from a second follow-up study, in which a different 3D scenario (but the same 2D scenario) was employed. The discrepancies in user performance and cognitive load experienced between the two 3D scenarios and the 2D scenario are subsequently reported in Sect. 12.9, which is titled "Results". Finally, in Sect. 12.10, we provide a discussion on our interpretation of the results.

12.2 Cognitive Load Theory

Cognitive Load Theory (CLT), a widely accepted theory in the field of educational psychology, aims to provide an account of the information processing mechanisms of the human cognitive system and its implications for instructional design. The primary objective of cognitive load theory is to elucidate how an increase in the cognitive demands associated with learning tasks, not only in educational settings but also in work environments, can impact individuals' ability to assimilate new information and store it in long-term memory [18–22]. It is important to note that in order for individuals to acquire new knowledge, they must be able to process new information, a process that depends on the capacity of their working memory, which is known to be limited [23–25]. However, when learners exceed these limitations and are faced with excessively demanding requirements, they experience cognitive overload [26]. Cognitive overload occurs when the cognitive load surpasses a certain threshold, leading to a failure in the acquisition and application of knowledge. This can arise due to ineffective presentation strategies employed during instruction or the presence of external distractions [20]. To design and create appropriate learning and work environments, it is essential to have a comprehensive understanding of the different types of cognitive load and their effects.

12.2.1 Categories and Effects of Cognitive Load

Sweller and his colleagues identified 10 cognitive load-related effects [19]:

- *The goal-free effect* refers to the phenomenon whereby problem-solving tasks without explicit goals yield superior learning outcomes compared to tasks with specific goals. This is because the absence of explicit goals allows for greater flexibility in exploring the problem, which in turn enhances the comprehension of its underlying structure. It is worth noting that goal-free tasks can still involve achieving the same precise objectives as traditional problems, but without placing exclusive focus on those goals.
- *The redundancy effect* is a phenomenon observed when the same information is presented through multiple sources and in various formats, such as text, images, or graphs. Unlike the split attention effect, where mental integration is required, the redundancy effect allows for understanding of the information even without any cognitive integration.
- *The expertise reversal effect* is closely related to the level of expertise in a particular domain. For novice learners, encountering information from multiple modalities can facilitate a deeper understanding and knowledge acquisition. However, for experts in a field, the redundancy of information may demand additional cognitive resources, as they are often able to comprehend the message from a single modality.
- *The guidance fading effect* involves gradually reducing the level of instructional guidance provided during problem-solving tasks as learners gain expertise. This approach, compared to alternating between worked examples and problems, leads to a more sustainable and effective instructional method.
- *The imagination effect* pertains to the reduction of cognitive load that can be achieved through mental practice of specific procedures. It is important to note that this technique is more effective for experienced learners compared to novices.

- *The self-explanation effect* illustrates that engaging in a mental dialogue about a working example of the material being learned helps learners construct a mental framework and comprehend the example. This process is particularly beneficial for expert learners rather than novices.
- *The element interactivity effect* posits that the impact of additional cognitive load is contingent upon the level of intrinsic cognitive load. When the intrinsic load is low, any additional load becomes inconsequential, whereas when the intrinsic load is high, the additional load becomes more significant.
- *The worked example effect/Problem completion effect* refers to the phenomenon wherein students who are provided with worked examples to study perform better on future tests compared to those who are required to independently solve the same problem. Van Merrienboer and Crook conducted research on a computer programming course and demonstrated that students who received a partial solution to a problem experienced lower cognitive load due to the reduced extraneous cognitive load associated with searching for examples independently.
- *The Split-Attention effect* arises when learners are required to divide their attention between two or more related information sources that cannot be comprehended without mental integration. Such types of learning necessitate the user to mentally integrate the provided information in order to effectively learn the material. Instructional strategies and formats that present the materials in an integrated format have been shown to facilitate more effective learning compared to scenarios where the information is presented in a split-attention format.
- *The modality effect* is closely associated with the split-attention effect and relies on the structure of the working memory, which processes visual and auditory information through separate channels. This effect can be advantageous in reducing extraneous cognitive load if textual information, pictures, and animations are presented in an auditory form.

The last three effects hold particular relevance to the chapter, as optimizing cognitive load is of significant importance in the design of digital environments and the use of digital content.

12.2.2 Measuring Cognitive Load

Various methods have been developed over the past four decades to measure cognitive load and its related effects. In the 1980s, cognitive load was treated as an assumed concept rather than a directly measured construct.

- Indirect measures
  Sweller and his colleagues [27] developed computational models that allowed for quantitative distinctions to serve as proxies for cognitive load. These indirect measures involve analyzing knowledge acquisition or learning performance. However, these models are not suitable for continuous measurement of cognitive load during learning activities [21, 22, 28, 29].

- Direct measures
  Subjective measures, such as Paas's [30] subjective rating scale and the NASA TLX [31], are commonly used in the literature to assess cognitive load. One advantage of these approaches is their applicability across various learning contexts, encompassing diverse subject matter and participant groups, and they provide valid information about the experience of the individuals being measured. However, there are limitations in terms of validity, reliability, and the availability of data in some cases [28, 32–34].
  Objective methods offer a range of possibilities. Such measurements are capable of continuous behavioral and physiological assessment via very detailed data acquisition. Some examples of objective measures include:

  - secondary task performance [35–37]
  - functional magnetic resonance imaging [38, 39]
  - electrodermal activity [40, 41]
  - electroencephalography [42, 43]
  - heart-rate measurement [44, 45]
  - eye-tracking and pupillometry analysis [46–51].

In the subsequent section, our focus will shift from the examination of human cognitive abilities to the fundamental cognitive theories that are relevant in contemporary technology-driven contexts. Therefore, while the objective of this chapter was to explore the characteristics of the human visual modality, the subsequent chapter will redirect its emphasis towards the practical implementation of these theories in the domains of Cognitive Infocommunications (CogInfoCom) and Cognitive Aspects of Virtual Reality (cVR). These domains are intricately interconnected with our comprehension of how information is organized, experienced, and navigated within 2D and 3D interfaces. This transition effectively connects the theoretical foundation with real-world, technology-centered outcomes, thereby reflecting the seamless integration of cognitive theories into modern digital contexts.

12.3 Cognitive Pupillometry

Research related to pupillometry dates back nearly six decades. Kahneman and Beatty demonstrated that increasing the length of a string of digits to be remembered is directly associated with a heightened load on memory, which subsequently results in an enlargement of pupil size [52]. Kahneman's [53] explanation of the pupillary response as a measure of "attentional capacity load" remains pertinent, as subsequent investigations have consistently shown a clear correlation between pupil dilation and the demands imposed by executive load or working memory load [54–56]. To gain a deeper understanding of why pupil dilation increases with increased task demands, it is crucial to have knowledge about the underlying biological and cognitive processes. The pupil demonstrates size variations in response to three specific types of stimuli, which can offer valuable insights into the underlying processes of perception, attention, and cognitive engagement. Specifically, it constricts when exposed to bright light (pupil light response) and when focusing on nearby objects (pupil near response); conversely, it dilates when there is heightened cognitive activity, such as increased arousal or mental effort (psycho-sensory pupil response), triggered either by an external stimulus or spontaneously [57].

The process of pupil dilation is controlled by the iris dilator muscle. This particular muscle is regulated by the sympathetic nervous system, which is a component of the autonomic nervous system responsible for functions such as arousal, wakefulness, and the fight-or-flight response. The connection between pupil dilation and the sympathetic nervous system elucidates why pupils tend to be larger when an individual is experiencing heightened arousal [58, 59]. The diameter of the human pupil typically ranges from approximately 2 to 8 mm [58, 60].

Currently, the measurement of pupil dilation has become relatively straightforward. The availability of affordable eye trackers has made it possible to achieve sufficient temporal resolution and accuracy for detecting even slight changes in pupil diameter.

12.4 Eye-Tracking

The historical lineage of eye-tracking devices and the associated methodology itself is not steeped in a long-standing tradition. Nonetheless, it has played a substantial role in the final decade of the 20th century and continues to exert its influence in the present day. As an instrument of inquiry, eye tracking is steadily gaining traction among researchers and finds practical application across a diverse range of scientific disciplines. The direction and duration of our visual fixation are not solely governed by attention alone, but rather are also influenced by cognitive processes encompassing perception, memory, language, and decision-making. While the connection between ocular movements and mental processes is not absolute, it is generally valid to assert that the eyes serve as a reflection of the cerebral processing of visual stimuli at any given moment [61]. Consequently, eye tracking is widely employed in studies that scrutinize cognitive operations. By virtue of its superior temporal sensitivity, the eye tracking modality has the capacity to furnish moment-to-moment insights into the temporal progression of cognitive processes, thereby transcending the mere presentation of the final outcome [62]. Furthermore, ocular movements are predominantly governed by autonomous mechanisms and are not easily or frequently susceptible to conscious control. Although individuals possess the ability to selectively determine what and when to visually attend to, the finer nuances of ocular movements are primarily reflexive in nature, rendering them largely unremembered. Consequently, the eye-tracking methodology emerges as a valuable tool for delving into non-conscious processes. Throughout the course of eye-tracking measurements, various types of ocular movements can be discerned, encompassing fixation, saccades, smooth pursuit, vergence, and pupillary dilation, the latter pertaining to the dilation and constriction of the pupil, regulated by the sympathetic and parasympathetic nervous systems [63]. Alongside this wealth of data, the so-called regions of interest also furnish invaluable information. These regions are designated by researchers to examine the frequency and duration of participants' focus on specific elements within an image, text, or even an individual's countenance. In such instances, predetermined sectors, referred to as regions of interest, are identified [62].

12.5 Cognitive Load in Virtual Spaces

The cognitive load theory posits that there exists a finite threshold to the amount of information the brain can effectively process within a given timeframe. Once this threshold is surpassed, performance is compromised and mental strain escalates [18–21]. Despite the several advantageous aspects of virtual reality, it often necessitates active engagement and engenders a dynamic and immersive encounter, which inadvertently augments cognitive processing demands. However, through judicious design, it is plausible to exert control over the intricacy and cognitive load associated with a virtual environment.

In recent years, there has been a growing interest among researchers in quantifying the cognitive load associated with learning, task-solving, and problem-solving in virtual and augmented reality. Some studies have demonstrated a decrease in cognitive load in these environments [26, 64–66], while others have shown an increase in cognitive load in virtual spaces [67]. Additionally, certain measurements have failed to show a significant difference in cognitive load between real-life tasks and their virtual reality counterparts [68]. It is important to note that these findings are particularly valuable in situations where it is crucial to assess the impact of a real place or event on individuals before it actually takes place [65, 68]. This research has significant implications in today's world, where the number of remote learners and workers is on the rise and there is a need for solutions that can reduce cognitive load.

The following section presents a series of experimental investigations that aim to determine the extent to which the transition from 2D interfaces to 3D spaces can impact users' performance and cognitive load. Moreover, this chapter identifies new cognitive abilities that are naturally facilitated by the utilization of 3D virtual reality spaces instead of traditional 2D interfaces for specific tasks.

12.6 Framework for the Experiments Conducted

To evaluate the effectiveness of user workflows and the cognitive load associated with them in 2D digital interfaces and various types of 3D environments, we developed a specific workflow and a set of experimental guidelines that can be replicated in different settings.

12.6.1 Details of the Workflow to Be Carried Out

The workflow assigned to the participants involved reading and/or viewing educational materials related to four different subtopics in the field of astronomy ("*Universe*", "*Planets*", "*Satellites*" and "*Space Research*"), followed by answering a questionnaire for each subtopic. The questionnaires included true-or-false questions, multiple-choice questions, and short-answer questions. Three examples of typical questions are:

- True or false?—Black holes can be observed based on the gravitational effects they have on surrounding gases, dust and stars;
- What are the rings of Saturn made of (select all that apply)?—ice, rocks, space debris, gases, asteroids, and/or moons.
- Why were Hubble's mirrors polished at night?

12.6.2 2D Setup

For the 2D condition, we utilized a conventional Google Sites page (https://sites.google.com/view/2deyetracking-egyoldalas/f%C5%91oldal, accessed on 14 January, 2023) loaded in a Chrome Browser. This page contained all the learning materials embedded in a sequential manner, interspersed with the corresponding questionnaires (which were in the form of embedded Google Forms within the Google Sites page). The documents and questionnaires for each subtopic were strictly ordered, allowing users to study three documents (PDF files, images, or YouTube videos) and then complete the associated questionnaire before moving on to the next subtopic (Fig. 12.1).

12.6.3 3D Setup

In the 3D condition, the same documents and questionnaires were presented in different 3D spaces within the MaxWhere platform (https://maxwhere.com). MaxWhere is a desktop 3D platform that enables the creation of 3D spaces with freely arranged display panels known as "smartboards" in MaxWhere terminology. Importantly, these smartboards can contain any type of document that a desktop browser can normally display, such as webpages, PDF files, images, and audio-video files.

In the case of the two-dimensional (2D) scenario, the learning materials were integrated into a Google form and users were required to scroll up and down in order to access different documents and questionnaires. Conversely, in the three-dimensional (3D) scenario, users were allowed and expected to freely navigate through the content.

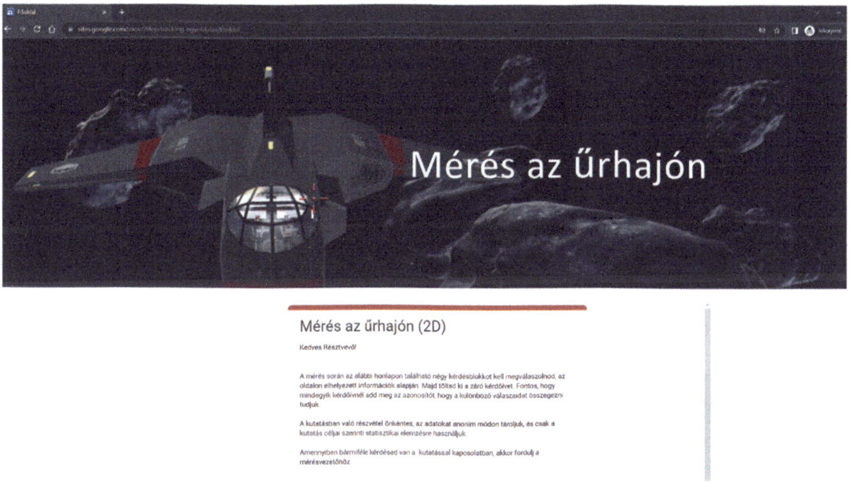

Fig. 12.1 Layout and design of the '2'D webpage using which the 2D scenario experiments were conducted

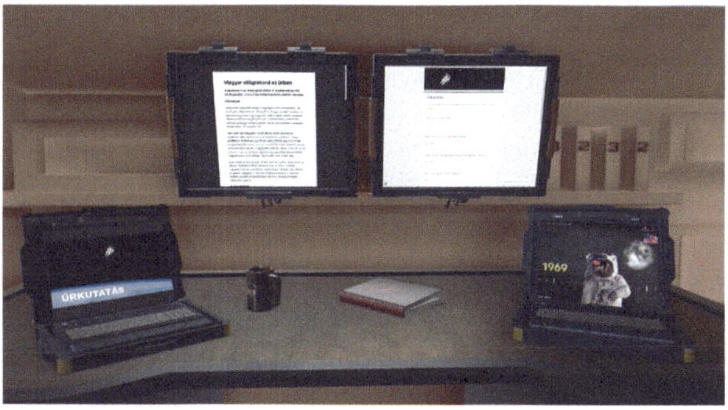

Fig. 12.2 Spatial arrangement of 2D content in a 3D virtual space

Figures 12.2 and 12.3 offer two examples of how documents were organized in a 3D layout.

It is important to highlight that the same exact documents and questionnaires that were embedded in the Google Sites page in the 2D scenario were also presented to users in the 3D scenario, without any additional content in either case. The only distinction was in the presentation of the content: in the 2D scenario, it appeared sequentially from top to bottom, while in the 3D scenario, it was displayed in a spatial arrangement. The relationship between these two scenarios can be observed in Fig. 12.4.

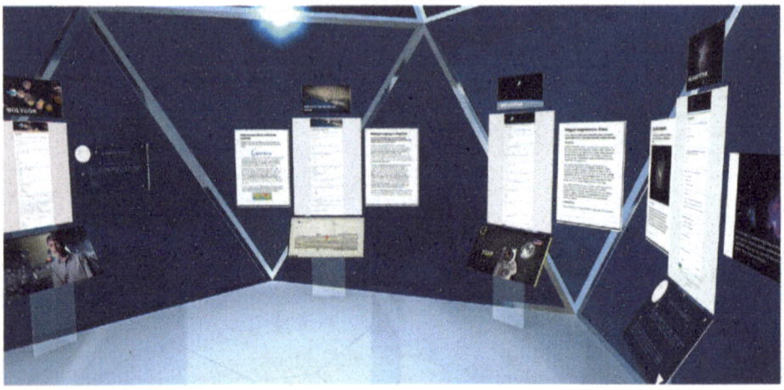

Fig. 12.3 Spatial arrangement of 2D content in a second 3D virtual space

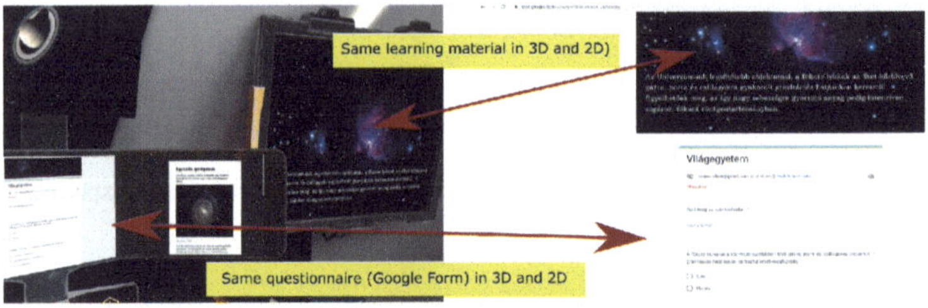

Fig. 12.4 This figure serves as an example to demonstrate that identical documents and questionnaires were utilized in both the 3D scenario (on the left-hand side) and the 2D scenario (on the right-hand side). No materials or questionnaires were added or removed from the experiment in either scenario. Here, we can observe that the PDF document related to the topic of the "Universe" is presented on the tilted panel at the back in the 3D scenario, as well as in the upper half of this particular view in the 2D scenario. Furthermore, while the questionnaire appears directly below the learning material in the 2D scenario, it is located on the left-hand side of the screenshot in the 3D space

12.6.4 Measurement Data Collected

The results obtained from the questionnaires were compared based on the percentage of correctly answered questions and the time required to complete all the questionnaires. It should be noted that, in the 3D scenario, the time spent on navigation was excluded from the latter metric, in order to directly compare the actual time spent on content exploration and questionnaire completion.

Meanwhile, data on eye gaze and pupil dilation were collected to evaluate the subjects' focus of attention and cognitive load. The measurements pertaining to the latter were acquired utilizing the EyeTribe eye tracker and its corresponding software (https://theeyetribe.com/dev.theeyetribe.com/dev.theeyetribe.com/general/index.html, accessed on 14 January, 2023).

In addition, the frequency of certain interaction patterns was recorded, including when:

- Users viewed content on a group of smartboards simultaneously from some distance ("holistic overview mode")
- Users alternated focus between different smartboards while remaining in a stationary position but frequently changing their camera orientation ("alternating mode");

Finally, in a post-experiment questionnaire, participants were requested to provide demographic information, answer questions regarding their digital leisure habits, and provide subjective assessments and evaluations of their sense of immersion in the virtual space (specifically in the 3D scenario). Participants were also asked to rate the difficulty of the topic blocks. For those who completed the tests on the two-dimensional interface, the final questionnaire did not include any questions related to the 3D space.

12.6.5 Procedure

Prior to commencing the experiments, written consent was acquired from all participants to utilize the gathered data in the following manner. The collected data was rendered anonymous and employed solely for the purpose of conducting the statistical analyses disclosed in this chapter. The experiments were conducted in accordance with the ethical principles laid out in the Declaration of Helsinki.

At the commencement of the session, the individuals performing the requisite task in three dimensions indicated their familiarity with the MaxWhere software and, if so, provided an approximate estimate of their software usage duration. Those participants who were not acquainted with the software underwent a learning period of approximately thirty minutes to acquaint themselves with it and acquire fundamental user skills. Basic proficiency entailed confident navigation within the software, activation and deactivation of display panels, and interaction with the displayed content. The evaluators assessed the mastery of proficient software utilization.

Subsequent to this, the participants were positioned in a tranquil room before a laptop computer. The room was dimly illuminated without any direct light source to prevent the introduction of undesired artifacts into the eye tracking data. The eye-tracker was calibrated prior to the commencement of the experiment for each test subject.

Following the calibration, the test administrator explicated the task to be performed. The participants were given the freedom to determine the order in which they completed the questionnaires, including the selection of the first and last questionnaire.

12.7 Preliminary Masurement Results

In an initial experiment, we contrasted the measurement data acquired from the two-dimensional scenario with that obtained from a specific three-dimensional space, as documented in [17].

The experiment was conducted in a space with a spaceship-themed environment, as shown in Fig. 12.5.

12.7.1 Test Subjects

In the case of the spaceship-themed virtual reality space, a total of 14 individuals partook in the experiment, but video data was corrupted for one person. The outcomes from an additional four subjects had to be discarded due to interruptions in pupil dilation measurements occurring midway through the experiment. The remaining nine subjects (three women, six men) had ages ranging from 17 to 55 years, with an average age of 32.5 years (Standard Deviation: 14.15).

A total of seven individuals (four women, three men) participated in the two-dimensional measurement and had ages ranging from 25 to 33 years, with an average age of 27.83 (Standard Deviation: 2.93).

The overall mean age of all participants was 30.84 (11.39) years. All participants were neurotypical Hungarian native speakers who volunteered to partake in the experiment. Informed consent was obtained from all participants before the commencement of the experiment, which was conducted in accordance with the institutional endorsement of the authors' affiliation. All of the data collected during the experiment was anonymized and exclusively utilized as input for the subsequent statistical analyses expounded upon in later sections of this chapter.

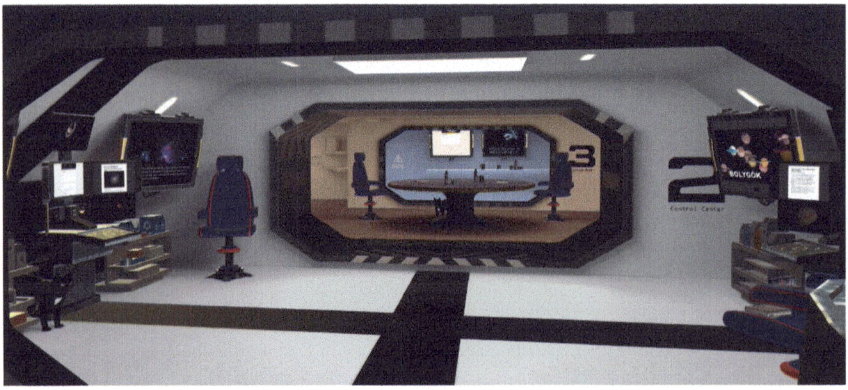

Fig. 12.5 Layout and design of the 3D space in which the first experiment was conducted, as reported in [17]

12.7.2 Preliminary Results

When comparing the outcomes derived from the 2D scenario and the 3D scenario, the collected data demonstrated a remarkable similarity in the accuracy of answers provided by participants in both the 2D and 3D situations, without any statistically significant discrepancies. Concurrently, within the domain of Satellites, participants completed the questionnaire in a significantly shorter period of time in the 3D setting compared to the 2D setting (while there were no notable variations in completion times within the other areas of inquiry). These findings, with the exception of the completion time in the case of the Satellites topic, did not substantiate the hypothesis that participants would exhibit superior performance in the 3D environment as compared to the 2D environment. It is worth noting, however, that participants certainly did not perform worse.

The presence of a significant amount of visual intricacy in the 3D environment led to the inference that reducing this intricacy—potentially by employing a more minimalist 3D space that necessitates minimal navigation (apart from view rotation)—could potentially improve performance in the 3D scenario.

In terms of subjective evaluations of difficulty and measurements of pupil diameter, it was observed that there was no discernible correlation between the two variables. This result was surprising. With the exclusion of the possibility that the tasks were actually more challenging when participants perceived them to be easier, this counterintuitive finding may also be attributed to the presence of visual complexity and consequently a certain level of excitement experienced by the test subjects.

12.7.3 Motivations Behind Follow-Up Study

Based on these points and our previous experience in designing three-dimensional spaces with customized layouts (see also [12]), our objective was to replicate the same experiment utilizing a distinct virtual reality (VR) space and a larger number of participants. Our aim was to employ a different virtual space in the subsequent study that would adequately address the following principles:

- The cognitive load should be reduced by limiting the need for navigation and minimizing visual clutter
- In virtual spaces with numerous spatial elements and accessories, heightened arousal may result in dilated pupils
- The importance of documents should be indicated by their size
- The arrangement of documents into clusters should enable users to perceive them as cohesive content units
- By ensuring a consistent layout for each block, with only the topic being different, users should be able to comprehend semantic relationships more efficiently and with less cognitive load.

- The digital content inside the virtual space should be displayed vertically or on slightly inclined panels, as users have demonstrated a preference for horizontal surfaces to a lesser extent [12].

Therefore, for our subsequent study described in later sections of this chapter, we selected a virtual space that lacked unnecessary distractions, was considerably smaller, required less navigation, and had more organized displays.

12.8 Follow-Up Experiment in a Second VR Environment

Our second experiment was conducted using the same two-dimensional scenario (albeit with a larger number of participants) and a different three-dimensional VR space that featured a circular arrangement of content clusters (Fig. 12.6).

In the three-dimensional virtual space, a total of 21 smartboards were utilized to place the contents. The questionnaires were positioned in the middle of each cluster, with the title slide above them and the documents containing the answers to the questionnaire questions positioned to the right, left, and bottom.

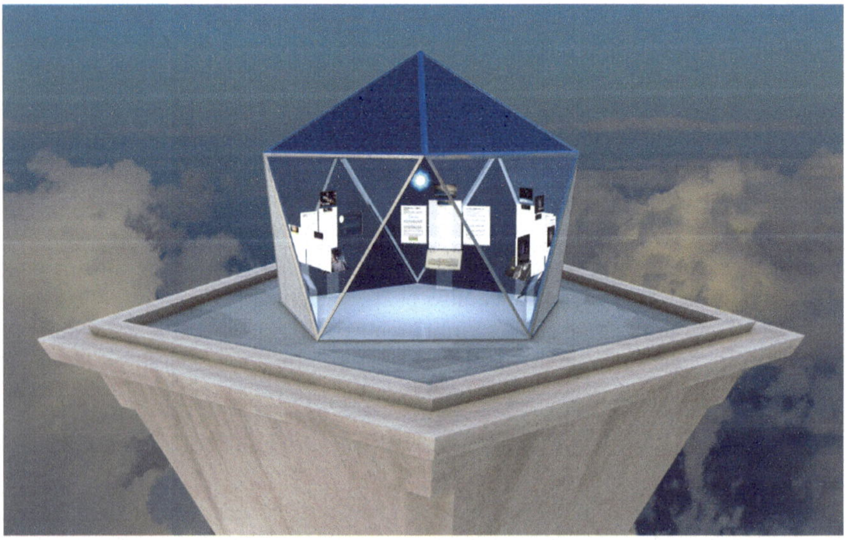

Fig. 12.6 Exterior of the 3D space in which the 3D case of the second experiment was conducted

12.8.1 Subjects

A total of 40 participants took part in the research, with an average age of 26.57 (SD: 6.96), consisting of 20 women and 20 men.

All participants were Hungarian native speakers and individuals with neurotypical development who volunteered to participate in the experiment. Informed consent was obtained from all participants prior to the experiment, which was conducted in accordance with the institutional approval of the authors' affiliation. All data collected during the experiment was anonymized and solely used for statistical analyses as outlined in subsequent sections of this chapter.

In the research study, the participants were divided into two distinct groups. A total of 20 individuals, consisting of 14 women and 6 men, with an average age of 27.35 (standard deviation: 6.72), took part in the measurement conducted in a 3D virtual space. Among the 20 subjects, 5 of them wore glasses during the experiment. Additionally, a separate group of 20 individuals, comprising of 14 men and 6 women, participated in the 2D measurement, with an average age of 25.8 (standard deviation: 7.29). Out of this group, 2 individuals wore glasses during the measurement.

12.8.2 Methods

The methods employed in the second experiment mirrored those of the first experiment, which were previously described in Sect. 12.6 concerning the framework utilized in our experiments. Specifically, the subjects engaged in both the 2D and 3D measurements were required to provide feedback on the block they retrospectively found to be the most challenging, as well as provide some demographic information such as gender, age, and highest level of education. In addition to the aforementioned data, the participants in the 3D measurement were also required to complete a series of questions related to their navigation experiences. Furthermore, they had to evaluate their General Presence, Spatial Presence, Involvement, and Experienced Realism experiences using a 7-point Likert scale in an IPQ questionnaire consisting of 14 questions.

12.8.3 Key Hypothesises

Prior to conducting our experiment, we formulated the following key hypotheses:

1. The 3D case would be characterized by subjective assessments of lower cognitive load and lower pupil dilation compared to the 2D case;
2. Questionnaires would generally be completed at a faster rate and with more accurate answers in the 3D measurement compared to the 2D measurementD.

3. When compared to the previous 3D group, the second 3D group would exhibit lower pupil dilation;
4. Subjective assessments made by the test subjects regarding the difficulty of the questionnaires would correlate with pupil dilation and the accuracy of their answers;
5. Lower cognitive load would be correlated with a shorter completion time.

12.9 Results

12.9.1 Score-Based Performance in 2D and 3D

The results of the Mann-Whitney test indicated that there was no statistically significant difference in the final scores of the four tasks between the 3D group (M = 22.387, SD = 2.167) and the 2D group (M = 22.125, SD = 1.879).

12.9.2 Completion Times in 2D and 3D

Furthermore, the Independent sample t-test revealed no statistically significant difference in the completion times of the four tasks between the 3D group (M = 36.7, SD = 6.959) and the 2D group (M = 37.8, SD = 9.059).

12.9.3 Pupil Dilation Between 2D and 3D

The group's descriptive statistics and the differences between the two groups are presented in Table 12.1 and Fig. 12.7. An independent sample t-test was conducted to compare the pupil dilation between the group that underwent 2D measurement (M = 22.204, SD = 3.971) and the 3D group (M = 19.464, SD = 2.902), ($t(34.792) = -2.492$, $p < 0.018$), (Cohen's d = 0.788).

Table 12.1 Descriptives of the pupil dilation data of the 2D and 3D groups

	Group	N	Mean	SD	SE	Coefficient of variation
Pupil Dilatation	3D	20	19.464	2.902	0.649	0.149
	2D	20	22.204	3.971	0.888	0.179

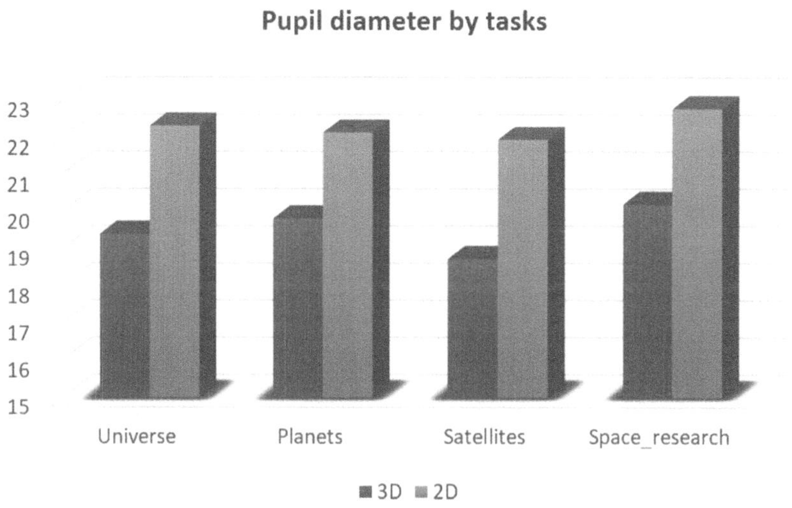

Fig. 12.7 Pupil diameter sizes grouped by tasks

Table 12.2 Descriptives of the pupil dilation data of the previous 3D and current 3D groups

	Group	N	Mean	SD	SE	Coefficient of variation
PupilTasks	3D sublimus	20	19.464	2.902	0.649	0.149
	3D spaceship	9	26.203	4.371	1.457	0.167

The results of the group who filled out the questionnaire in the 3D space (N = 20) showed that they have had smaller pupil dilation with the mean rank of this group being 11.30, while the other group who—for the same task- used the Spaceship 3D space for the same task (N = 9) had a mean rank of 22.23. A Mann-Whitney test revealed statistically significant results, U = 16, $p < 0.001$, r = 0.65 (Table 12.2 and Fig. 12.8).

The Spearman coefficient analysis did not reveal any correlation between the difficulty of the questionnaires and pupil dilation. Similarly, there was no correlation observed between pupil dilation and completion time.

12.10 Discussion

Among the cognitive load effects mentioned in the theoretical introduction, split-attention and modality effect are commonly observed in online teaching materials. This is because teachers frequently present information through separate documents

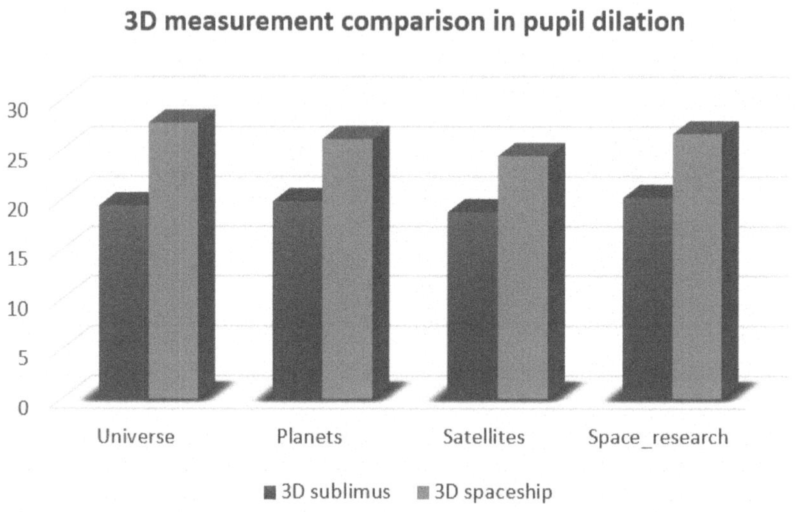

Fig. 12.8 Comparison of the pupil diameter size between the previous 3D research results and the current ones

that are related in content. The aim of the study was to develop a virtual environment that could alleviate the cognitive load for students and mitigate the impact of these two aforementioned effects.

The pupillometry measurements yielded significant differences between the two groups, favoring the second virtual reality (VR) scenario. Thus, it can be concluded that by utilizing previous research findings, a VR space was successfully created that reduces cognitive load during task-solving in the 3D desktop virtual environment. In this virtual space, despite the utilization of multiple modalities and documents, the capabilities of VR enable the interpretation and overview of the information as a unified element or block. During the measurements, the subjects were observed using an intermediate view, where individual documents were enlarged and blocks or specific elements within a block appeared together in front of the user. I suggest referring to this capability as the "holistic overview" capability. Additionally, when the user remains in the same position but rotates to view different parts of a block, the term "alternating mode" aptly describes this user behavior. These capabilities allow users to obtain an overview of multiple documents, similar to real-life situations where documents are placed beside each other on a table or a book stand. Due to the simplicity and cleanliness of the created virtual space, users can focus their attention solely on the content and task solution.

There was no significant difference in completion time and overall score between the two groups. Although the group working in the 3D environment consistently achieved better results, further studies with a larger sample size are required to determine the outcomes more conclusively. Furthermore, there is evidence of modalities blending, particularly with videos in the virtual space. When the questionnaire window is active in front of users, they can respond in real-time based on the audio information from a previously initiated video. While this can also be accomplished on two-dimensional (2D) surfaces, the natural navigational methods in 3D environments provide users with the sensation that they have simply turned their heads to view other content.

The subsequent analysis of correlation did not uncover any noteworthy disparity between the perceived complexity of the given sections and the expansion of the pupil, nor the duration taken to finish tasks, for either group. The initial measurement employed a three-dimensional virtual space, which encompassed a greater variety of colors and spatial objects, and necessitated more extensive navigation due to its larger dimensions. This was juxtaposed with a more simplified and optimized virtual environment with fewer distractions. The statistical analysis demonstrated a significant variation in pupil expansion between the two groups, which implies that the distractors, which were perceived as established elements at the time of measurement, augmented the cognitive burden-increasing influence of the split-attention effect.

Further measurements are essential to ascertain whether, despite acting as disruptive elements in the completion of short-term tasks and the comprehension of specific documents requiring text understanding, such distractors can still facilitate long-term memorization based on the principles of the memory palace. To explore this, it would be beneficial to employ not only eye-tracking but also EEG measurements.

Nevertheless, these outcomes enable me to infer that virtual reality (VR) environments on desktop platforms can function as a feasible alternative and a prospective avenue for distance learning when compared to conventional two-dimensional (2D) interfaces. Considering their manifold proven advantages, it is worthwhile to contemplate their implementation in the broader society.

References

1. Radianti, J., Majchrzak, T.A., Fromm, J., Wohlgenannt, I.: A systematic review of immersive virtual reality applications for higher education: Design elements, lessons learned, and research agenda. Comput. & Educ. **147**, 103778 (2020)
2. Checa, D., Bustillo, A.: A review of immersive virtual reality serious games to enhance learning and training. Multimed. Tools Appl. **79**, 5501–5527 (2020)
3. Tham, J., Duin, A.H., Gee, L., Ernst, N., Abdelqader, B., McGrath, M.: Understanding virtual reality: Presence, embodiment, and professional practice. IEEE Trans. Prof. Commun. **61**(2), 178–195 (2018)
4. Javaid, M., Haleem, A.: Virtual reality applications toward medical field. Clin. Epidemiol. Glob. Health **8**(2), 600–605 (2020)
5. Loureiro, S.M.C., Guerreiro, J., Eloy, S., Langaro, D., Panchapakesan, P.: Understanding the use of virtual reality in marketing: a text mining-based review. J. Bus. Res. **100**, 514–530 (2019)
6. Coban, M., Bolat, Y.I., Goksu, I.: The potential of immersive virtual reality to enhance learning: a meta-analysis. Educ. Res. Rev. **36**, 100452 (2022)
7. Makransky, G., Petersen, G.B.: The cognitive affective model of immersive learning (camil): a theoretical research-based model of learning in immersive virtual reality. Educ. Psychol. Rev. 1–22 (2021)

8. Huang, C.L., Luo, Y.F., Yang, S.C., Lu, C.M., Chen, A.-S.: Influence of students' learning style, sense of presence, and cognitive load on learning outcomes in an immersive virtual reality learning environment. J. Educ. Comput. Res. **58**(3), 596–615 (2020)
9. Hamilton, D., McKechnie, J., Edgerton, E., Wilson, C.: Immersive virtual reality as a pedagogical tool in education: a systematic literature review of quantitative learning outcomes and experimental design. J. Comput. Educ. **8**(1), 1–32 (2021)
10. Wenk, N., Penalver-Andres, J., Buetler, K., Nef, T., Müri, R.M., Marchal-Crespo, L.: Effect of immersive visualization technologies on cognitive load, motivation, usability, and embodiment. Virtual Real. **27**(1), 307–331 (2023)
11. Makransky, G., Borre-Gude, S., Mayer, R.E.: Motivational and cognitive benefits of training in immersive virtual reality based on multiple assessments. J. Comput. Assist. Learn. **35**(6), 691–707 (2019)
12. Sudár, A., Csapó, Á.B.: Elicitation of content layout preferences in virtual 3D spaces based on a free layout creation task. Electronics **12**(9), 2078 (2023). [Online]. https://www.mdpi.com/2079-9292/12/9/2078
13. Daeijavad, P., Maurer, F.: Layouts of 3d data visualizations small multiples around users in immersive environments. In: IEEE International Symposium on Mixed and Augmented Reality Adjunct (ISMAR-Adjunct), vol. 2022, pp. 258–261. IEEE (2022)
14. Luo, W., Lehmann, A., Widengren, H., Dachselt, R.: Where should we put it? layout and placement strategies of documents in augmented reality for collaborative sensemaking. In: Proceedings of the 2022 CHI Conference on Human Factors in Computing Systems, pp. 1–16 (2022)
15. Dozio, N., Marcolin, F., Scurati, G.W., Ulrich, L., Nonis, F., Vezzetti, E., Marsocci, G., La Rosa, A., Ferrise, F.: A design methodology for affective virtual reality. Int. J. Hum. Comput. Stud. **162**, 102791 (2022)
16. Glaser, N., Schmidt, M.: Systematic literature review of virtual reality intervention design patterns for individuals with autism spectrum disorders. Int. J. Hum. Comput. Stud. **38**(8), 753–788 (2022)
17. Sudár, A., Csapó, Á.B.: Descriptive markers for the cognitive profiling of desktop 3d spaces. Electronics **12**(2), 448 (2023)
18. Sweller, J., Van Merrienboer, J.J., Paas, F.G.: Cognitive architecture and instructional design. Educ. Psychol. Rev. 251–296 (1998)
19. Sweller, J.: Cognitive Load Theory. Psychology of Learning and Motivation, vol. 55, pp. 37–76. Elsevier (2011)
20. Sweller, J., van Merriënboer, J.J., Paas, F.: Cognitive architecture and instructional design: 20 years later. Educ. Psychol. Rev. **31**, 261–292 (2019)
21. Plass, J.L., Moreno, R., Brünken, R. (eds.): Cognitive Load Theory. Cambridge University Press (2010)
22. Paas, F., Renkl, A., Sweller, J.: Cognitive load theory and instructional design: recent developments. Educ. Psychol. **38**(1), 1–4 (2003)
23. Cowan, N.: Working Memory Capacity: Classic Edition. Psychology Press (2016)
24. Turner, M.L., Engle, R.W.: Is working memory capacity task dependent? J. Mem. Lang. **28**(2), 127–154 (1989)
25. Unsworth, N., Engle, R.W.: The nature of individual differences in working memory capacity: active maintenance in primary memory and controlled search from secondary memory. Psychol. Rev. **114**(1), 104 (2007)
26. Thees, M., Kapp, S., Strzys, M.P., Beil, F., Lukowicz, P., Kuhn, J.: Effects of augmented reality on learning and cognitive load in university physics laboratory courses. Comput. Hum. Behav. **108**, 106316 (2020)
27. Sweller, J.: Measuring cognitive load. Perspect. Med. Educ. **7**, 1–2 (2018)
28. Brunken, R., Plass, J.L., Leutner, D.: Direct measurement of cognitive load in multimedia learning. Educ. Psychol. **38**(1), 53–61 (2003)
29. Brünken, R.E., Plass, J.L., Moreno, R.E.: Current Issues and Open Questions in Cognitive Load Research, pp. 253–272. Cambridge University Press (2010)

30. Paas, F.G.: Training strategies for attaining transfer of problem-solving skill in statistics: a cognitive-load approach. J. Educ. Psychol. **84**(4), 429 (1992)
31. Hart, S.G., Staveland, L.E.: Development of NASA-TLX (task load index): Results of empirical and theoretical research. In: Advances in Psychology, vol. 52, pp. 139–183. Elsevier (1988)
32. Anmarkrud, Ø., Andresen, A., Bråten, I.: Cognitive load and working memory in multimedia learning: conceptual and measurement issues. Educ. Psychol. **54**(2), 61–83 (2019)
33. Van Gog, T., Kirschner, F., Kester, L., Paas, F.: Timing and frequency of mental effort measurement: Evidence in favour of repeated measures. Appl. Cogn. Psychol. **26**(6), 833–839 (2012)
34. Cierniak, G., Scheiter, K., Gerjets, P.: Explaining the split-attention effect: Is the reduction of extraneous cognitive load accompanied by an increase in germane cognitive load? Comput. Hum. Behav. **25**(2), 315–324 (2009)
35. Brünken, R., Plass, J.L., Leutner, D.: Assessment of cognitive load in multimedia learning with dual-task methodology: auditory load and modality effects. Instr. Sci. 115–132 (2004)
36. DeLeeuw, K.E., Mayer, R.E.: A comparison of three measures of cognitive load: Evidence for separable measures of intrinsic, extraneous, and germane load. J. Educ. Psychol. **100**(1), 223 (2008)
37. Park, B., Knörzer, L., Plass, J.L., Brünken, R.: Emotional design and positive emotions in multimedia learning: An eyetracking study on the use of anthropomorphisms. Comput. & Educ. **86**, 30–42 (2015)
38. Van Dillen, L.F., Heslenfeld, D.J., Koole, S.L.: Tuning down the emotional brain: an fMRI study of the effects of cognitive load on the processing of affective images. Neuroimage **45**(4), 1212–1219 (2009)
39. Whelan, R.R.: Neuroimaging of cognitive load in instructional multimedia. Educ. Res. Rev. **2**(1), 1–12 (2007)
40. Buchwald, M., Kupiński, S., Bykowski, A., Marcinkowska, J., Ratajczyk, D., Jukiewicz, M.: Electrodermal activity as a measure of cognitive load: A methodological approach. In: Signal Processing: Algorithms, Architectures, Arrangements, and Applications (SPA), vol. 2019, pp. 175–179. IEEE (2019)
41. Li, P., Li, Y., Yao, Y., Wu, C., Nie, B., Li, S.E.: Sensitivity of electrodermal activity features for driver arousal measurement in cognitive load: the application in automated driving systems. IEEE Trans. Intell. Transp. Syst. **23**(9), 14954–14967 (2021)
42. Anderson, E.W., Potter, K.C., Matzen, L.E., Shepherd, J.F., Preston, G.A., Silva, C.T.: A user study of visualization effectiveness using EEG and cognitive load. Comput. Graph. Forum **30**(3), 791–800 (2011). Wiley Online Library
43. Friedman, N., Fekete, T., Gal, K., Shriki, O.: EEG-based prediction of cognitive load in intelligence tests. Front. Hum. Neurosci. **13**, 191 (2019)
44. Solhjoo, S., Haigney, M.C., McBee, E., van Merrienboer, J.J., Schuwirth, L., Artino, A.R., Battista, A., Ratcliffe, T.A., Lee, H.D., Durning, S.J.: Heart rate and heart rate variability correlate with clinical reasoning performance and self-reported measures of cognitive load. Sci. Rep. **9**(1), 1–9 (2019)
45. Vanneste, P., Raes, A., Morton, J., Bombeke, K., Van Acker, B.B., Larmuseau, C., Depaepe, F., Van den Noortgate, W.: Towards measuring cognitive load through multimodal physiological data. Cogn. Technol. & Work **23**, 567–585 (2021)
46. Joseph, A.W., Murugesh, R.: Potential eye tracking metrics and indicators to measure cognitive load in human-computer interaction research. J. Sci. Res. **64**(1), 168–175 (2020)
47. Zagermann, J., Pfeil, U., Reiterer, H.: Measuring cognitive load using eye tracking technology in visual computing. In: Proceedings of the Sixth Workshop on Beyond Time and Errors on Novel Evaluation Methods for Visualization, pp. 78–85 (2016)
48. Gavas, R., Chatterjee, D., Sinha, A.: Estimation of cognitive load based on the pupil size dilation. In: 2017 IEEE International Conference on Systems, Man, and Cybernetics (SMC), pp. 1499–1504. IEEE (2017)
49. van der Wel, P., Van Steenbergen, H.: Pupil dilation as an index of effort in cognitive control tasks: a review. Psychon. Bull. & Rev. **25**, 2005–2015 (2018)

50. Sibley, C., Coyne, J., Baldwin, C.: Pupil dilation as an index of learning. Proc. Hum. Factors Ergon. Soc. Annu. Meet. **55**(1), 237–241 (2011). SAGE Publications Sage CA: Los Angeles, CA
51. Perkhofer, L., Lehner, O.: Using gaze behavior to measure cognitive load. In: Information Systems and Neuroscience: NeuroIS Retreat, vol. 2019, pp. 73–83. Springer (2018)
52. Kahneman, D., Beatty, J.: Pupil diameter and load on memory. Science **154**(3756), 1583–1585 (1966)
53. Kahneman, D.: Attention and Effort, vol. 1063. Citeseer (1973)
54. Van Gerven, P.W., Paas, F., Van Merriënboer, J.J., Schmidt, H.G.: Memory load and the cognitive pupillary response in aging. Psychophysiology **41**(2), 167–174 (2004)
55. Võ, M.L.-H., Jacobs, A.M., Kuchinke, L., Hofmann, M., Conrad, M., Schacht, A., Hutzler, F.: The coupling of emotion and cognition in the eye: introducing the pupil old/new effect. Psychophysiology **45**(1), 130–140 (2008)
56. Laeng, B., Sirois, S., Gredebäck, G.: Pupillometry: a window to the preconscious? Perspect. Psychol. Sci. **7**(1), 18–27 (2012)
57. Mathôt, S.: Pupillometry: Psychology, physiology, and function. J. Cogn. **1**(1), (2018)
58. Wang, C.-A., Munoz, D.P.: A circuit for pupil orienting responses: implications for cognitive modulation of pupil size. Curr. Opin. Neurobiol. **33**, 134–140 (2015)
59. Samuels, E.R., Szabadi, E.: Functional neuroanatomy of the noradrenergic locus coeruleus: its roles in the regulation of arousal and autonomic function part i: principles of functional organisation. Curr. Neuropharmacol. **6**(3), 235–253 (2008)
60. McDougal, D., Gamlin, P.: 1.26—pupillary control pathways. In: Masland, R.H., Albright, T.D., Albright, T.D., Masland, R.H., Dallos, P., Oertel, D., Firestein, S., Beauchamp, G.K., Catherine Bushnell, M., Basbaum, A.I., Kaas, J.H., Gardner, E.P. (eds.) The Senses: A Comprehensive Reference. Academic Press, New York (2008, pp. 521–536. [Online]. https://www.sciencedirect.com/science/article/pii/B9780123708809002826
61. Reichle, E.D., Reineberg, A.E., Schooler, J.W.: Eye movements during mindless reading. Psychol. Sci. **21**(9), 1300–1310 (2010)
62. Carter, B.T., Luke, S.G.: Best practices in eye tracking research. Int. J. Psychophysiol. **155**, 49–62 (2020)
63. Rayner, K.: Eye movements and attention in reading, scene perception, and visual search. Q. J. Exp. Psychol. **62**(8), 1457–1506 (2009)
64. Küçük, S., Kapakin, S., Göktaş, Y.: Learning anatomy via mobile augmented reality: effects on achievement and cognitive load. Anat. Sci. Educ. **9**(5), 411–421 (2016)
65. Frederiksen, J.G., Sørensen, S.M.D., Konge, L., Svendsen, M.B.S., Nobel-Jørgensen, M., Bjerrum, F., Andersen, S.A.W.: Cognitive load and performance in immersive virtual reality versus conventional virtual reality simulation training of laparoscopic surgery: a randomized trial. Surg. Endosc. **34**, 1244–1252 (2020)
66. Lai, A.-F., Chen, C.-H., Lee, G.-Y.: An augmented reality-based learning approach to enhancing students' science reading performances from the perspective of the cognitive load theory. Br. J. Educ. Technol. **50**(1), 232–247 (2019)
67. Andersen, S.A.W., Mikkelsen, P.T., Konge, L., Cayé-Thomasen, P., Sørensen, M.S.: The effect of implementing cognitive load theory-based design principles in virtual reality simulation training of surgical skills: a randomized controlled trial. Adv. Simul. **1**(1), 1–8 (2016)
68. Armougum, A., Orriols, E., Gaston-Bellegarde, A., Joie-La Marle, C., Piolino, P.: Virtual reality: a new method to investigate cognitive load during navigation. J. Environ. Psychol. **65**, 101338 (2019)

Part IV
Augmented Communication Channels in VR

In this part, we focus on novel forms of meaning representation supported by VR, and present a brief comparison with respect to existing media.

Chapter 13
A Taxonomy of Representations of Meaning in VR

Abstract This chapter explores ways in which VR environments can enhance existing methods for representing information and for guiding interactions compared to other kinds of interfaces. In particular, we relate the theory of cognitive infocommunication channels to virtual reality spaces, and investigate how it applies to the design of VR spaces and user experiences therein. Examples are provided from the MaxWhere VR platform.

Keywords Virtual Reality (VR) · VR-based learning · VR memory palace

13.1 Introduction

Virtual and augmented reality (VR/AR) is becoming one of the fastest growing industrial sectors. According to Fortune Business Insights, the VR market size reached almost 20 billion USD in 2022 at a global level, and can be expected to reach 165 billion USD by 2030.[1] Consequently, producers are focused on enhancing hardware capabilities for 3D to meet the rising demand. In addition, an increasing number of new companies are developing methods to efficiently handle both 2D and 3D content for virtual reality (VR) and augmented reality (AR) applications.

Considering these developments, it is important to explore some of the fundamental ideas behind VR and AR, and how they came about. Also, a key question is how these technologies, when integrated into platforms, may shape the future of computing. This chapter primarily concentrates on VR, with examples taken from desktop VR, though many of our insights are applicable to other forms of 3D technologies as well.

Some of the key ideas that are of particular interest to this chapter can be linked to the Memory Palace Method (MPM), also known as the method of loci, which dates back to the classical antiquities. To understand the principles of this technique

---

[1] https://www.fortunebusinessinsights.com/industry-reports/virtual-reality-market-101378.

I. Horváth et al., *Cognitive Aspects of Virtual Reality*, Studies in Big Data 156,
https://doi.org/10.1007/978-3-031-68130-1_13

and its connection to VR, it is useful to start by examining how humans acquire new knowledge and subsequently retrieve what they've learned before–processes that are fundamental to MPM.

Generally speaking, the human learning process can be characterized through the following points:

- Knowledge can be seen as a capability not only to recall insights, but also to relate them to a web of already existing insights, often in more than one way;
- Such relationships can form the basis of further, novel insights;
- Because insights are by definition derived rather than ready-made, first-principle concepts, learning is by necessity a multi-faceted process that revolves both around a set of constantly changing (active) concepts as well as a set of relatively fixed conceptual components;
- Hence the process of effective learning consists of more than just memorization: it often includes a step of **chunking** followed by **memorization**. Chunking helps to divide larger-scale, composite concepts into smaller ones, leading to a more reliable formation of memories [1];
- As a result, it is possible to talk about **knowledge systems** and **thinking systems**, which describe the ways in which the fixed conceptual components and the active set of concepts are stored, linked to each other and recalled (usually as a function of their relationships) [2].

The key idea is that for learning to truly take hold, it is essential not only to memorize the fundamental principles but also to understand and remember the significant connections among concepts derived from these principles, often called insights. Nevertheless, due to their inherent complexity, generating insights and then distilling them into digestible parts for memory retention is generally a challenging endeavor. This challenge is further amplified in the current era, where we are constantly bombarded with a vast amount of information in our daily lives.

Over time, humans have developed numerous strategies to tackle the challenges involved in memorizing new insights. Due to the associative nature of the human brain, often the most effective strategies are those that create links between new ideas and those that are already firmly embedded in our minds. A significant number of these strategies are verbal, utilizing the structure of language to establish connections through acronyms, key phrases, or other mnemonic devices. These devices are particularly memorable because of their rhythmic or musical qualities, or because they have a meaningful connection to other words or concepts we already understand [3–5].

Just as the auditory aspects of language facilitate memory, visual strategies also play a crucial role in learning and memory formation. This effectiveness stems from the human brain's evolutionary background, where recognizing and remembering spatial relationships and locations was critical—far more so than memorizing abstract, non-spatial information [6, 7]. Consequently, the Memory Palace Method (MPM), which leverages mental visualization within a three-dimensional space, remains one of the most potent techniques for securing long-term memories [7]. MPM is particularly suitable for associating seemingly mundane or unrelated pieces of information with a comparatively more significant body of insights, thereby making the insights easier to recall. It accomplishes this by forming chains of both natural and constructed associative links across a sequence of imaginary locations within a visualized space.

When considered separately, VR and MPM may appear to address distinct challenges: the former offers a technological platform for engaging digital experiences, while the latter provides a cognitive strategy for memorization. However, the intrinsic spatial orientation of both methods, coupled with their shared ability to transcend the limitations of physical reality, raises the intriguing possibility that VR could enhance or even revolutionize the traditional Memory Palace, especially when applied with similar objectives in mind, albeit on a potentially grander scale. Historical research has already highlighted VR's capacity to bolster the effectiveness of MPM [8], as well as to enrich the user experience with more, or more accessible and memorable, content [9–11]. Consequently, this chapter aims to explore the extent to which VR environments can further advance the acquisition, communication, and retention of new information, positioning them as a versatile tool in the realms of education and knowledge preservation.

The chapter is structured as follows. Section 13.2 revisits earlier discussions on the feasibility of conceptual mapping within the scope of cognitive infocommunication (CogInfoCom) channels [12]. The objective here is to shed light on the ways human cognition can create connections between newly encountered concepts and those that are already well-established. Subsequently, Sect. 13.3 delves into an exploration of how the entirety of the conceptual mapping theory within CogInfoCom channels, along with the methodologies incorporated by the Memory Palace Method, could become seamlessly integrated into the architecture of Virtual Reality (VR) environments. In this regard, we posit that VR spaces are not only useful for effective communication but also serve as a modern expansion of the classical Memory Palace Method. Through a series of illustrative examples and subsequent discussions in Sect. 13.4, we make two basic claims:

- Firstly, that VR environments are inherently equipped to encompass the diverse spectrum of conceptual mapping strategies utilized in the Memory Palace Method, as further detailed within the framework of CogInfoCom channels.
- Secondly, that when thoughtfully designed, VR environments can function both as an innovative memory enhancement tool and a communicative medium, offering a more evocative and efficient alternative to the traditional methodologies of the Memory Palace Method.

13.2 Conceptual Mapping in CogInfoCom Channels

Previous research has revealed several methods through which meaning can be assigned to various sensory channels in humans [12]. According to this theoretical framework, it is useful to differentiate at the outset between **direct** and **indirect** communication methods. This distinction is based on whether the concepts being conveyed are represented directly in the messages themselves, or if, alternatively, analogies or spatio-temporal correlations are employed to establish connections between concepts that are, on the surface, unrelated. This approach allows for a nuanced understanding of how information can be communicated in ways that either straightforwardly utilize the intended concepts or creatively link them to other ideas through metaphorical or relational mappings.

On the indirect communication side, we can further differentiate between structural mapping, corroborative (co-)stimulation, and scenario-based mapping. **Structural mapping** involves aligning the dimensions and their respective scales of one concept with those of another, distinct concept. For instance, signaling a train's approach with a specific color light that turns on or off, and indicating the train's nearness by the light's blinking rate, exemplifies structural mapping. **Corroborative stimulation**, or **co-stimulation**, leverages the spatial and temporal closeness between the concept being communicated and the delivery mechanism of the message to create an associative link, regardless of their dissimilarity in form. **Scenario-based mapping**, on the other hand, arranges the message's components in a spatial and temporal sequence that mirrors a real or hypothetical scenario. This scenario-based arrangement enables the creation of links that can be easily recognized and remembered, making the information conveyed more relatable and retainable.

13.3 Links Between CogInfoCom Channels, the Memory Palace and VR Spaces

The traditional Memory Palace Method (MPM), understood as a technique for facilitating the mapping between memory storage and retrieval, fundamentally operates as a scenario-based instrument. It establishes a spatio-temporal structure that aids in the memorization of individual concepts by associating them with a spatial environment that is mentally visualized and, consequently, assists in their sequential recall. This scenario-based approach ensures that information is remembered in a specific sequence, enhancing the orderliness and efficiency of recollection. It is crucial to recognize that the individual concepts placed within this memory framework can be encoded using a variety of methods–encompassing both direct and indirect strategies—based on their characteristics such as shape, color, size, tactile sensations, auditory cues, or even olfactory elements. This flexibility in the encoding process allows for a rich and multi-sensory memorization experience, leveraging the full range of human perceptual capabilities.

Given the discussion thus far, the Memory Palace Method (MPM) can be viewed as a practical application of conceptual mapping theory within the context of CogInfoCom channels. Similarly, it can be argued that today's Virtual Reality (VR) environments have the capacity to encompass all forms of conceptual mapping discussed earlier. Consequently, a significant correlation between MPM and VR spaces is evident, particularly in their shared emphasis on visual engagement and the ability to abstractly represent concepts. In the following, we illustrate these connections through different perspectives, citing examples from the MaxWhere VR platform[2] These examples aim to showcase how VR can serve not only as a dynamic extension of the Memory Palace Method but also as a versatile platform for conceptual mapping, highlighting VR's potential to enhance and expand upon traditional memorization and learning techniques.

13.3.1 VR Spaces Can Encapsulate All Direct and Indirect Forms of Conceptual Mapping

To highlight this point, let us consider a few examples from among the MaxWhere spaces.

In the layout of the X-Podium space, shown in Fig. 13.1, the utilization of color coding across different sections of the platform for various applications is a prominent feature. This color-coded layout facilitates a clear distinction between different groups of content, embodying a form of structural mapping where specific colors are associated with particular use cases. Such an approach leverages color as a means to organize and categorize information, effectively aiding the learning process by making distinctions between content groups more intuitive and immediately recognizable.[3] In addition, the integration of MaxWhere's story feature enables a guided digital tour through a sequence of preset viewpoints within the 3D environment, tailored to the specific sequence necessitated by the user's objectives. This guided progression through the digital landscape exemplifies scenario-based mapping, with the ordered series of viewpoints guiding the interactions dictated by the given use case, thereby facilitating a structured and immersive learning experience within the virtual space.

Similarly, let us explore the environment named Mars, illustrated in Fig. 13.2 through two screenshots (see also[4]). This virtual space employs digital guidance, thereby incorporating scenario-based mapping. The effectiveness of this mapping is amplified when the content integrated into the environment pertains to subjects such as the physics, geography, or geology of space. Establishing a vivid connection between the subject matter and the visual design of the surroundings creates a blend of scenario-based mapping and co-stimulation. This combination is particularly suitable for reinforcing key concepts, immersing users in a simulated location that feels convincingly real. An intriguing aspect of the Mars environment is its embodiment of "initiation"–akin to a rite of passage. Users are required to traverse a long and

---

[2] MaxWhere VR: http://www.maxwhere.com.

[3] Incidentally, the segmentation of content into distinct groups significantly enhances the efficacy of learning processes, as discussed in [13].

[4] https://youtu.be/B4PksATlI-c.

Fig. 13.1 X-Podium in MaxWhere's VR platform.

ascending path (ideally lined with educational content such as lecture slides) before reaching their ultimate goal: a glass hub where they can engage in lab activities using cloud-based services. This feature of VR holds significant educational value, not only in terms of content delivery but also by enriching the psychological experience of learning, effectively marrying the acquisition of knowledge with the learner's journey through a simulated, yet impactful, environment.

Given the ability of VR to replicate fully animated environments, it is evident that the spectrum of CogInfoCom mapping techniques not showcased in the aforementioned examples–specifically, low-level and high-level direct mapping–can be seamlessly integrated into VR spaces as well. VR's versatility allows for a wide array of mapping techniques to be employed, enhancing the learning and engagement experience. Additional illustrations of such techniques in action within various MaxWhere environments further underscore this point (refer to Fig. 13.3):

Fig. 13.2 Mars in MaxWhere's VR platform.

- the use of 2D display panels of different sizes, as well as different colors to highlight the relative importance of the different panels, such as in the Team Head, Team Leader and Team Member spaces (structural mapping)
- the use of visual animations and potentially even sound effects to drive home key concepts more effectively, such as in the Control Lab spaces which use visual animations (high-level direct mapping and corroborative stimulation)
- the use of modifiable animations, so that the structure of modifiability helps students to recall key features of the subject matter, as in case of the control theory related Control Lab spaces (structural mapping)
- the use of various modalities (even within one human sensory modality, such as vision) to better highlight key concepts, such as in the Control Lab spaces, where users can simultaneously visualize the system under control (the pendulum), different kinds of graphs of its performance and the controller parameters (low-level direct mapping and corroborative stimulation)

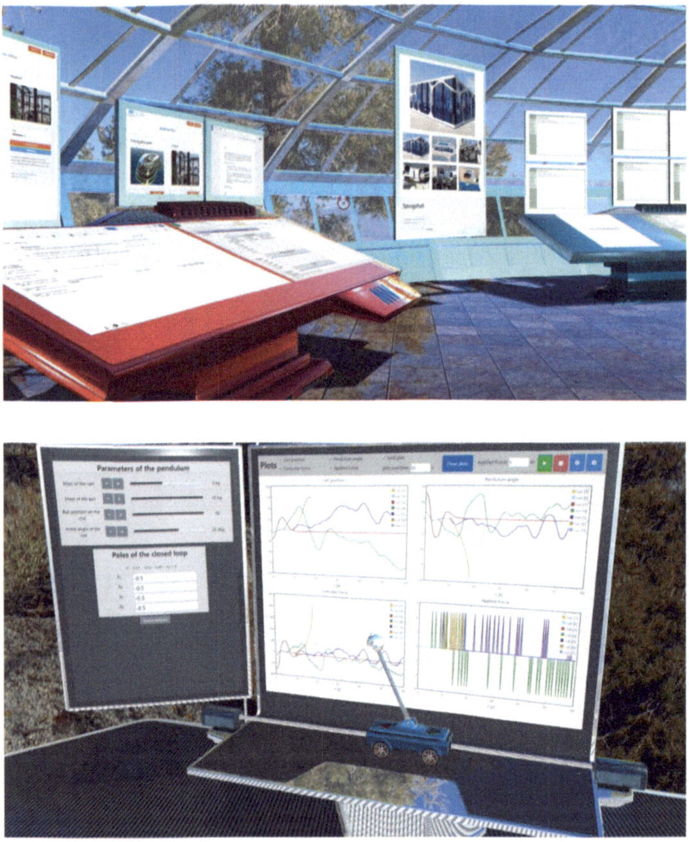

Fig. 13.3 The Team Head and Control Lab Pendulum spaces in MaxWhere's VR platform.

Table 13.1 Materialization of different forms of conceptual mapping in VR spaces.

Level	Method	Analogy in VR
Direct	High-level	3D objects
	Low-level	details of form and animations
Indirect	Structural	layout, digital guidance and interactive controls
	Co-stimulation	multi-modal feedback
	Scenario-based	the VR environment

Based on the above it can be concluded that all of the mapping techniques derived from the theory of CogInfoCom channels, as well as from MPM can be used to practical effect in VR spaces. Table 13.1 summarizes this conclusion.

13.3.2 VR Spaces Can Serve as a Memory Management Tool and a Communication Medium That Is Even More Effective Than the Memory Palace Method

The examples mentioned previously suggest a broader range of applications for VR spaces compared to MPM. The focal aspect of this discussion is VR's capacity to encapsulate the psychological dimensions of learning as a dynamic process. This capability extends beyond merely presenting information; it involves engaging learners on a deeper, more interactive level, thereby leading to a more immersive and embodied learning experience. Indeed, even if a VR learning platform almost exclusively supports the visual (as opposed to auditory, tactile, haptic, vestibular etc.) modality, the ability to freely navigate within the space and to consume the content within a structured process can enhance understanding and retention. This aligns with the cognitive and emotional aspects of learning.

For example, as hinted at by the example of the Control Lab spaces (which fit well into the concept of virtual laboratories [14]), VR can provide constant visual or other feedback to ensure that misunderstandings can be brought to light as early in the learning process as possible. It can also use different modalities to capture the key phases of learning as a process–as hinted at in the example on the Mars space as a process of initiation.

The final example considered in this chapter is that of the Lecture Space SZIU, developed by the Széchenyi István University (Fig. 13.4). As reflected by its name, this is primarily a lecture space, but it also includes a small working area, with a table of 5 display panels. The design of the space conveys a significant message to students, emphasizing the importance of proactive exploration and selection in the learning process. It essentially advises, "Start by observing your surroundings and identifying resources. Next, sift through these materials to find what's relevant, gather it on your worktable, and use it to create something innovative." This methodology implicitly promotes a project-based learning strategy, urging students to engage in independent thought, discernment, and prioritization of information. By advocating for this approach, the space not only guides students towards self-directed learning but also underscores the value of critical thinking and the skill of distilling essential information from a broader context.

Fig. 13.4 Lecture Space SZIU in MaxWhere's VR platform

13.4 Discussion

In earlier discussions, it was established that VR spaces function effectively as tools for memory management and communication. They encompass the entirety of analogy-based mapping techniques identified within the realm of CogInfoCom channels and engage deeply with the psychological facets of how humans learn. Despite these insights, one might still ask whether VR offers advantages beyond traditional learning methodologies like MPM and conventional auditory and textual learning approaches. The question arises: Is the primary value of VR its ability to furnish learners, particularly those who may struggle with visual imagination, with a more captivating and thus more compelling learning environment? Or does VR extend its benefits further by enhancing comprehension, engagement, and retention in ways that traditional methods cannot, leveraging its immersive and interactive capabilities to transform and elevate the educational experience beyond the superficial appeal of visual stimulation?

Considering how orators of Cicero's caliber might have dedicated immense effort to crafting the ideal mental landscapes for their discourses, ensuring these cognitive constructs didn't conflict with previously established mental images, it is clear that VR spaces might be able to provide users with significant short-cuts to achieve a similar result. Yet, it is essential to recognize that the challenge today may not lie solely in a diminishing capacity or willingness to engage in such intensive memorization efforts.[5] In addition, the challenge may also include the ability to manage an increasingly overwhelming amount of complex information that contemporary society is exposed to (see [16–18], which suggest that information itself is an evolving entity). From this perspective, VR emerges as a viable tool for embedding the principles of the memory palace into modern learning practices, aiding in the organization, memorization, and retrieval of crucial knowledge. Additionally, the external visualization capabilities of VR allow for the detailed crafting and continuous refinement of virtual spaces, making it possible to encapsulate a wealth of information in a cohesive manner.

A distinctive benefit of VR over conventional MPM lies in the persistent nature of VR environments–their capacity for prolonged modification and interaction suggests a future where artificial intelligence (AI) tools might autonomously generate and customize VR spaces for specific communicative and educational purposes. Current developments within platforms like MaxWhere, which already automate digital guidance based on the spatial design, hint at future capabilities of AI in this domain. Looking forward, it is conceivable that AI will advance to not only tweaking existing environments in terms of object quantity, placement, and organization but also crafting entirely new VR spaces from scratch.

This anticipates a dynamic educational landscape where the unique characteristics of lectures, seminars, or collaborative projects–each with its distinct patterns of focus and structural needs (e.g., main arguments, supporting information, supplementary content)–can be automatically translated into VR configurations. Such innovation could revolutionize the customization of learning environments, ensuring they align precisely with the pedagogical objectives and content structure. The promise of AI in facilitating these transformations speaks to a future where VR not only enhances the learning experience through immersion and interactivity but also adapts fluidly to the diverse and evolving educational requirements of both instructors and learners, tailoring experiences that are as unique as the subjects they aim to teach.

13.5 Conclusion

The examples presented in the chapter support the notion that well-designed 3D VR spaces are extremely useful generic tools for supporting learning and memory management. At the same time, VR spaces that are well designed, i.e. contain noticeable and interesting components to focus one's attention, can serve as a basis for the

[5] Refer to [15] for a thorough discussion on the shifting paradigms of human learning styles.

"algorithmification" of human learning, given that they are capable of using a diverse range of conceptual mappings and scenarios to help the formation of powerful memories. Importantly, the effectiveness of VR spaces in achieving these goals can be measured, and thus improved upon using increasingly automated tools. As a result, VR spaces can be considered as a memory management tool, and can ultimately provide a unique platform for the blending of human and AI capabilities towards the creation of knowledge.

References

1. Powers, W.G., Bodie, G.D., Fitch-Hauser, M.: Chunking, priming and active learning: Toward an innovative and blended approach to teaching communication-related skills. Interact. Learn. Environ. **14**(2), 119–135 (2006). [Online]. https://doi.org/10.1080/10494820600800182
2. Habók, A.: Evaluating a concept mapping training programme by 10 and 13 year-old students. Int. Electron. J. Elem. Educ. **4**, 07 (2012)
3. Manalo, E., Bunnell, J.K., Stillman, J.A.: The use of process mnemonics in teaching students with mathematics learning disabilities. Learn. Disabil. Q. **23**(2), 137–156 (2000)
4. Knott, D., Thaut, M.H.: Musical mnemonics enhance verbal memory in typically developing children. Front. Educ. **3**, (2018). [Online]. https://www.frontiersin.org/articles/10.3389/feduc.2018.00031
5. McPherson, F.: Mnemonics for Study, 2nd edn. Wayz Press (2018). [Online]. https://www.amazon.com/Mnemonics-Study-2nd-ed-Skills/dp/192716642X
6. Rodríguez, F., López, J., Vargas, J., Broglio, C., Gómez, Y., Salas, C.: Spatial memory and hippocampal pallium through vertebrate evolution: insights from reptiles and teleost fish. Brain Res. Bull. **57**(3), 499–503 (2002). [Online]. https://www.sciencedirect.com/science/article/pii/S0361923001006827
7. Logie, R.H.: Visuo-spatial processing in working memory. Q. J. Exp. Psychol. Sect. A **38**(2), 229–247 (1986)
8. Legge, E.L., Madan, C.R., Ng, E.T., Caplan, J.B.: Building a memory palace in minutes: Equivalent memory performance using virtual versus conventional environments with the method of loci. Acta Psychologica **141**(3), 380–390 (2012)
9. Horváth, I., Sudár, A.: Factors contributing to the enhanced performance of the MaxWhere 3D VR platform in the distribution of digital information. Acta Polytechnica Hungarica **15**(3), 149–173 (2018)
10. Lampert, B., Pongracz, A., Sipos, J., Vehrer, A., Horvath, I.: MaxWhere VR-learning improves effectiveness over clasiccal tools of e-learning. Acta Polytechnica Hungarica **15**(3), 125–147 (2018)
11. Berki, B.: 2D advertising in 3D virtual spaces. Acta Polytechnica Hungarica **15**(3), 175–190 (2018)
12. Baranyi, P., Csapo, A., Sallai, G.: Cognitive Infocommunications (CogInfoCom). Springer (2015)
13. Horváth, I.: Innovative engineering education in the cooperative VR environment. In: 2016 7th IEEE International Conference on Cognitive Infocommunications (CogInfoCom), IEEE, pp. 359–364. IEEE (2016)
14. Budai, T., Kuczmann, M.: Towards a modern, integrated virtual laboratory system. Acta Polytechnica Hungarica **15**(3), 191–204 (2018)
15. Kövecses-Gősi, V.: Cooperative learning in VR environment. Acta Polytechnica Hungarica **15**(3), 205–224 (2018)
16. Skyrms, B.: Signals: Evolution, Learning, and Information. Oxford University Press, Oxford, GB (2010)

17. Deacon, T.W.: Incomplete Nature: How Mind Emerged from Matter. Ser. Norton, A Norton paperback. W.W (2013)
18. Dennett, D.C.: Intuition Pumps and Other Tools for Thinking. W. W. Norton & Company, New York (2013)

Part V
VR Context Control

This part provides an overview of the innovative research conducted across three pivotal chapters, aiming to enhance user experience within 3D Virtual Reality (VR) environments through personalization and optimal cognitive load management.

The first chapter focuses on an examination of the current IT infrastructure through Human-Computer Interaction (HCI) analysis. Its objective is to identify which platformis most suitable for meeting user needs under optimal cognitive load conditions. This necessitated the development of a newmethodology that allows for the objective evaluation of digital interfaces. By focusing on operational complexity measurement, definitions and metrics were established to clearly determine the number and complexity of user operations, enabling an objective comparison of different interfaces. Subsequent validation of this measurement tool revealed that operations in 3D spaces could be performed with lower cognitive load, showing an increase in performance.

The second chapter explores whether the positive outcomes presented in the first chapter can be further enhanced through personalization. After profiling users, personalized work environments were created, and empirical studies confirmed their positive effects on both short-term and long-term memory and time efficiency. Through this research, significant contributions are made to the VR field by highlighting the advantages of personalization in VR workspaces and how it can enhance work efficiency by aligning with users' learning styles and work processes.

The third chapter targets IT implementation, aiming to develop a framework for VR context control based on the construction of cognitive profiles. This involves defining and formalizing a template language for unambiguously designing schematic layouts in 3D spaces for various digital content, tailored to individual learning styles as per Kolb's taxonomy. This new approach facilitates the presentation of 2D digital content in 3D layouts in a manner that respects the geometry of the 3D space and user preferences. It proposes a system for automatically customizing the user's work environment in the most optimal manner based on these indicators, without the need for human intervention. The concept of 'context control' and the introduced framework offer fresh perspectives on designing VR environments tailored to individual users, potentially leading to more effective and efficient virtual work experiences. This research holds practical implications for the development of personalized VR applications across various domains, including education, training, engineering, and corporate and professional workflows.

Chapter 14
Operational Complexity and Evaluating Digital Interfaces

Abstract The 21st Century has seen disruptive changes in technology, especially with the internet and the human-machine coevolution, which makes us wonder how these advancements help or hinder the human way we do tasks. This chapter presents an exploratory study contrasting traditional 2D interfaces with innovative 3D Virtual Reality (VR) platforms to present the evolving field of digital literacy and its operational effectiveness. Through a comprehensively designed experiment, it examines the number of user operations and the effectiveness of digital workflows across different technological environments. A novel conceptual framework, inclusive of both qualitative and quantitative metrics, underpins this investigation, aiming to refine the evaluation of digital workflow effectiveness. The experiment's 1 findings reveal significant benefits of using 3D interfaces over traditional 2D interfaces for digital workflows. Notably, the shift to 3D environments results in a 27% reduction in elementary user operations and a 54% decrease in the cost of accessing information, compared to traditional 2D Windows systems. In scenarios where 2D project management tools like Trello are integrated with immersive 3D VR environments, the efficiency gains become even more pronounced. Such hybrid arrangements lead to a 37% decrease in basic operations and a 41% reduction in the cost of accessing information when using desktop VR with icon-based interfaces, compared to standard 2D operating systems. Additionally, the integration of Trello with the 3D VR platform MaxWhere significantly enhances knowledge sharing and comprehension in project management, achieving a 65% improvement in the number of basic operations and a 66% enhancement in operational efficiency over conventional 2D collaborative tools.

Keywords 3D virtual reality · Cognitive infocommunications · Human-machine coevolution · Human computer interactions · User operation · Operational complexity · Digital workflow · Virtual reality

I. Horváth et al., *Cognitive Aspects of Virtual Reality*, Studies in Big Data 156,
https://doi.org/10.1007/978-3-031-68130-1_14

14.1 Introduction

In the fast-evolving realm of information technology, the 21st Century has been marked by a profound digital transformation that influences every facet of our daily lives. Central to this transformation is the emergence of a networked culture, prompting a reevaluation of traditional paradigms of human-computer interaction and digital literacy. This chapter presents an experiment designed to compare and contrast the operational effectiveness of traditional 2D interfaces with dynamic and immersive 3D VR platforms. By foregrounding the efficiency and complexity of digital workflows in these environments, the study seeks to unpack the nuances of digital operations within the context of contemporary technological advancements.

The focus of this investigation is the development of a comprehensive framework, encompassing a broad spectrum of concepts, metrics, and experimental procedures tailored to assess the effectiveness of digital workflows. This framework serves as the cornerstone for a systematic analysis that evaluates the intricacies of user operations within four specific digital work environments: the traditional 2D Windows operating system, the MaxWhere 3D desktop VR environment, the integration of 2D Trello with 3D MaxWhere, and the Immersed 3D Immersive VR solution. Through a meticulous comparison of user operation complexity, this chapter illuminates the tangible benefits of 3D technology in enhancing digital workflow efficiency. Empirical evidence from the study underscores a significant reduction in both user and machine operations when employing 3D interfaces over their 2D counterparts, affirming the potential of 3D VR platforms to streamline complex digital tasks. Notably, the integration of Trello within MaxWhere's 3D spaces was shown to dramatically decrease the number of elementary user operations and the time allocated to overview-related tasks, thereby enhancing productivity and operational oversight. These findings advocate for a paradigm shift towards the adoption of 3D platforms in everyday work and educational settings, heralding a new era of digital interaction that promises greater efficiency and reduced complexity in digital workflows. This chapter is an extended and revised version of the following publications: [1–3]. The chapter is structured into sections that address the research background, the main goals, and research results constituting a substantial area of inquiry within the overall research theme.

14.2 Operational Complexity

In the digital era, marked by significant paradigm shifts, the emergence of global connectivity both physically and virtually, and the surge in information flow, there has been a notable increase in the operational complexity within organizations. This complexity arises as organizations evolve into intricate systems, paralleled by the increasingly complex structure of information flow. It becomes imperative to identify and utilize information resources more efficiently to enhance information productivity, as emphasized in various studies [4]. A significant portion of the literature on operational complexity in work processes focuses on analyzing the 'system' itself, namely the working environment, and its capabilities in relation to human operational complexity. A key distinction made by Li [5] between perceived complexity and objective complexity underscores the crucial role of human cognitive factors in understanding operational complexity. The complexity of an operation step is defined by the number of operations in an operation unit, while the complexity of the logical structure of an operation encompasses the logical sequence of performing the activities within an operation unit [6]. Rasmussen [7] introduces a different perspective by presenting modeling related to performance representation at various levels: skill, rule, and knowledge. This approach illuminates the diverse human behaviors (skill-based, rule-based, and knowledge-based) encountered in different environments and situations, and the associated human performance, which can be characterized at multiple levels of abstraction—from the physical form to functional representation, and even up to intention or goal representation. Rasmussen also emphasizes the interdisciplinary nature of addressing the complexity of this topic, which requires research and analysis spanning across fields from engineering to optimization, and to cognitive science. In addition, Fritzsche et al. highlight the significance of ergonomics in digital environments and advocate for a design approach based on human modeling. Traditionally, human performance has been quantified based on the number of tasks (operations) that can be performed per unit of time [8]. This approach, however, is evolving with the changing digital landscape and the increasing recognition of the multifaceted nature of human interaction with complex systems. Further enriching this discussion, it's crucial to consider how cognitive load, perception, and human performance measurement impact VR environments. Cognitive load in VR refers to the mental effort required to process the virtual environment and its tasks. Effective VR design should aim to balance this load, providing intuitive interfaces and avoiding information overload. Perception in VR, involving sensory interpretations, is key to the immersive experience, and issues like motion sickness due to sensory mismatches must be addressed. Measuring human performance in VR involves metrics such as task completion time and error rates. This data can be used for real-time adjustments and customization of VR experiences, optimizing them for different learning styles and cognitive capacities. These aspects demonstrate the importance of human-centered design in virtual environments, especially in applications where user engagement and learning outcomes are paramount.

14.3 Human-Computer Interactions in the Digital Work

The widespread use of computer systems has greatly enhanced the efficiency and ease of daily activities. However, the increasing complexity of digital processes often leads to a higher mental and cognitive load as users navigate and interact with these systems. The study of human-computer interaction (HCI) is a critical aspect of the development of computer systems. HCI focuses on researching, designing, and evaluating how people interact with computers, aiming for effective and seamless exchanges. This field draws upon various disciplines, including social and cognitive psychology, linguistics, and theories of communication, as well as graphic and industrial design [9].

Understanding both the human user and the computer software system is crucial for successful interaction. Török [10] highlights four key domains in comprehending the human aspect: behavior, limitations, needs, and cognitive processes. Understanding human behavior allows systems to recognize the motivations behind actions. By closely monitoring activities, systems can gather nuanced, context-specific insights into user intentions, thereby fostering more collaborative and intuitive interactions between humans and computers. Traditional interfaces often require explicit user instructions, yet human cognition is not inherently command-driven but rather responsive to environmental signals and gestures. Gesture-based interactions, which are increasingly common and intuitive, exemplify this [10].

HCI involves four fundamental elements: affordance, collaboration, acknowledging limitations, and adaptable interfaces [11] Török [10] points out the interconnectedness of these components, where collaboration aids in understanding human limitations, leading to more dynamic interfaces. Effective interfaces leverage affordance to enhance smooth cooperation [10].

The ever-increasing volume of information that users manage has rendered screen size a limitation [12]. Information is often scattered across various windows and tabs, requiring users to constantly switch between tasks or software, and maintain mental track of the workflow without visual aids [12]. Unlike digital systems, which demand formal categorization, everyday organizations relies on visual recognition and grouping without specific names [13].

The capacity of a system to display information effectively is also crucial [14]. Users generally prefer an overview of information, as observed with large, high-resolution displays where users tend to distance themselves for a full view [14].

Virtual work environments offer various collaboration tools, including real-time communication via audio, video, or instant messaging, and collaborative document editing [15]. The integration of both session-centric and document-centric collaboration tools was an early goal in the design of virtual workspaces [15].

Attention and cognitive focus during work are also vital [16]. Interfaces should aid users in maintaining focus and staying within the context of their thought process. Direct interaction with information keeps users engaged in their task, whereas menu navigation can be distracting, requiring additional cognitive effort [16].

This chapter aims to examine work processes and human behaviors to inform more ergonomic system design. It involves breaking down these elements into manageable components, basic functions, and logical sequences, allowing for a comparative analysis of different systems.

14.4 Definitions and Metrics

In this section, the author outlines fundamental concepts and characteristics that distinctly encapsulate the intricacy of user interactions and the quantity of required actions across various 2D and 3D digital work settings. The explanations provided in this document draw upon the definitions initially introduced in the 2018 paper by Horvath and Sudar [3] and Horvath and Berki [2]. Yet, the swift advancements in technology have necessitated an update and refinement of these earlier definitions to align them with the evolving landscape of virtual workspaces. A thorough exploration of their multifaceted application potential is crucial. It is imperative that readers are able to accurately comprehend and interpret the metrics and terminologies employed. The proposed model is predicated on the notion that evaluating a system's functional intricacy enables its comparison with other digital platforms, given that these metrics are quantifiable and unbiased. Assessing a system's effectiveness involves more than just simplifying the total number of tasks into a set of basic operations and examining their interrelationships. It also necessitates attention to the effort involved in obtaining information, a concept referred to as the cost of information access.

Sections 14.4.1–14.4.6 presents the complexity of the user operations specified in the definitions and the method of converting them to elementary operations which carry the cognitive load weights in the values defined for the changes.

14.4.1 Elementary Operation—EO

An elementary operation is a simple interaction from the user that triggers the execution of a pre-defined process from the machine [3].

Single selection—select the digital item with a single click/tap or touch The most common selection function usage aims activation of digital content—transfer of control.

Double click/tap, touch—start a digital item, display full size, and activate a special function.

Key press—on a real or virtual device.

Press/push—long press of a key, a mouse button, or a controller—e.g. to maintain an operation.

Scroll up/down operation.

Swipe—on a touchpad, touchscreen, or virtual interface.

Head, - hand, - foot, - other body part movement.

From here on, each of these elementary operations will be regarded as having a complexity of 1 EO.

14.4.2 Complex Operation—CO

A complex operation is one that consists of more than 1, but at most 3 EOs that are performed in coordination.

For instance, executing a copy action using the "Ctrl+C" keyboard shortcut involves a click to select and the simultaneous pressing of two keys for copying and pasting. Similarly, the familiar drag-and-drop maneuver, applicable in both 2D and 3D digital environments, entails a selection elementary operation (EO), followed by a push action, and completed in an inverse click or release.

Drawing from these instances, we suggest establishing a correlation between the metrics of 1 Copy Operation (CO) and 3 Elementary Operations (EO).

14.4.3 Navigation-Based Elementary Operation—NBEO

An elementary operation is navigation-based if it is performed not with a keyboard but with a mouse, touchpad, touch screen, on-screen visualization (e.g. virtual keyboard), and hand- or other body part movement. In a 3D VR environment, this includes orbiting, rotation, positioning, changing direction, etc. Take, for instance, the 2D operating system environment, where selecting files from the file manager requires first navigating the cursor to the correct spot before any other actions can be taken. Hence, navigation is the starting point of the action sequence. Similarly, in a 3D environment, positioning oneself appropriately is crucial for interacting with a digital twin.

In the context of this study, we have categorized navigation-based elementary operations (NBEOs) as having a complexity of 1.5 times that of a standard elementary operation. Therefore, we equate 1 NBEO to 1.5 EOs.

14.4.4 Ordering Operation—OO

An example of an ordering operation can be seen when users organize multiple windows on their screen, allowing them to easily switch focus between them. This process involves both moving and resizing each window. Moving a window is done by dragging it to a desired spot, while resizing is achieved by pulling on the window's corner at least once to adjust both horizontal edges to their final position. Therefore, the complexity of this task is calculated as a minimum of 1 Elementary Operation (EO) for repositioning, plus 1 Non-Basic Elementary Operation (NBEO) for the initial resizing, and an additional 2 NBEOs for any further adjustments, totaling at least 5.5 EOs per window. However, in practical scenarios, this complexity often reaches around 7 EOs, especially when users opt to resize the window on all four sides or need to refine their overall arrangement.

In a 3D context, reordering involves arranging digital content in a useful workflow order sequence. This includes 2 EOs for head and hand movements, 1 EO for selecting or grasping, 1.5 EOs for navigation, and another EO for placing or releasing.

Based on this example, we use a conversion where 1 Ordering Operation (OO) is equivalent to 5.5 Elementary Operations (EOs) in our study.

14.4.5 High Distraction Operations—HDO

Operations that are initiated by users through an EO or CO, but which also incur a period of waiting time, are considered high-distraction operations due to the iteration of inhibition and update associated with waiting and the frequent switching operations. Examples in 2D of such operations include file downloads, file opening operations in specific applications, file compression, etc., or 3D, simulation start-up, evaluation, interference with the operation of the digital twin, etc. The weight of the cognitive load associated with an operation is assumed to be 2.5, based on the time required to perform the operation and user feedback we adopt the equivalence 1 HDO = 2.5 EO.

14.4.6 High Alternation Operations—HAO

High alternation operations require the user to switch back and forth between multiple windows to perform the necessary digital workflow. However, it is essential to note that HAO involves a complete change of the screen content, which results in a significant cognitive load. Related to this are the executive functions: inhibition (inhibiting dominant, prepotent responses), switching (detaching from an irrelevant task set and switching to a task set relevant to the current goals), and refreshing (updating the content of working memory).

This fact alone motivates using the 1 HAO metric by adopting the equivalence of 1 HAO = 5 EO.

The evaluation of operational efficiency using basic operations as a benchmark is shown in Fig. 14.1.

14.4.7 Monitoring Density—MD

Monitoring density refers to the quantity of information units (such as menus, icons, etc.) that can be simultaneously observed and understood by a user. It quantifies the amount of information presented at once and is measured by the count of these information units. It's crucial to recognize that having a high or low monitoring density isn't inherently beneficial or detrimental; rather, it depends on the context. A critical

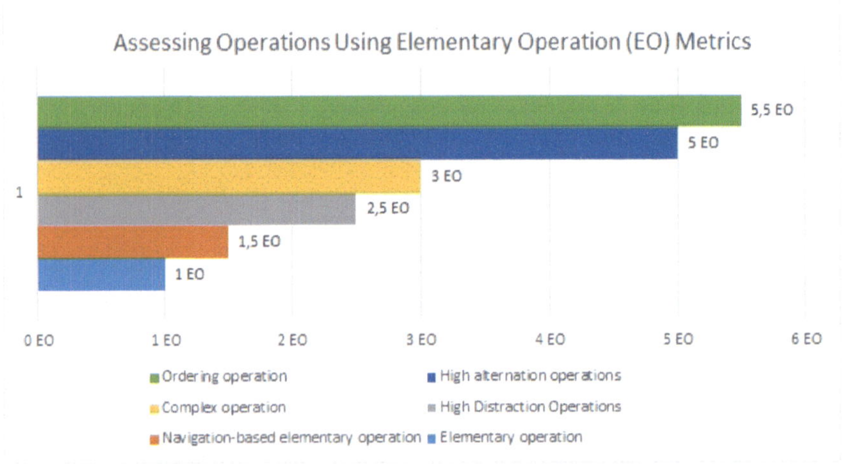

Fig. 14.1 Evaluation of operational efficiency using basic operations as a benchmark

aspect is that a higher monitoring density demands a more effective representation to aid the user in processing a greater volume of information elements efficiently.

14.4.8 Information Availability—IA

Indicates what percentage of the information (digital content) needed to execute a workflow is available in the digital work environment when executing up to 1.5 EO (1NBEO)

Remark: e.g. 100%, if all the information required to execute the workflow is available and can be accessed by a navigation operation.

14.4.9 Information Access Cost—IAC

The weighted sum of the time spent accessing information for each type of operation, where the weights are the number of elementary operations corresponding to that type of operation.

$$IAC = \sum_{i=1}^{n} O_i * t_i \tag{14.1}$$

Remark: this metric is high even when complex operations are performed in the same amount of time and when simple operations are performed for a long time. It encompasses the complexity of the task at hand as well as the user's proficiency and skill level.

14.4.10 Information Validity—IV

Indicates the percentage of the information presented in the digital work environment that is directly required for task execution.

IV = Number of Valid Information Units/Summa necessary information units number.

Remark: To calculate the Information Validity (IV) percentage, it's essential first to ascertain the specific type and quantity of digital information units necessary for the task, such as relevant documents, data tables, diagrams, timelines, etc. Subsequently, an analysis of the content within the digital workspace is conducted to identify what digital content exists and what is lacking. For instance, if 10 units of digital content are required but only 4 are available in the workspace, then the IV is calculated to be 40%.

IV reaches 100% when the digital workspace contains exclusively the information necessary for the task, without any excess or deficiency. Conversely, IV is 0% if the digital workspace fails to provide any relevant information for the task at hand.

14.4.11 Confuse Information Content—CIC

It indicates the percentage of the total information content in the workflow that is confusing (irrelevant) to task performance. Remark: IV + CIC = 100%, i.e. IV and CIC are disjoint sets.

14.4.12 Personalized Workflow Order Ability—PWO

This is an indicator of the facility of the digital work environment to provide users with the ability to arrange digital content in their own way. The value is 0 if the option is not provided, 1 if the user can set the layout himself, and 2 if the optimal layout is automated with AI support.

14.4.13 Personalized Information Overview—PIO

Indicates the ability of the digital work environment to provide users with the ability to set personalized information overview. The value is 0 if the option is not given, 1 if the user can set the layout himself, and 2 if it is automated with AI support.

14.4.14 Preference Point

In the virtual environment, preference points are specific locations that users frequently visit to find optimal vantage points for overseeing the space and accomplishing their tasks.

Arrangement patterns:

- Content—refers to the tendency of users to group content with similar themes or subjects together, typically in close-knit clusters.
- Type—indicates a preference for organizing similar types of content in proximity, forming clustered arrangements.
- Mixed—denotes a primary organizational principle based on content, where similar content is grouped together, but within these groupings, a secondary principle based on the type of content is applied.

14.4.15 Task Execution Tools—TET

The number of hardware tools needed to perform a task—shows how many independent tools the user needs to perform the task. The system is optimal when TET goes to 1.

For instance, executing a task in an immersive VR setting might require additional equipment, such as a computer, VR headset, a pair of controllers, a wireless adapter, or a cable for the headset, leading to a TET value of 5. Conversely, in a 2D workspace, the total value for a setup comprising a computer, a keyboard, a mouse, and a monitor would result in a TET of 4.

Should the work environment be operable with just a single laptop, tablet, or smartphone, the assigned value will be 1.

14.4.16 Degree of Interaction—DI

A measure of the ability of the digital work environment to provide interaction. The interaction can be human-human, human-machine (e.g. Digital Twin), and human-AI. The value of the metric is determined by the presence or absence of the following factors:

- Interaction with co-users
- 2D digital content management
- Real-time content management and change tracking
- Navigation (2D interface or 3D space)
- Interaction with 3D objects

- Interaction with digital twins
- Simulation management
- Adding and removing simulations
- Interaction with AI with human initiation
- Interaction with AI with both sides (AI and human) of the initiation.

Remark: The calculation of DI hinges on the specific weighting allocated to each element. To authenticate this metric, additional investigative approaches are needed, relevant to the task at hand. This might include conducting surveys to establish benchmark values and verifying these figures.

These metrics are designed to offer a quantifiable and foreseeable method for assessing the most efficient level of operational complexity while minimizing cognitive demand.

14.5 Investigating 2D and 3D Interfaces in Terms of Cognitive Load

This study presents a comprehensive analysis comparing user operations in various digital systems, focusing on workflows within classic 2D operating systems, mixed 2D project management and 3D desktop VR, and immersive VR solutions. The primary emphasis lies on user tasks and the cognitive load needed to execute a digital workflow or project. Various work settings call for varying types of tasks, which can be compared using the framework introduced earlier in this document. So the comparative examination takes into the number of user operations, account operational complexity, and cognitive load, during the use of common digital content like email clients, online meeting platforms, document editors, computational spreadsheets, and internet browsers. These contents, integral to office and business tasks, are organized around projects and are uniformly available across the four systems. For this analysis, Office 365 was utilized, enabling real-time content editing and tracking. The mixed 2D and 3D solution specifically, evaluates the efficiency of Trello collaborative project management software on both Windows 2D operating system (Fig. 14.2) and the MaxWhere 3D VR platform. The methodology includes a novel framework that offers both quantitative and qualitative insights into user interactions required for digital project completion.

14.5.1 Digital Content Sharing Across Platforms

All platforms shared the same digital contents, including email clients, online meeting platforms, document editors, spreadsheets, and internet browsers. In all systems, these contents were pre-set in the working environment, and the Office 365 suite was used for real-time content editing and tracking.

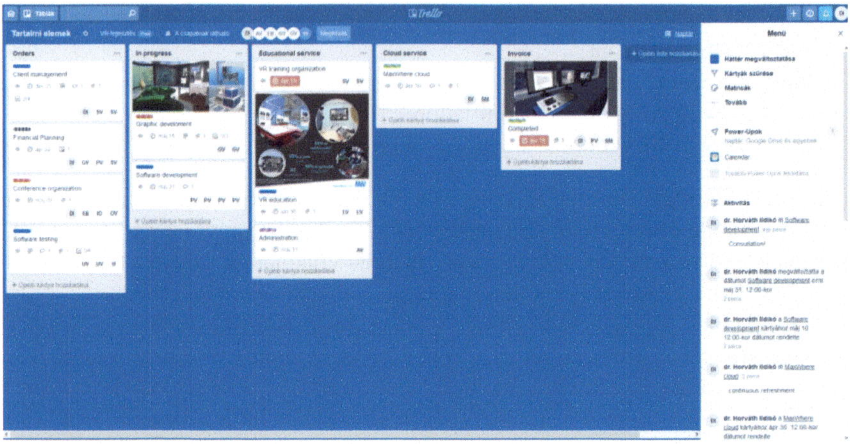

Fig. 14.2 The experiment's Trello board configuration

- 2D—Windows Operating System: The Windows operating system was used to assess the 2D interface. Users had access to one display and launched programs from desktop icons for lower operational complexity.
- 2D Trello, a 2D collaborative project management tool, was used for organizing projects into boards, lists, and cards. This system, based on the "going deeper" principle, involved navigating through a hierarchy of project components. Trello facilitated task organization and collaboration but required multiple user operations for accessing and closing cards and attachments.
- Desktop VR—MaxWhere MaxWhere, a desktop VR system, provides predesigned spaces with smartboards to display various digital contents. The smartboards functioned as a comprehensive platform for project content management. The Let's Meet Extra space, equipped with five smartboards, was optimized to display the required digital content with minimal distractions (Fig. 14.3).
- Mixed 2D Trello and 3D Desktop VR integrated system combines Trello's project management with MaxWhere's immersive 3D visualization, enhancing user comprehension and efficiency. It allows for spatial organization of project elements, making it easier for users to understand and remember relationships between different digital contents. By moving beyond Trello's hierarchical navigation, it leverages 3D spaces for more intuitive workflow management, reducing cognitive load and the number of user operations needed. This approach merges the organizational strengths of Trello with the natural, direct engagement of 3D environments, streamlining digital workflows and project guidance. 3D explains what 2D cannot [1], as 3D objects are capable of explaining relationships among different bits of digital content, offering a deeper understanding and quicker comprehension of complex information that 2D representations cannot match.

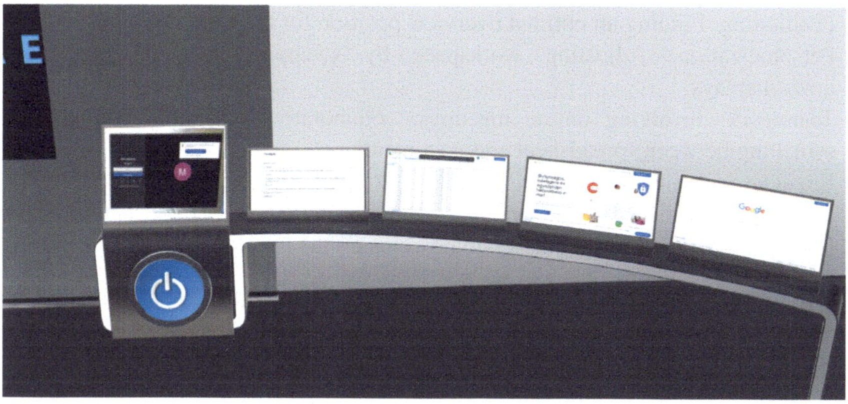

Fig. 14.3 Screenshot of the workflow optimized Let's Meet Extra MaxWhere space

- Immersive VR—Immersed Immersed VR offered a more enveloping virtual office experience with up to five virtual monitors, hand tracking, and multi-screen sharing for collaborative work. This platform provided a heightened sense of social presence and efficient work management through advanced VR technologies.

Traditional 2D Windows Setup: One segment of users engaged with digital workflows through a conventional method, receiving an email. This email detailed the workflow in text and included necessary digital resources as links within the email's content.

For the 2D Trello, the combined Trello and MaxWhere, and the Immersive VR, the mechanisms for sharing digital workflows were similarly structured. Participants were provided with access to a Trello board or a VR space. They received a Trello board link via email, while the MaxWhere space required installation on their laptops. The assessment of user operations commenced once all necessary applications were launched and operational.

Note: The focus of this investigation was on user operations linked to processing information. Consequently, the initial setup of content within emails, Trello, MaxWhere, or Immersive VR environments was considered outside the scope of this study.

14.5.2 General Workflow

The workflow used for this analysis aimed to represent typical online collaborative tasks, allowing for personalization based on individual habits and information acquisition preferences. The general workflow included:

- Positioning: Finding an optimal overview position for the task.
- Personalization: Adjusting workspace by swapping two program windows/displays.
- Teamwork: Involving online meetings, collaborative document editing, and switching between spreadsheet and document.
- Email Communication: Sending emails to external contacts.
- Browser Utilization: For information search and data transfer to documents.
- Interface Closure: Ending the session by closing the interface.

This unified methodology and workflow analysis across different platforms aimed to understand the efficiency, effectiveness, and operational complexity of various productivity systems in managing digital workflows and collaborative projects.

This quantitative assessment was complemented with an empirical, exploratory survey. The survey aimed to evaluate user satisfaction. This quantitative assessment was enhanced through a practical, investigative survey and we applied a 10-point Likert scale. A total of fifty-two volunteers took part in the survey.

14.6 Evaluation of Examination Results

14.6.1 Survey on User Satisfaction and Interface Usability

A total of 52 individuals participated in our web-based survey, consisting of 32 men, 17 women, and 3 who either did not respond or provided unusable data due to incorrect usage of the 10-point Likert scale. Among these, 34 were identified as belonging to Generation Z with an average age of 20.82 years (SD: 2.22), while 17 were from older generations, averaging 38.5 years in age (SD: 11.1).

Analysis of the data showed no notable differences in feedback across genders or age groups. Interestingly, 42% of the Generation Z participants and 77% of those from older generations reported having experience with VR applications, likely reflecting the prevalence of IT and engineering fields among distance learning students. Participants generally felt comfortable using an average of five different types of digital tools and collaborating with up to five individuals on online projects.

In querying participants on the ease of performing specific activities related to user satisfaction, including optimal screen positioning, personalization through adjusting their workspace, engaging in teamwork via online meetings and document sharing, communicating through email, utilizing browsers for research and data integration, and closing interfaces after use, the survey aimed to garner insights into user interface preferences. Figure 14.4 provides detailed insights into user feedback.

Respondents rated various features of digital work environments on a 10-point scale, from least to most important, revealing key priorities as well: the ability to easily overview information (Mean: 9.31, SD: 1.104) and user-friendliness (Mean: 9.31, SD: 1.157) topped the list. Immediate access to necessary information for task execution was highlighted as crucial for selecting a working environment. However,

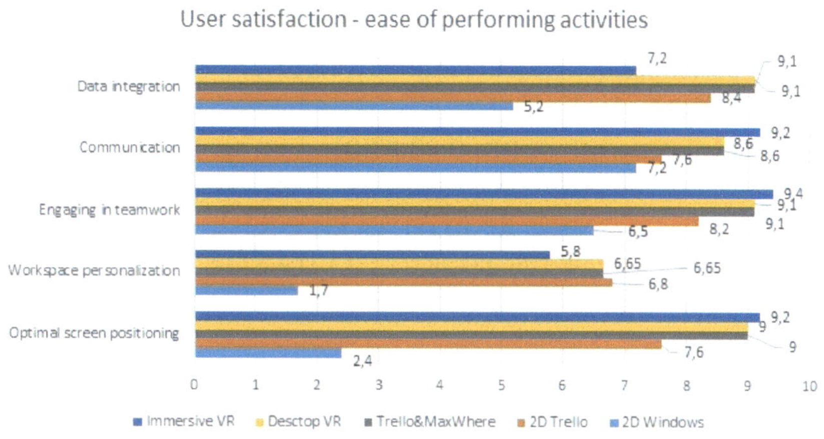

Fig. 14.4 User satisfaction feedback connecting to the ease of performing activities

the potential for artificial intelligence support was deemed less critical, ranking lowest among the features considered.

14.6.2 Results of Quantitative Investigations

14.6.2.1 Operations in 2D Windows Operation System [2]

1. *Positioning:* Finding a position to have an optimal overview for the task.
    a. Open online meeting.—**1 NBEO = 1.5 EO**
    b. Minimize down to taskbar.—**1 NBEO = 1.5 EO**
    c. Open the document.—**1 NBEO = 1.5 EO**
    d. Minimize down to taskbar.—**1 NBEO = 1.5 EO**
    e. Open email client.—**1 NBEO = 1.5 EO**
    f. Minimize down to taskbar.—**1 NBEO = 1.5 EO**
    g. Open browser.—**1 NBEO = 1.5 EO**
    h. Minimize down to taskbar.—**1 NBEO = 1.5 EO**
    i. Open spreadsheet.—**1 NBEO = 1.5 EO.**
2. *Personalization:* Swapping two program windows/displays.—**1 OO = 5.5 EO**
3. *Teamwork:* online meeting, editing documents together.
    a. Online meeting for discussing the tasks.
        i. Select online meeting to participate.—**1 NBEO = 1.5 EO.**
    b. Making a note of what was discussed in the document.
        i. Alternating between document and online meeting.—**1 HAO = 5 EO.**

c. Back-and-forth switching between spreadsheet and document.
    i. Loading the document.—**1 NBEO = 1.5 EO**
    ii. Loading the spreadsheet.—**1 NBEO = 1.5 EO.**

4. *email:* Sending an email.—**1 NBEO = 1.5 EO.**
5. *Browser:*
    i. Search for information.—**1 NBEO = 1.5 EO**
    ii. Transfer information to the document.—**1 CO = 3 EO.**
6. Closing the applications.—**5 * 1 NBEO = 7.5 EO.**

Total: 42 EO

Operations in 2D Trello.

1. *Positioning situating oneself to gain a comprehensive perspective for efficient task execution*
    a. Loading Trello by clicking on the link.—**1 NBEO = 1.5 EO.**
2. *Operations*
    a. View the list of team members: Open list **1 NBEO = 1.5 EO**
    b. Open cards **5 x 1HDO = 12,5 EO**
    c. Open attachments **5 x 1HAO = 7,5 EO + CL**
    d. Close attachments **5 x 1HAO = 7,5 EO + CL**
    e. Scroll to the bottom and back of card **5 x 2 x 1 EO = 10 EO**
    f. Closing application **1 NBEO= 1,5 EO.**

Total in Trello 42 EO

14.6.2.2 Operations in 3D Desktop VR Platform

Operations in MaxWhere desktop virtual reality.

1. *Positioning*: Finding a position to have an optimal overview for the task.
    a. Loading space by click on link.—**1 NBEO = 1.5 EO**
    b. Positioning with mouse.—**1 NBEO = 1.5 EO.**
2. *Personalization*: Swapping two program windows/displays.
    a. **In desctop VR—MaxWhere** reloading two content, no real window moving
        i. Smartboard selection by one click.—**1 EO**
        ii. Copy browser address.—**1 CO = 3 EO**
        iii. Click on other smartboard.—**1 EO**
        iv. Paste browser address.—**1 CO = 3 EO.**

3. *Teamwork*: online meeting, editing documents together.
    a. Online meeting for discussing the tasks.
        i. Click to enter the meeting.—**1 EO**
    b. Making a note of what was discussed in the document.
        i. Select the note.—**1 HAO = 5 EO**
    c. Back-and-forth switching between spreadsheet and document.
        i. Loading the document.—**1 EO**
        ii. Loading the spreadsheet.—**1 EO**.

4. *email*: Sending an email.
    a. Click on email window.—**1 EO**.

5. *Browser*:
    i. Search for information.
        i. Double click for better visibility and activation.—**1 EO**
    ii. Transfer information to the document.
        i. Copy and paste the information to the document.—**1 CO = 3 EO**
        ii. Click on fullsize button to return to the original position.—**1 EO**.

6. Closing the interface.—**1 NBEO = 1.5 EO**

    Total in Desktop VR: 26.5 EO.

14.6.2.3 Operations in Immersed VR Platform

1. *Positioning*: Finding a position to have an optimal overview for the task.
    a. Opening VR space.—**1 NBEO = 1.5 EO**
    b. Switch on VR headset.—**1 NBEO = 1.5 EO**
    c. Enable head tracking (unlock screen).
        i. Raising the left hand.—**1 EO**
        ii. Button press.—**1 NBEO = 1.5 EO**
    d. Enable hand tracking: pinch ring finger and thumb together—**1 EO**.

2. *Personalization*: Swapping two program windows/displays.
    By default, the monitors are grouped together in a snap grid.
    a. Ungroup the monitors.—**1 EO**
    b. Grab the selected monitor (pointing to the monitor by touching index finger and thumb together), and move/navigate with fingers together—move the arm out of "snap gird" and move to the new position.—**1 CO = 3 EO**
    c. Select the second monitor and mot to a new position. **1 CO = 3 EO**
    d. Lock the screen.—**1 EO**.

3. *Teamwork*: online meeting, editing documents together.
   a. Online meeting for discussing the tasks.
      i. Turn to enter the meeting.—**1 EO**
   b. Making a note of what was discussed in the document.
      i. Select the note.—**1 HDO = 5 EO**
   c. Back-and-forth switching between spreadsheet and document.
      i. Loading the document.—**1 EO**
      ii. Loading the spreadsheet.—**1 EO.**
4. *email*: Sending an email.
   a. Click on email window.—**1 EO.**
5. *Browser*:
   a. Search for information.
      i. Double click for better visibility and activation.—**1 EO**
   b. Transfer information to the document.—**1 CO = 3 EO.**
6. Closing the interface.—**1 NBEO = 1.5 EO.**

Total: 29 EO

The findings reveal a significant reduction of 27% in elementary operations when tasks are performed within a 3D immersive virtual work environment, as opposed to traditional 2D Windows systems. Additionally, the immersive VR environment results in a 54% decrease in the cost of accessing information. With desktop VR that utilizes icons, there's a 37% drop in elementary operations and a 41% reduction in information access cost compared to the conventional 2D operating systems (Fig. 14.5).

3D VR spaces enhance concentration and drastically reduce the effort required for tasks such as searching for documents, opening them, or altering their layout on the 2D Trello desktop, which organizes digital content according to workflow needs rather than alphabetical order. While effective, Trello has its limitations for workflow execution. The findings indicate that although the Trello knowledge management system does not decrease the volume of fundamental user actions, the Elementary operations relative to the traditional Windows-based work environment, facilitate a notable 11% enhancement in the efficiency of accessing information. In contrast, 3D provides immediate access to digital content in a spatial workspace.

The comparative analysis indicates that the 3D VR platform enhances the speed of information sharing and comprehension, outperforming the 2D Trello system. The integration of Trello with MaxWhere significantly expands the capacity for knowledge sharing and understanding in project management, surpassing current popular 2D collaborative tools. This synergistic use of both software tools results in a 65% improvement in the number of elementary operations conducted and a

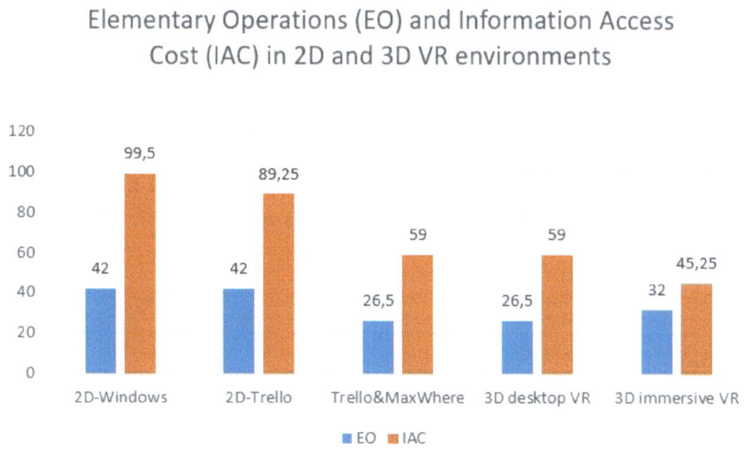

Fig. 14.5 Number of Elementary Operations (EO) and Information Access Cost (IAC) in 2D and 3D VR environments

66% enhancement in operational efficiency compared to 2D alternatives. Increasing the amount of digital content and performing more complex tasks lead to further improvements in these metrics [1].

14.7 Conclusions

This chapter provides a comprehensive analysis of an exploratory study that contrasts traditional 2D interfaces with innovative 3D virtual reality (VR) platforms, specifically examining their operational efficiency within user workflows. Through the introduction of a novel conceptual framework, the study successfully identified a set of qualitative and quantitative metrics designed to improve the evaluation of digital work process efficiency. The experimental results highlight the significant advantages of utilizing 3D interfaces over their 2D counterparts for enhancing digital workflows.

The comparative analysis reveals that 3D VR platforms significantly enhance the speed of information sharing and comprehension, clearly outperforming traditional content management systems like 2D Trello. Furthermore, integrating Trello with MaxWhere—a 3D collaborative tool—markedly boosts knowledge sharing and comprehension in project management, offering superior capabilities compared to widely used 2D collaborative tools. This combined use of both tools leads to a substantial 65% increase in the number of elementary operations conducted and a 66% improvement in operational efficiency in comparison to using 2D solutions alone.

The study highlights the escalating benefits of handling more digital content and undertaking more complex tasks, further elevating these performance metrics. A key takeaway from the research is the profound impact of 3D spatial content display on workflow execution. The flexibility and efficiency provided by 3D interfaces facilitate easy content sharing and arrangement tailored to the user's workflow needs, significantly reducing the number of user operations required and thereby enhancing workflow efficiency. Particularly, neither the concurrent use of traditional 2D Windows-based software nor the application of the 2D Trello project management system, despite its utility in other areas, matches the capabilities offered by 3D spatial content display.

References

1. Horváth, I.: MaxWhere 3D capabilities contributing to the enhanced efficiency of the Trello 2D management software. Acta Polytechnica Hungarica **16**(6), 55–71 (2019)
2. Horváth, I., Berki, B.: Investigating the operational complexity of digital workflows based on human cognitive aspects. Electronics **12**(3), (2023). [Online]. https://www.mdpi.com/2079-9292/12/3/528
3. Horváth, I., Sudár, A.: Factors contributing to the enhanced performance of the MaxWhere 3D VR platform in the distribution of digital information. Acta Polytechnica Hungarica **15**(3), 149–173 (2018)
4. Cekuls, A.: The impact of complexity to collaboration and ability of businesses to share information across teams. In: 21st International Multidisciplinary Scientific GeoConference SGEM 2021. STEF92 Technology (2021)
5. Li, K., Wieringa, P.: Structured operation environment and perceived complexity in human supervisory control. Asian J. Control. **3**(3), 181–189 (2001). [Online]. https://api.semanticscholar.org/CorpusID:110441579
6. Zhang, Y., Li, Z., Wu, B., Wu, S.: A spaceflight operation complexity measure and its experimental validation. Int. J. Ind. Ergon. **39**(5), 756–765 (2009). [Online]. https://api.semanticscholar.org/CorpusID:14132608
7. Rasmussen, J.: Skills, rules, and knowledge; signals, signs, and symbols, and other distinctions in human performance models. IEEE Trans. Syst. Man Cybern. SMC-13(3), 257–266 (1983). [Online]. https://api.semanticscholar.org/CorpusID:1525146
8. Fritzsche, L., Jendrusch, R., Leidholdt, W., Bauer, S., Jäckel, T., Pirger, A.: Introducing ema (editor for manual work activities)—a new tool for enhancing accuracy and efficiency of human simulations in digital production planning. In: Interacción (2011). [Online]. https://api.semanticscholar.org/CorpusID:15956498
9. Card, S.K., Moran, T.P., Newell, A.: The psychology of human-computer interaction. In: Card, S.K. (ed.). CRC Press (1983)
10. Torok, A.: From human-computer interaction to cognitive infocommunications: a cognitive science perspective. In: 2016 7th IEEE International Conference on Cognitive Infocommunications (CogInfoCom), pp. 000433–000438 (2016). [Online]. https://api.semanticscholar.org/CorpusID:22735468
11. Shneiderman, B., Plaisant, C., Cohen, M., Jacobs, S., Elmqvist, N.: Designing the User Interface Strategies for Effective Human-Computer Interaction. Pearson Higher Education & Professional Group (2017)
12. Card, S., Henderson Jr, D.A.: A multiple, virtual-workspace interface to support user task switching. In: Proceedings of the CHI'87 Human Factors in Computing Systems Conference. ACM Press (1987)
13. Henderson, D.A., Jr., Card, S.: Rooms: the use of multiple virtual workspaces to reduce space contention in a window-based graphical user interface. ACM Trans. Graph. **5**(3), 211–243 (1986)

14. Endert, A., Bradel, L., Zeitz, J., Andrews, C., North, C.: Designing large high-resolution display workspaces. In: Proceedings of the International Working Conference on Advanced Visual Interfaces - AVI'12, pp. 58–65. ACM Press (2012)
15. Spellman, P.J., Mosier, J.N., Deus, L.M., Carlson, J.A.: Collaborative virtual workspace. In: Proceedings of the International ACM SIGGROUP Conference on Supporting Group Work: The Integration Challenge the Integration Challenge—GROUP '97, pp. 197–203. ACM Press (1997)
16. Green, T.M., Ribarsky, W., Fisher, B.: Building and applying a human cognition model for visual analytics. Inf. Vis. **8**(1), 1–13 (2009)

Chapter 15
Performance-Enhancing Impact of User Profiling

Abstract The chapter examines the integration of user profiling in the design of digital environments, both 2D and 3D, highlighting its role in optimizing spatial arrangements to facilitate enhanced interaction and accessibility. The methodology presented underscores the importance of the topological arrangement of digital content, where the connectivity and semantic value of content nodes are leveraged to create environments that fit with the user's learning preferences and behaviors. This comprehensive analysis aims to highlight the critical role of user profiling in fostering performance improvements across various sectors, emphasizing its significance in engineering personalized experiences and automation in the digital age. Through a detailed examination of practical applications and strategies in user profiling, the chapter contributes to a deeper understanding of how personalized digital environments can be effectively constructed and optimized, paving the way for future advancements in personalized learning and digital interaction.

Keywords Cognitive infocommunications · User profiling · Personalized interactions · Information acquisition habits · Learning styles · 3D VR

15.1 Introduction

In the field of computer science, the identification for delivering personalized user experiences is continually shaped by the advancement of user profiling techniques. As technology advances, our capacity to provide tailored recommendations, streamline processes, and deeply understand user behavior improves. This chapter explores the multifaceted impact of user profiling on performance enhancement, delving into its role in generating personalized recommendations and facilitating automation processes.

User profiling is characterized by its integration of artificial intelligence (AI) and machine learning [1], which stand as a cornerstone the understanding and detecting the complexities of user preferences and actions. Thanks to the advent of real-time

I. Horváth et al., *Cognitive Aspects of Virtual Reality*, Studies in Big Data 156,
https://doi.org/10.1007/978-3-031-68130-1_15

profiling, systems are now capable of adjusting to user interactions on the fly, facilitating experiences that are both dynamic and personalized. The practice of collaborative profiling methodologies further enriches user understanding by tapping into social interactions and community dynamics.

The success of user profiling depends on the effective combination of varied and rich data sources. Strategies like profile extraction [2] gather crucial user data, while the importance of maintaining data integrity is emphasized to overcome issues like data duplication and inaccuracies.

Furthermore, user profiling extends beyond conventional web interactions, encompassing a myriad of user behaviors such as text selection, mouse movements, and scroll speeds. This holistic approach to profiling enables systems to capture nuanced user preferences, facilitating tailored experiences across various digital platforms.

As we navigate through the intricacies of user profiling, we also explore its applications in educational and working contexts. By leveraging perceptual preference questionnaires and user-centric profiling techniques, personalized educational materials can be crafted to align with individual learning styles and preferences.

This chapter aims to unpack the complex mechanisms of user profiling and underscore its critical role in fostering performance improvements across a spectrum of sectors. By examining the strategies and practical applications of user profiling, we intend to showcase the engineering behind personalized experiences and automation in the contemporary digital age. We introduce the notion of personalized digital environments in both 2D and 3D spaces and discuss Kolb's learning style taxonomy as a foundational element for further analysis within this chapter.

This chapter builds upon and significantly revises content from previous publications [3, 4], including analyses on personalized learning in 3D virtual reality and the motivations and tools for personalized workspaces in VR settings.

15.2 Personalizing Interactions: The Impact of User Profiling on Digital Experiences

In the field of computer science, the method of personalization has seen remarkable advancements through sophisticated user profiling techniques. These techniques stand at the forefront of creating personalized digital experiences, transforming how users interact with technology across various platforms. The integration of artificial intelligence (AI) and machine learning into user profiling has revolutionized our understanding of user behavior and preferences [5], opening the way for dynamic and responsive digital environments.

15.2.1 The Role of User Profiling in Enhancing Digital Experiences

User profiling, through the lens of AI and machine learning, offers an understanding of individual behaviors and preferences. This deep insight is critical in the development of personalized recommender systems, decision support systems, and tailored digital environments. AI-supported real-time profiling, for instance, allows systems to adapt instantly to user interactions, offering personalized content and recommendations that support it on a personal level. It enables the provide the PWO (Personal Workflow Order) level 2. This immediate adaptability not only enhances user engagement but also elevates the overall user experience.

The methodologies behind user profiling have evolved, embracing collaborative and community-driven approaches. These methods leverage the power of social interactions and relationships, enriching the profiling process with a broader perspective of user preferences and behaviors. Moreover, the integration of comprehensive data from varied sources addresses key challenges such as redundancy and cleanliness, ensuring the accuracy of user profiles.

Techniques like profile extraction aid in gathering relevant user information, including web browsing, and social media data mining. Various profiling approaches cater to different user needs, such as explicit profiling based on static user characteristics or implicit profiling focusing on dynamic user behavior analysis. Hybrid profiling combines both static attributes and behavioral information for more comprehensive profiles. Additionally, user profiling extends beyond traditional web interactions to include tracking behaviors such as text selection, mouse movements, and scroll speeds. It also encompasses actions like bookmarking pages, saving HTML documents, or printing pages. Profile integration involves identifying and resolving issues with data cleanliness, such as identifying unique versus duplicate data. Understanding user interests is crucial, achieved through techniques like Interest exploration [6] which categorizes users based on their behaviors, content-based, and collaborative filtering, which group users with similar interests.

15.2.2 Bridging User Profiling with Interface Design: From 2D to Immersive VR

The impact of user profiling extends beyond traditional web interactions, influencing the design and functionality of user interfaces, including the transition from 2D to immersive 3D virtual reality (VR) environments. This transition signifies a shift towards more natural and intuitive user interactions, facilitated by an understanding of user preferences and behaviors gleaned from profiling.

In 2D interfaces, personalization involves organizing and customizing graphical elements based on user preferences. Here the value of Personalized Workflow Order ability and the value of Personalized Information Overviews is 1 these manipulations require a high level of cognitive load. The evolution into desktop virtual reality and immersive VR systems represents a paradigm shift where user profiling plays a crucial role in creating intuitive and engaging experiences. In this context, user profiling can be conducted through a knowledge-based approach, utilizing initial assessments, questionnaires, or via a machine learning-based approach, where the creation of profiles is grounded in attributes inferred from the user's actions. The option of Context Control Based on cognitive profile construction methods allows increasing the PWO and PIO level 2.

Immersive VR systems by aligning closely with human instincts, these systems facilitate a swift shift from actions governed by explicit instructions to those based on intuitive skills, enhancing user engagement and ease of use in the virtual realm. In this platform user profiling based on an AI-based approach is common, so the PWO and PIO are level 2. For instance, VR environments, designed with user profile data, can adapt to individual learning styles in educational applications and may offer personalized content that maximizes engagement and effectiveness.

15.2.3 The Cognitive Aspect of Personalized Interactions

The cognitive load on users is significantly influenced by how well digital environments align with their natural behaviors and preferences. In 2D operational systems personalizing and organizing the user interface demands the execution of operations based on knowledge, which places a higher cognitive load on users compared to operations based on rules. However basic movements, actions, and choices are governed by precise rules, yet years of practice have evolved these into skills-based operations. Typically, desktop VR systems are compatible with devices like a physical keyboard, mouse, touchpad, or touchscreen. Additionally, certain situations permit the use of hand-tracking for more intuitive gesture controls. However, in the case of the MaxWhere desktop VR application, while cognitive navigation technology facilitates navigation and manipulation within a 3D context, it does not support hand-tracking or gesture-based controls. Profiling techniques help in designing interfaces that reduce this cognitive burden, facilitating a seamless transition from rule-based to skill-based execution of actions. This is particularly evident in immersive VR systems, where systems are designed to mirror the natural movements and actions we perform daily, such as turning our heads, lifting our hands, grasping objects, and walking. This design approach leverages our innate behaviors to facilitate operations within the virtual environment, significantly minimizing the cognitive effort required to interact with the system. Immersive VR applications, which require additional devices like headsets and hand tracking, demonstrate the cost implications of creating more interactive and engaging experiences. However, the personalized nature of these interactions, informed by user profiling, significantly enhances the user experience by offering a higher degree of interaction possibilities and direct manipulation of information. Moreover, the effectiveness of digital environments, as indicated by metrics like Task Execution Tools (TET) value, Information Availability (IA) [7], and Personalized Workflow Order (PWO) ability underscores the importance of personalization.

15.2.4 The Future of Personalized Digital Experiences

The integration of user profiling into the design and implementation of digital environments marks a significant advancement in personalizing user interactions. By understanding and anticipating user needs, digital platforms can offer more relevant, engaging, and intuitive experiences. As we move forward, the continued evolution of profiling techniques and their application in interface design, especially in immersive technologies like VR, will play a pivotal role in shaping the future of digital experiences. The challenge lies in balancing the benefits of personalization with the ethical considerations of data privacy and security, ensuring that personalization enhances digital experiences in a responsible and user-centric manner.

This revised version aims to more cohesively integrate the discussion of user profiling with its applications in digital experience design, particularly emphasizing the transition to immersive VR environments and the cognitive aspects of personalized interactions.

15.3 Personalization Concepts in Virtual Reality

IT developments entail a degree of standardization, a natural progression in the field. However, concurrently, there is a growing user demand for heightened customization and personalization of 2D digital, and 3D VR environments. In this section, we present a concise overview of personalized digital workspaces in both 2D and 3D environments, alongside Kolb's taxonomy of learning styles, which will serve as a foundational reference point in subsequent sections of this chapter. By summarizing this research background, our objective is to underscore the distinctive contributions of our work, which diverge from and, in certain cases, complement existing studies.

15.3.1 Personalized 2D Digital Workspaces

In the past few years numerous IT solutions have emerged in scientific literature and industry aimed at appliance personalized digital workspaces utilizing conventional 2D graphical user interfaces. These include:

- Cloud-Enabled Productivity Suites: Cloud-based productivity suites like Microsoft Office 365 and Google Workspace empower employees to access work documents and applications seamlessly across multiple devices via internet connectivity. They also facilitate collaboration and communication through features such as real-time editing and video conferencing.

- Employee Engagement Platforms: Platforms such as Workday and SAP SuccessFactors offer personalized experiences to employees based on their job roles, preferences, and competencies. They serve as conduits for targeted learning and development initiatives tailored to individual needs.
- Digital Workplace Solutions: Digital workplace platforms such as Microsoft SharePoint, Slack, and Workplace by Facebook serve as centralized hubs for employees to access information, collaborate with peers, and communicate effectively.
- Task Management Tools: Platforms like Trello, Asana, and Monday.com offer personalized task management experiences, allowing users to organize and prioritize tasks according to their preferences. These tools often feature customizable boards, lists, and task cards, enabling users to tailor their workflow to suit their individual needs.
- Note-Taking Applications: Apps such as Evernote, OneNote, Notion, Obsidian, or Roam Research provide personalized note-taking experiences, allowing users to capture and organize their thoughts, ideas, and information in a way that best suits their workflow. These applications often offer customizable templates, tags, and organizational structures to accommodate different note-taking styles.
- Virtual Desktop Infrastructures (VDIs): VDI technology enables organizations to furnish virtual desktop environments to employees, granting them access to personalized desktop settings and applications from any device with an internet connection.
- Digital Assistant Solutions: Integrating digital assistant solutions like Amazon Alexa for Business and Microsoft Cortana into organizational digital workspaces offers personalized assistance and streamlines routine tasks. With advancements in large language models (LLMs), the widespread adoption of such solutions is anticipated in nowadays.
- Customizable Dashboard Solutions: Dashboard tools such as Google Data Studio, Tableau, and Microsoft Power BI provide personalized data visualization experiences, allowing users to create customized dashboards and reports tailored to their specific data analysis needs. These platforms offer a range of visualization options, filters, and interactive features to enable users to explore and analyze data in depth.
- Personal Finance Management Apps: Applications like Mint, YNAB (You Need a Budget), and Personal Capital offer personalized financial management experiences, allowing users to track expenses, set budgets, and plan for financial goals based on their individual financial situations and priorities. These apps often offer personalized insights and recommendations to help users make informed financial decisions.

These examples illustrate the diverse range of personalized 2D digital workspaces available across various domains, catering to different user needs and preferences.

15.3.2 *Customization and Personalization of 3D VR Environments*

Personalization in VR platforms is primarily built upon the fusion of user data and feedback, working symbiotically to tailor individual experiences based on distinct needs and preferences. This subsection will delineate the prevalent directions for personalization within 3D VR solutions.

VR platforms gather insights into user preferences, encompassing elements like preferred color schemes, avatars, or virtual settings, which are then leveraged to tailor the experience. For instance, platforms like VR Chat empower users to customize avatars and fashion their own virtual domains.

Platforms analyze user behavior data, such as meeting attendance frequency or engagement in specific activities, to personalize experiences. This data informs recommendations for relevant events or activities and enables customization of the user's virtual workspace. For example, the Oculus/Meta platform utilizes behavior and preference data to suggest VR apps and experiences aligned with user interests, while also dynamically adjusting the virtual environment based on users' gaze and hand movements in real-time.

Feedback mechanisms allow users to provide input on their experiences, which informs future personalization efforts. Platforms like Mozilla Hubs facilitate user ratings of virtual rooms, aiding in recommending similar spaces in the future. Similarly, the Oculus/Meta platform enables users to rate apps, share reviews, and provide feedback via social media channels, offering valuable insights for platform improvement.

Some VR platforms integrate with traditional 2D digital platforms, such as social media or email, to personalize experiences based on existing user data. For instance, Immersed VR enables users to import LinkedIn profiles to suggest relevant events or networking opportunities.

Recently, an increasing number of VR platforms have introduced support for integrating 2D digital content, such as PDFs, images, videos, and websites, into 3D environments. This feature enables users to interact with 2D content seamlessly within 3D settings. Platforms like AltspaceVR, Spatial, VRChat, ENGAGE, VirCA, and MaxWhere VR offer functionalities for importing and displaying various types of 2D content within virtual environments, enhancing collaboration and productivity. A focus on the user characteristics and user habits of digital content placement and usage in virtual space [8].

In the field of VR education, adaptive learning solutions like zSpace and Cerego harness adaptive learning algorithms to tailor learning experiences based on individual performance and preferences. These platforms adjust difficulty levels, pacing, and content to accommodate diverse learning styles and optimize learning outcomes.

While there may not be a VR platform exclusively personalizing digital content based on learning style, numerous educational VR platforms employ adaptive learning methodologies to create bespoke learning experiences tailored to individual needs.

15.4 The Concept of Personalization in VR Education

Scientific literature extensively explores various aspects pertinent to VR workspaces, including the cognitive benefits of 3D environments, the design and development of VR training systems, and opportunities for personalization. In the area of effectiveness:

- In 3D VR environments, digital workflows exhibit a reduction of at least 50 percent, as evidenced by Lampert et al. [9].
- According to Horváth and Sudár (2018), workflows in 3D VR show a decrease of 30 percent in user operations and 80 percent in machine operations [10].
- The rate of information recall is 50% higher [11].
- The MaxWhere 3D platform increases the effectiveness of online collaborative project management software–requiring 72 percent less elementary user operations, and 80 percent less time spent on overview-related tasks [12].

In the area of visual attention and memory:

- VR advertisements have demonstrated superior effectiveness compared to traditional web-based ads in a 3D environment, based on a questionnaire-based free recall test, as outlined by Berki [11].
- The equipment of 3D Vr space and the location of control questions relative to the theoretical content influence both the short and long- term memorization [13].
- Sudár and Csapó conducted research on the connection between visual attention, fixation and cognitive load, in which descriptive markers were identified [14].

15.5 Information Acquisition and User Requirements

This session summarizes various types of information acquisition tests and explores the significance of aligning information types with user preferences. Understanding how individuals acquire and process information is crucial in designing effective systems, interfaces, and content that cater to their needs and preferences.

Information acquisition refers to the process of gathering, interpreting, and internalizing information from various sources. Different individuals have unique ways of acquiring and processing information, influenced by factors such as cognitive abilities, learning styles, and preferences. By employing information acquisition tests, we can assess and understand how users interact with different types of information and interfaces.

The following tests are commonly used in psychology, neuropsychology, and cognitive neuroscience to evaluate various aspects of cognitive function and can provide valuable insights into an individual's cognitive abilities and potential areas of impairment as well.

- Corsi Block-Tapping Test (CBT) [15, 16] The goal of the Corsi Block-Tapping Test is to measure visuospatial working memory capacity. In this test, participants are presented with a series of blocks or cubes placed in a random order and asked to reproduce the sequence by tapping the blocks in the same order they were presented. The length of the sequences is progressively increased to assess the participant's memory span.
- Digit Span Test [17] The Digit Span Test aims to assess short-term memory and attention span. Participants are read a sequence of digits at a rate of one digit per second and asked to repeat the sequence immediately afterward. The length of the digit sequences is gradually increased until the participant can no longer recall them accurately.
- Stroop Test [18, 19] The Stroop Test evaluates cognitive processing speed, attention, and the ability to inhibit automatic responses. In this test, participants are presented with words printed in colored ink. The words themselves may name colors, but the ink color may be different from the word's meaning (e.g., the word "red" printed in blue ink). Participants are asked to name the ink color while ignoring the word's meaning, which often leads to interference and slower response times.
- Trail Making Test [20, 21] The Trail Making Test assesses visual attention, task-switching ability, and executive functioning. The test consists of two parts. In Part A, participants are instructed to connect a sequence of numbers in ascending order. In Part B, participants must alternate between connecting numbers and letters in ascending order and in a specific pattern (e.g., 1-A-2-B-3-C, etc.). The time taken to complete each part and any errors made are recorded and used to evaluate cognitive performance.
- Raven's Progressive Matrices [22, 23] Raven's Progressive Matrices are non-verbal tests of fluid intelligence and abstract reasoning. Participants are presented with a series of visual pattern completion problems and asked to identify the missing piece that completes each pattern. The difficulty of the problems increases progressively throughout the test, allowing for the assessment of cognitive abilities across different levels of complexity.

In the context of daily digital tasks and learning, individuals interact with a variety of digital content. It's beneficial for them to recognize which form of digital information (text, image, or sound) they find most memorable, essentially pinpointing their preferred method of information consumption. To ascertain users' preferences in information acquisition, we can utilize standard psychological assessments designed for learning. For instance, Horváth, in his exploration of personalized virtual reality (VR) educational experiences [24], employed a gamification-based test based on a multisensory learning approach to gauge students' information acquisition preferences.

To assess visual memory, students were shown a sequence of 10 images without taking notes. This was immediately followed by a text-based logical puzzle, designed to shift their focus away from the images they had just viewed.

Upon completing the puzzle, they were asked to write down any words they could recall from the images, with a time limit of one minute for this recall task.

The auditory memory test followed a similar format, where students listened to 10 words instead of viewing images. This was also succeeded by a logical task to divert their attention, after which they needed to recall and write down the words they had heard.

For the reading comprehension test, participants were shown 10 words, and then given a logical question to answer: "If yesterday was the third day after Monday, what day is the day after tomorrow?" After responding, they were required to list the words they had memorized.

The count of accurate responses across the different categories (text, image, sound) offered insights into the students' preferred methods of information absorption. This information was instrumental in establishing user personas and categorizing users based on their information acquisition preferences.

The study did not explore kinesthetic learning styles, as the focus was on digital interaction. The assumption is that users navigating in a three-dimensional space using a mouse or touchpad, with its intuitive control over spatial movement, inherently engage with 3D interfaces and metaphors, making specific tests for kinesthetic learning unnecessary for of this research.

The subsequent step in preliminarily identifying user expectations is linked with understanding their learning style. Various theories highlight the significance of learning styles, with some being particularly pivotal in educational strategies. The shift towards an active, experiential-based education that embraces diverse learning styles has been described as transformational learning. This approach, championed by Jesus et al. [25], is presented as a superior alternative to conventional teaching methods, as argued by Jacoby in 1996.

Coffield and colleagues, in their 2004 study [26], cataloged over 70 different models of learning styles, indicating a wide array of approaches to understanding how individuals learn best. Among these, Kolb's Experiential Learning Model [27, 28] emerges as particularly influential, as noted by Kayes and others between 2005 and 2011. The model's strength lies in its emphasis on the process of experiential learning, focusing on dynamic learning activities over static learning characteristics, as discussed by Turesky and Gallagher [29].

15.5.1 Understanding Kolb's Learning Styles: Accommodator, Assimilator, Divergent, and Convergent

David Kolb's Experiential Learning Theory [27] outlines four distinct learning styles based on a four-stage learning cycle. These styles are:

- Accommodator: Prefers hands-on experience, relies on intuition rather than logic, and tends to excel in practical tasks. Accommodators are good at adapting to specific circumstances and solving problems in innovative ways. They learn best by doing and feeling, thriving in situations that require action and risk-taking.

- Assimilator: Leans towards abstract conceptualization and logical analysis. Assimilators are adept at understanding wide-ranging information and organizing it in a clear, logical format. They prefer thinking and watching, valuing ideas and theories over practical application. This style is more comfortable with a deductive reasoning approach, focusing on forming theoretical models rather than practical implementation.
- Divergent: Emphasizes concrete experience and reflective observation. Individuals with a divergent style are imaginative and can see things from multiple perspectives. They are strong in brainstorming and generating creative ideas. Divergents prefer to watch rather than do, drawing upon their broad interests to gather information. They excel in situations that require generating alternative ideas and solutions.
- Convergent: Combines abstract conceptualization with active experimentation. Convergent learners solve problems by applying theories to practical scenarios. They prefer technical tasks and are less concerned with interpersonal and social aspects. This style is characterized by finding practical uses for ideas and theories,-and excelling in situations where there is a single correct answer or solution to a problem.

Each of these learning styles reflects the different ways individuals prefer to engage with and process information, highlighting the importance of a diverse educational approach to cater to varying needs and strengths.

15.6 Enhancing VR Experience Through Personalized Space Design Automatic VR Space Construction

15.6.1 Optimization of Spatial Arregement

This section presents the methodology for creating a 3D environment that seamlessly integrates 2D digital content, optimized to the unique preferences of each user. The process involves several key steps aimed at creating a VR space that not only reflects individual user profiles but also optimizes their interaction with digital content.

Step 1: User Profiling for Custom VR Environments

User profiling is the foundational step, executed through two distinct approaches within the digital environment:

- Knowledge-Based Profiling: Utilizes preliminary assessments and questionnaires to gather user data. This approach facilitates a semi-automated creation of VR spaces, blending user-provided information with system-generated insights.
- Machine Learning-Based Profiling: Relies on analyzing user behavior to construct profiles. This method enables a more dynamic and automated VR space construction, adapting to user interactions in real-time. This approach facilitates a fully automated creation of VR spaces, however, it enables the user to control the space.

Step 2: Workflow Analysis
This step involves examining how users with varied preferences engage with digital content. By analyzing the interactions of users' different learning styles, identified and validated by the author [24], insights into content engagement patterns are obtained.
Step 3: Constructing Workflow Graphs
Workflow graphs are developed for each identified learning style, illustrating the flow of interaction with digital content: Graph Construction: Utilizes access paths to map out the workflow, with content types represented as nodes and the transitions between them as edges.
Step 4: Content Significance Analysis
Through graph analysis, it becomes possible to pinpoint central and peripheral content elements. This distinction is crucial for understanding user engagement and content relevance.
Step 5: Spatial Arrangement Development
Leveraging the insights gained from the workflow graphs, a spatial layout scheme is devised. An illustrative example (Fig. 15.1) demonstrates how central content, such as online meeting tools for divergent thinkers, is positioned within the VR space to reflect its significance and frequent access.

The methodology concludes that the optimal learning styles are informed by the topological arrangement of digital content, factoring in the connectivity and semantic value of each node. This approach is encapsulated in the Context Control Based on Cognitive Profile Construction method, detailed further in Chap. 16.

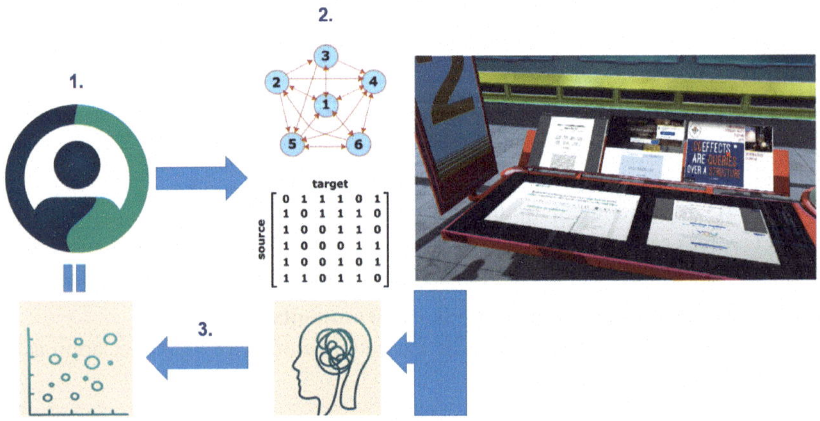

Fig. 15.1 Process of automatic VR space construction

15.7 Conclusion

The chapter presented a new approach to the integration of user profiling in the design of digital environments, emphasizing its crucial role in optimizing spatial configurations to improve interaction and accessibility in both 2D and 3D contexts. The methodology outlined demonstrates the significance of carefully considering the topological arrangement of digital content, leveraging the connectivity and semantic value of content nodes to tailor environments that align with user learning preferences and behaviors. This thorough analysis underlines the crucial role of user profiling in driving performance enhancements across diverse sectors, highlighting its importance in crafting personalized experiences and enabling automation in the digital era.

Through an in-depth exploration of practical applications and user profiling strategies, the chapter contributes significantly to our understanding of how personalized digital environments can be efficiently designed and optimized. The process of Automatic VR Space Construction, detailed in steps from user profiling to the development of spatial arrangement, exemplifies a methodical approach to creating VR spaces that not only reflect the unique profiles of users but also enhance their engagement with digital content by optimizing content accessibility and relevance. This personalized space design enhances user interaction with VR environments, making digital content more accessible and engaging according to individual learning preferences and behaviors.

References

1. Kanoje, S., Girase, S., Mukhopadhyay, D.: User profiling trends, techniques and applications (2015). arXiv:1503.07474
2. Tang, J., Yao, L., Zhang, D., Zhang, J.: A combination approach to web user profiling. ACM Trans. Knowl. Discov. Data (TKDD) **5**(1), 1–44 (2010)
3. Horváth, I.: An analysis of personalized learning opportunities in 3d VR. Front. Comput. Sci. **3**, 673826 (2021)
4. Horváth, I., Csapó, Á.B.: Motivations and tools relevant to personalized workspaces in VR environments. Electronics **12**(9), 2059 (2023). [Online]. https://www.mdpi.com/2079-9292/12/9/2059
5. Fawcett, T., Provost, F.J.: Combining data mining and machine learning for effective user profiling. In: KDD, vol. 96, pp. 8–13 (1996)
6. Middleton, S.E., Shadbolt, N.R., De Roure, D.C.: Capturing interest through inference and visualization: Ontological user profiling in recommender systems. In: Proceedings of the 2nd International Conference on Knowledge Capture, pp. 62–69 (2003)
7. Horváth, I., Berki, B.: Investigating the operational complexity of digital workflows based on human cognitive aspects. Electronics **12**(3), (2023). [Online]. https://www.mdpi.com/2079-9292/12/3/528
8. Sudár, A., Csapó, Á.B.: Elicitation of content layout preferences in virtual 3D spaces based on a free layout creation task. Electronics **12**(9), 2078 (2023). [Online]. https://www.mdpi.com/2079-9292/12/9/2078

9. Lampert, B., Pongracz, A., Sipos, J., Vehrer, A., Horvath, I.: MaxWhere VR-learning improves effectiveness over clasiccal tools of e-learning. Acta Polytechnica Hungarica **15**(3), 125–147 (2018)
10. Horváth, I., Sudár, A.: Factors contributing to the enhanced performance of the MaxWhere 3D VR platform in the distribution of digital information. Acta Polytechnica Hungarica **15**(3), 149–173 (2018). Mar
11. Berki, B.: Better memory performance for images in MaxWhere 3D VR space than in website. In: 2018 9th IEEE International Conference on Cognitive Infocommunications (CogInfoCom), pp. 000281–000284. IEEE (2018)
12. Horváth, I.: Maxwhere 3d capabilities contributing to the enhanced efficiency of the trello 2d management software. Acta Polytechnica Hungarica **16**(6), 55–71 (2019)
13. Horváth, I.: Behaviors and capabilities of generation CE students in 3D VR. In: Baranyi, P. (ed.) Proceedings of the 10th IEEE International Conference on Cognitive Infocommunications: CogInfoCom 2019, pp. 491–494. IEEE (2019)
14. Sudár, A., Csapó, Á.B.: Descriptive markers for the cognitive profiling of desktop 3D spaces. Electronics **12**(2), 448 (2023). Jan
15. Corsi, P.M.: Human memory and the medial temporal region of the brain. Ph.D. Dissertation, McGill University, Montreal (1972)
16. Milner, B.: Interhemispheric differences in the localization of psychological processes in man. Br. Med. Bull. (1971)
17. Wechsler, D.: Wechsler adult intelligence scale-revised manual. Psychological Corporation, New York (1981). (No Title)
18. Stroop, J.R.: Studies of interference in serial verbal reactions. J. Exp. Psychol. **18**(6), 643 (1935)
19. MacLeod, C.M.: Half a century of research on the Stroop effect: an integrative review. Psychol. Bull. **109**(2), 163 (1991)
20. Reitan, R.M.: Validity of the trail making test as an indicator of organic brain damage. Percept. Mot. Ski. **8**(3), 271–276 (1958)
21. Tombaugh, T.N.: Trail making test a and b: normative data stratified by age and education. Arch. Clin. Neuropsychol. **19**(2), 203–214 (2004)
22. Raven, J.C.: Progressive Matrices: a Perceptual Test of Intelligence, vol. 19, p. 20. HK Lewis, London (1938)
23. Manual for raven's progressive matrices and vocabulary scales. In: Standard Progressive Matrices (1983)
24. Horváth, I.: An analysis of personalized learning opportunities in 3d VR. Front. Comput. Sci. **3**, 673826 (2021)
25. Pedrosa de Jesus, H.T., Albergaria Almeida, P., Joaquim Teixeira-Dias, J., Watts, M.: Students' questions: Building a bridge between Kolb's learning styles and approaches to learning, vol. 48, no. 2/3, pp. 97–111. Education+ Training (2006)
26. Coffield, F., Moseley, D., Hall, E., Ecclestone, K.: Learning Styles and Pedagogy in Post-16 Learning. Learning and Skills Research Centre (2004)
27. Kolb, D.A.: Learning styles and disciplinary differences. Mod. Am. Coll. **1**(January 1981), 232–235 (1981)
28. McLeod, S.: Kolb-Learning Styles. Simply Psychology (2017)
29. Turesky, E.F., Gallagher, D.: Know thyself: Coaching for leadership using Kolb's experiential learning theory. Coach. Psychol. **7**(1), 5–14 (2011)

Chapter 16
Adaptive Content Layout in 3D Spaces

Abstract This chapter presents a systematic framework for personalizing workspace layouts in 3D Virtual Reality (VR) environments through a structured template language, enabling dynamic adaptation of content and layout to individual user preferences and task requirements. It introduces a hierarchical template system for content layout specification and transformation, complemented by an associative semantic search to enhance workspace relevance and efficiency by linking semantically related content. The research hypothesizes that such personalized VR workspaces significantly improve task performance in speed and accuracy. An innovative service is proposed for efficient digital content management, tailored to workflow demands and user preferences, focusing on optimizing content selection, layout creation, and the alignment of user preferences with the workflow's temporality and the 3D space's geometry. The objective is to explore the impact of personalized workspace reconfiguration on cognition within VR, aiming to contribute to a deeper understanding of this dynamic and guide future VR technology developments.

Keywords Virtual reality · Virtual workspace · Desktop VR · Digital content synchronization

16.1 Introduction

Technological advancements, notably the progression from basic command-line interfaces (1D) and graphical user interfaces (2D) to the realms of Virtual and Augmented Reality (VR/AR) and metaverse innovations, promise to revolutionize the volume and quality of digital interactions performed by users. Such evolutions facilitate previously unimaginable or heavy modes of collaboration, communication, and information exchange, breaking away from the confines of conventional computing architecture. For example, VR/AR solutions allow geographically dispersed teams to collaborate within three-dimensional spaces, thereby enhancing productivity and streamlining communication through the adoption of visual symbols and spatial metaphors [1].

I. Horváth et al., *Cognitive Aspects of Virtual Reality*, Studies in Big Data 156,
https://doi.org/10.1007/978-3-031-68130-1_16

Considering the limitations of using a "linear" interface system like MS-DOS in today's digital age, it's apparent that such an environment would severely restrict our digital task capabilities. However, these technological shifts, while expanding current operational capacities, introduce challenges that necessitate adaptability and skill development among individuals and organizations alike. This adaptation is crucial as the integration of new technologies may redefine professional roles, with automation and artificial intelligence supplanting conventional tasks, thereby altering the skill sets valued in the workforce and increasing the demand for managing huge amounts of information effectively. VR may also be the ideal navigation tool for large or complex databases [2].

In the field of education, the transition to immersive, spatially oriented learning within three-dimensional digital environments marks a significant departure from the traditional, linear educational methodologies. This chapter explores the cognitive advantages offered by virtual reality, emphasizing its potential to enhance a broad spectrum of human faculties, including psychological, cognitive, and social skills, particularly those related to understanding complex spatial, temporal, and conceptual relationships [3].

Asserting that three-dimensional virtual spaces surpass their two-dimensional counterparts in educational and professional efficacy, this study investigates how the personalization of VR workspaces could further amplify these benefits. VR environments are adaptable enough to be customized for various purposes, including the improvement of spatial skills [4], the enhancement of visual memory [5, 6], Immersion, presence, and episodic memory [7], the sharpening of attentional abilities [8, 9], and the support of individuals with cognitive challenge0s [10]. These adaptable layouts ensure that users can maximize their cognitive capabilities [11]. By introducing 'context control' for the customization of digital content in three-dimensional layouts to align with user-specific cognitive and learning styles, a multi-faceted approach is proposed. This includes developing diverse layout schemas tailored to different styles, identifying each user's distinct style for schema selection, and adapting the schema to optimize the virtual workspace environment [12, 13].

This strategy overrides traditional static content arrangement, advocating for dynamic and personalized creation of VR workspaces. It aims to enhance work efficiency by reducing the time and cognitive effort needed for users to configure their optimal working conditions. The framework's adaptability extends its applicability beyond learning and professional tasks to a broad spectrum of domains, offering a versatile solution for designing tailored VR experiences [14, 15].

Moreover, the chapter emphasizes the importance of organizing digital content within virtual environments to streamline workflows, enhance information accessibility [16], and improve overall project comprehension. This spatial organization not only facilitates memory recall but also enables smooth transitions between different project phases or data sources. Through the exploration of a structured template language, the study seeks to enable the personalization and adaptable reconfiguration of VR workspaces, leveraging principles from cognitive inforcommunications and communicative VR to optimize the human-machine interface within these digital spaces. Established in the emerging field of VR workspace personalization, this research contributes to the advanced understanding of enhancing human-computer interactions for more efficient and intuitive virtual work and learning environments.

Through these focused sections, this chapter aims to contribute significantly to the field of human-computer interaction, specifically in the context of VR. The ultimate goal is to build a bridge for more intuitive, efficient, and personalized digital environments that align closely with individual cognitive profiles and preferences. This chapter is an extended and revised version of the following publications [12, 17].

16.2 Digital Adaptive Content Layout for Personal Learning Styles

The novelty of the context control framework presented in this chapter is that it operates based on users' cognitive profiles (e.g., learning styles) and also offers a semi-automated way to adapt 3D content layouts to the individual preferences of users, within the constraints of a wide range of possible 3D environments. Authors propose that a strategic framework for developing personalized VR workspaces, encompassing both content and spatial arrangement, could significantly improve the efficiency and precision of digital task execution, beyond simply aligning content or tasks with individual learning styles.

16.2.1 Methodology for Generating Personalized Content Layouts

To begin our investigation, we define context control in VR as the set of algorithmic solutions and user-interaction designs that curate digital content and present it in a 3D layout corresponding to users' working and learning styles while adhering to physical constraints. This definition underscores the dual aspects of content selection and layout creation, highlighting the need for adaptable frameworks.

Definition 1. Context control in VR is a set of algorithmic solutions and user-interaction designs that enable digital content to be curated, and presented in a 3D layout in such a way that corresponds to users' working and learning styles while also adhering to the physical constraints of the 3D environment [12].

It is crucial to emphasize that context control entails both the selection of content and the creation of its layout. Moreover, the layout must accommodate the user's preferences and the physical dimensions of the virtual space. For instance, if a table's surface is inclined at a 30-degree angle, rather than flat, the chosen layout schema, although generally appropriate, would require adjustments for that particular setting. Herein lies the necessity for a formal framework capable of structured modifications (customization), highlighting why the concept of context control as defined here encompasses both computational and design elements in interaction.

16.2.1.1 Understanding User Interaction with Content

Initial exploration involves observing individuals with diverse learning preferences interacting with different document types within specified workflows. This observation leads to the creation of workflow graphs, depicting navigation patterns among various content types.

In a previous study conducted by one of the authors, university students underwent an entire semester immersed in a customized 3D desktop VR educational setting. Details of this investigation, including the pathways between various content elements related to assigned tasks, are documented in [18]. These pathways served as the foundation for constructing workflow graphs, where different content types are depicted as nodes and frequent transitions between them are represented by links (Fig. 16.1).

16.2.1.2 Analysis and Layout Schema Design

The subsequent phase involves analyzing workflow graphs to identify central and marginal content elements within tailored workflows. This analysis informs the design of spatial layout schemas reflective of user interaction patterns and learning style preferences. Visual representations, such as Fig. 16.2, illustrate the schematic design process tailored to specific learning styles, providing clarity for readers The

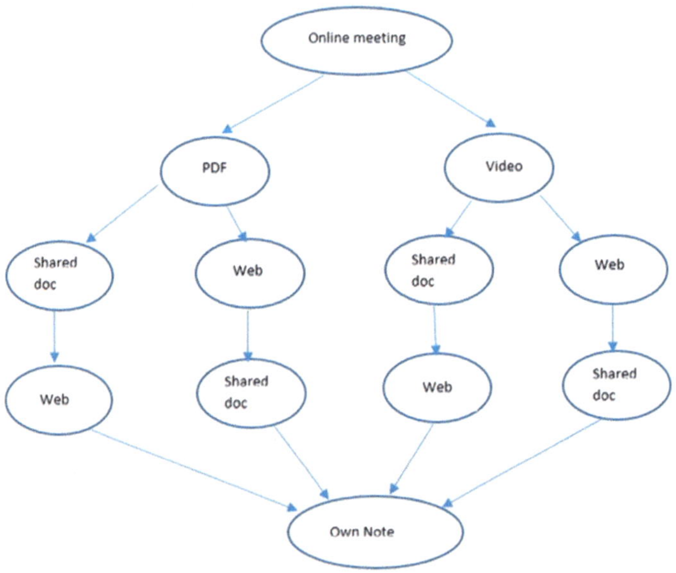

Fig. 16.1 The workflow of a Divergers learning style user. Workflow: Observe—Understand—Discuss—Prepare (describe, draw, etc.)

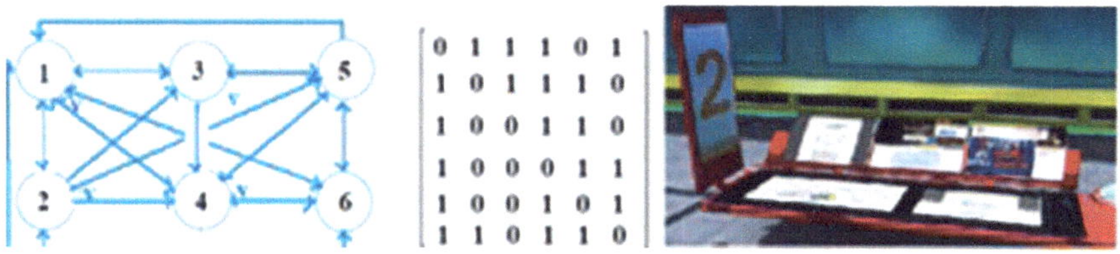

Fig. 16.2 Diverger's learning style graph, adjacency matrix and digital layout

workflow graph in Fig. 16.2 shows which content elements assume a central position and which remain peripheral within a workflow tailored to the divergent learning style at hand.

16.3 Methods for Input User Profiling

The journey of user profiling has been deeply intertwined with advancements in data mining and machine learning [19], marking a significant evolution from basic information retrieval to sophisticated, predictive systems that understand and anticipate user needs. The concept of a user profile has been defined in several ways. Initially, the practice of user profiling was grounded in the simple collection and analysis of user data, primarily focusing on explicit information provided directly by users through surveys or online forms. This approach, while foundational, was limited by its static nature and reliance on user participation, which often led to issues of data accuracy and privacy concerns. User Profiling can be defined as the process of identifying the data about a user interest domain. This information can be used by the system to understand more about the user and this knowledge can be further used for enhancing the retrieval for providing satisfaction to the user. User profiling has two important aspects efficiently knowing the user and based on those recommending items of interest. El Allioui et al. [20], in their study, explained user profiles as a set of data structures that describe the environment for human-computer interaction. The integration of data mining techniques signified a pivotal shift in user profiling methodologies. Data mining, as described in the knowledge data discovery (KDD) model [21], involves extracting valuable information from large datasets, a principle that resonates with the core of user profiling. The adaptation of the KDD process to User Data Discovery (UDD) [22] highlighted a nuanced approach to profiling, where the system begins with minimal information about the user and progressively learns and adapts based on data acquired over time. This dynamic process underscores a more refined and responsive method of understanding user preferences and behaviors.

Machine learning has further revolutionized user profiling by enabling systems to automatically learn and improve from experience without being explicitly programmed. This approach is particularly effective in recommendation systems, where machine learning algorithms analyze patterns in user behavior to predict and suggest items of interest. Such systems evolve with each interaction, becoming increasingly adept at tailoring recommendations to individual users. The application of machine learning in cognitive infocommunication systems extends beyond recommendation, facilitating behavior modeling and sentiment analysis and the evolution of user needs. Moreover, the evolution of user profiling has embraced the complexity of human-computer interaction, acknowledging that each individual has unique and consistent preferences for perceiving, organizing, and retaining information. Learning styles, as identified by Coffield et al. [23] and exemplified by Kolb's Experiential Learning Model [24], play a significant role in tailoring digital environments to match the varied ways people engage with content and processes. This recognition of individual differences further exemplifies the depth and sophistication of modern user profiling techniques, which are now capable of accommodating a broad spectrum of cognitive and behavioral nuances.

16.4 Personalized Layout Adjustment

Learning preferences act as dependable indicators for how individuals engage with and adapt to their educational and professional surroundings. Coffield and his colleagues have cataloged over 70 distinct models of learning styles [23], among which Kolb's Experiential Learning Model [24] stands out due to its emphasis on learning through experience rather than predetermined cognitive traits [25]. This model, which highlights the importance of experiential learning in daily tasks and knowledge acquisition, identifies four primary learning styles: Diverger, Assimilator, Converger, and Accommodator [24].

Divergers excel in generating creative ideas and thrive in team settings where brainstorming and expansive thinking are required. Their learning is enhanced by engaging in activities that allow for reflective observation and concrete experience. Such individuals are well-suited to tasks that involve collaboration, as well as the creation of reports and summaries due to their broad interests and information-gathering skills [26].

Assimilators prefer a structured and logical approach to learning, excelling in tasks that demand the organization of a wide array of information into coherent, logical frameworks [26]. Their strength lies in dealing with theoretical models, making them ideal for roles in design and planning.

Convergers, who favor abstract conceptualization, are naturally inclined towards technical rather than interpersonal challenges, showing a propensity for problem-solving in fields like management and product development.

Accommodators are characterized by their hands-on approach to learning, often relying on active experimentation and concrete experiences. This group of learners is inclined towards taking risks and frequently depends on instinctive feelings rather than analytical reasoning, making them adept at roles that require testing and hands-on experimentation.

Beyond these learning styles, it's crucial to acknowledge the diverse cognitive, emotional, and physical facets that make up an individual's approach to learning [27]. People exhibit varied preferences in how they absorb information, whether through auditory, visual, or textual means, reflecting the diverse ways they interact with their educational environments.

Utilizing these insights, the spatial layout framework can be devised to reflect the observed patterns.

16.4.1 Enhancing Personalized Workspaces: Introducing a Hierarchical Template Language

Direct application of the generated layout schemas isn't always feasible due to variations in scale and the availability of suitably arranged 3D objects across different VR spaces. Additionally, diverse tasks and workflows may necessitate varying numbers and types of documents, even when the geometric layout of a workspace is known. This calls for adaptability in terms of digital content display.

To address these challenges, we propose a formal hierarchical template language to specify content layout schemas in 3D and outline how they may evolve over time. Concurrently, we suggest a canonical set of operations empowering users to manipulate these layout schemas and tailor them to any specific 3D environment.

16.4.1.1 Introducing a Hierarchical Template Language Framework

The research presented in the preceding study by Horvath and Csapo demonstrates [12] that the layout schemas devised are highly compatible with the diverse learning styles categorized within Kolb's framework and users' information acquisition habits. However, the direct application of these schemas is occasionally impeded by variances in scale and the availability of appropriately configured 3D objects across distinct virtual reality (VR) environments.

Moreover, considering that various tasks and processes might necessitate differing quantities of documents and a diverse array of content, the geometric configuration of a workspace, even when established, demands flexibility regarding the assortment and classification of documents presented.

To address these complexities, the authors introduce an advanced hierarchical template language designed for defining content layout schemas in three dimensions, as well as their dynamic evolution over time. Concurrently, we suggest a comprehensive suite of operations enabling users to modify and adapt these layout schemas to meet the unique requirements of any specific 3D setting.

Within this framework, it's assumed that the strategic manipulation of context within these VR environments allows for the availability of an extensive selection of layouts and configurations. These predetermined layout sets are engineered to offer customized spatial solutions for digital workspaces in VR, tailored to meet users' specific needs and preferences. This customization ensures the optimal arrangement, size, and spatial relationship of digital displays (smartboards).

Furthermore, acknowledging the unique digital content demands of each project and workflow is essential. Thus, there's a provision for integrating 2D digital content, which can significantly enhance the utility and versatility of these virtual workspaces. Digital assets or documents are primarily obtained from an existing database, potentially repurposed across various projects, or by importing entirely new digital materials onto designated display units or smartboards. This dual approach facilitates a seamless integration of digital content, enriching the user experience within the VR environment by accommodating a wide range of project-specific requirements and preferences.

This template language was introduced in a prior scientific publication authored by the researchers [12].

- A **work area** illustrates a comprehensive content arrangement and is defined by the following:
  - A *work area type* (for example:convergent, divergent, assimilator, accommodator or other preferences);
  - A 3-dimensional *position* vector;
  - A 4-dimensional (quaternion) *orientation*;
  - A *width* and *height* (in centimeters);
  - A *current scale*, with an initial value of 1.0;
    as well as the following parameters, the meaning of which are detailed later in this list:
  - A list of *work area parts*;
  - A partitioning of work area parts identifying work area parts that can be moved together, referred to as *move groups*;
  - A partitioning of work area parts identifying work area parts that can be rotated together, referred to as *rotate groups*;
  - A *centroid*, expressed as a 3-dimensional vector, that points from the location of the work area to a point in space which helps determine the global directions in which the work area can be shifted (translated);
  - A list of so-called *progressions*.
- A **work area part** is a planar region in space belonging to a work area, that is characterized by the following:
  - A unique *index* to identify the work area part inside the work area;
  - A *work area part type* (pdf, video, collaborative document, email, online meeting, table, diagram, domain-specific software, browser);
  - A *relative position vector*, the coordinates of which are expressed as a percentage of the width and height of the work area in the x and y dimensions, and in nominal terms (in centimeters) in the z dimension;

  - A *relative orientation*, expressed as a quaternion specifying a rotation with respect to the orientation of the work area;
  - A *relative width and height*, expressed as a percentage of the width and height of the work area, respectively;
  - A list of *panels* (to be detailed later in this list).

- A **panel** is a rectangular display panel characterized by the following:

  - A *relative position vector*, the coordinates of which are expressed as a percentage of the width and height of the work area part in the x and y dimensions, and in nominal terms (in centimeters) in the z dimension;
  - A *relative width and height*, expressed as a percentage of the width and height of the work area part, respectively;
  - A *url*, specifying the content displayed inside the panel;
  - A *physical panel id*, linking the panel with a physical entity (a canvas) inside the 3D space.

- A **progression** is a state within the lifecycle of a work area that contextually specifies the set of panels that are contained in the work area relative to a previous and a subsequent progression. Progressions are characterized by the following:

  - A *work area part index* specifying the work area part that is affected by the modifications in the progression;
  - A list of *ids of deleted physical panels*, i.e., panels that are removed from the prior progression;
  - A list of *panels to create*;
  - A list of *panels to delete*, which correspond to the ids of deleted physical panels, and can be useful when stepping backwards in the list of progressions.

Further details on these parameters and their meaning can be found in the following subsections. When the values of each of these parameters—in the context of work areas, work area parts, panels and progressions—are specified as part of a JSON file, the resulting description can be parsed automatically, and the resulting physical layout constructed based on the derivations in Sect. 16.4.1.2.

16.4.1.2 Calculating the Parameters of Individual Panels

The architecture delineated by the template language is organized in a rigorous hierarchical manner, where workspaces consist of components of the workspace, which in turn include panels. Transformations at each tier are determined in relation to the tier directly above, with the positioning and alignment of entities being specified based on the dimensions and orientation of the entity at the preceding tier. In particular, to ascertain the location, direction, and dimensions of a distinct physical entity (such as a canvas) that aligns with a given panel, the subsequent formulas must be employed:

$$position_P = position_{WA} + \underbrace{Q_{WA} \cdot \left[\begin{pmatrix} width_{WA} \\ height_{WA} \\ 1 \end{pmatrix} \cdot \begin{pmatrix} x_{rposWAP} \\ y_{rposWAP} \\ z_{rposWAP} \end{pmatrix} \right]}_{\text{shiftvec for work area part}} + \underbrace{Q_{WA} \cdot Q_{WAP} \cdot \left[\underbrace{\begin{pmatrix} width_{WA} \\ height_{WA} \\ 0 \end{pmatrix} \cdot \begin{pmatrix} width_{rWAP} \\ height_{rWAP} \\ 0 \end{pmatrix}}_{\text{width and height of work area part}} \cdot \begin{pmatrix} x_{rposP} \\ y_{rposP} \\ 0 \end{pmatrix} \right]}_{\text{shiftvec for individual panel}} \tag{16.1}$$

$$Q_P = Q_{WA} \cdot Q_{WAP} \tag{16.2}$$

$$\begin{aligned} height_P &= height_{WA} \cdot rheight_{WAP} \cdot rheight_P \\ width_P &= width_{WA} \cdot rwidth_{WAP} \cdot rwidth_P \end{aligned} \tag{16.3}$$

where the subscripts *pos* and *rpos* stand for position and relative position, *WA* stands for work area, *WAP* stands for work area part, *P* stands for panel, and *Q* stands for quaternion (orientation).

16.4.1.3 Positioning Panels

The method for determining the placement of a panel within a workspace leverages the concept that the x and y coordinates of a workspace component's position vector ($x_{rposWAP}$ and $y_{rposWAP}$) are expressed as proportions of the workspace's overall width and height. Consequently, the product of these proportions in the second component of the equation generates a scale-adjustable translation vector. This means the translation vector will vary in tandem with any changes in the workspace's dimensions. In the equation's third component, the dimensions of a workspace component—defined as percentages of the workspace's width and height—when multiplied, provide the actual dimensions of that workspace component (Fig. 16.3).

Therefore, the process of calculating a canvas's placement involves several key steps:

- Initially, it considers the baseline position of the workspace;
- Adds to it a shift vector obtained by multiplying the relative position of the work area part of the panel in terms of the width and height of the work area (and the nominal scale in the z-dimension) and rotating it according to the orientation of the work area;

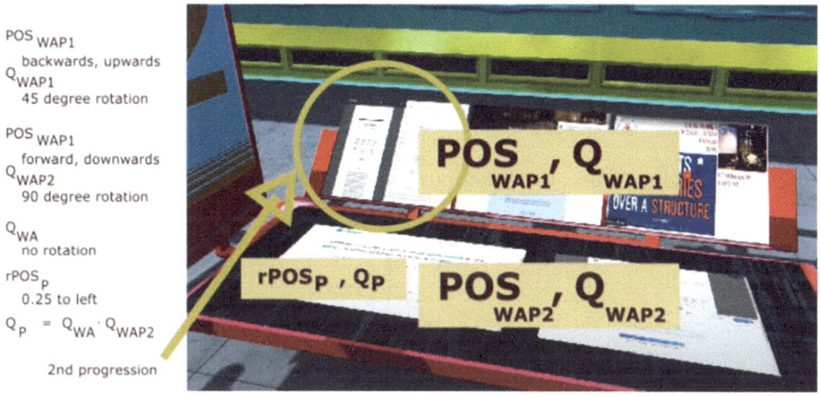

Fig. 16.3 Key parameters of dynamic positioning and scaling

- Finally, adds to it a second shift vector obtained by multiplying the relative position of the panel inside the work area part by nominal width and height of the work area part, and rotating it according to the orientation of the work area part (obtained by multiplying the quaternion of the work area with that of the work area part).

16.4.2 Modifying Layouts Based on the Proposed Template Language

In this section, we examine the support provided by the suggested template language for resizing, repositioning, and reorienting single panels, components of the workspace, and the entire workspace itself. We also explore the capability and mode to alter the number of panels in any specific section of the workspace following a template-based approach.

16.4.2.1 Scaling Work Areas

Scaling is only allowed at the level of the entire work area, given that the relative scales of the different constituent work area parts can be regarded as fixed. However, whenever the current scale of the work area is modified, the only operation that needs to be carried out is to divide the height and width of the work area by the previous scale, and multiply it by the new current scale. Next, all the panels can be regenerated based on Eqs. (16.1)–(16.3) above, with the modified values of $width_{WA}$ and $heightWA$.

16.4.2.2 Moving Work Areas or Move Groups Within Them

When shifting the location of panels, one can choose to either shift the entire work area, or to move panels belonging to a subset of work area parts defining a move group.

Regardless of this choice, the direction of the forward vector is always computed as

$$fwdvec = Q_{WA} \cdot \begin{pmatrix} x_{centroidWA} \\ y_{centroidWA} \\ z_{centroidWA} \end{pmatrix} \tag{16.4}$$

For instance, consider when the centroid vector is defined as (0; 0; 1)—meaning it points straight at the virtual camera. If we combine this vector with the workspace's orientation, we get a vector that points in the opposite direction of the workspace. This is helpful because all parts of the workspace move together in one direction when pushed forward. Since each panel can be oriented differently, using the panel orientations as a reference for movement would quickly change the overall shape of the workspace.

When the workspace or any part of it is moved up or down, every panel shifts along the global vertical (y) axis, keeping the angle (pitch) of all panels unchanged.

By calculating the cross-product of these vectors, we also determine the left/right direction, thus covering all three directions of movement. Using this method, one can adjust either the entire workspace or specific parts of it. Similar to resizing, specific equations can be used to determine the new positions and orientations of all panels.

16.4.2.3 Rotating Work Areas or Rotate Groups Within Them

Adopting a comparable strategy, one can likewise rotate the workspace or specific sections within it. Here, one must adjust either the overall orientation of the workspace or that of its distinct segments, and then refresh both the position and orientation of the separate panels.

The sole limitations are as outlined below:

- In case the complete work area is rotated, only rotations along the vertical (yaw) axis are allowed—since the pitch axes of rotation for each individual work area part will usually be parallel to one another, but will not coincide;
- In case individual work area parts are rotated, only rotations along the horizontal (pitch) axis are allowed—this helps guarantee that the yaw orientations of the individual parts do not change relative to each other;
- Finally, rotations along the outward-pointing axis (roll) are never allowed, as all panels are required to intersect the globally horizontal plane along a line (not a single point).

16.4.2.4 The Logic of Progressions and Regressions

As defined by the template language, the layout of work areas is organized through a sequence known as progressions. Each progression represents a specific layout configuration, allowing for a shift forward to the subsequent progression (progressive transition) or backward to the prior one (regressive transition). Consequently, defining a progression involves outlining the incremental adjustments required for movement in either direction.

In detail, the framework treats moving forward as the norm, detailing both the identifiers of removed physical panels and those of panels to be created (with the latter encompassing full details necessary for the fabrication of these panels). For a step back, the procedure is inverted: panels slated for creation become the reference for which panels to eliminate, and a "panels to delete" roster (including details for the panels equivalent to those removed) serves as the reference for creating panels.

This mechanism conceptually mirrors cell division (mitosis) in biology, albeit with the inclusion of a hypothetical reverse process. Within a workspace, panels may divide, relocate, rotate, and/or combine as part of the progressive or regressive transitions.

In Fig. 16.4, the blue frames indicate Work Area Part 1 and Work Area Part 2. The blue frames also indicate the Rotate Groups. The orange frames indicate the Move Groups. These can be moved separately or together depending on the selection. Progression and regression can be performed for each panel.

Fig. 16.4 Interactive work areas and group dynamics: Visualization of rotate and move functions

16.4.3 Associative Semantic Process for 2D Content Search and Insertation

After the successful generation and modification of automated spatial configurations, this subsection introduces a solution that highlights the use of selectable and dynamically adaptable two-dimensional entities, such as documents, spreadsheets, PDFs, and other content. The goal of this solution is to optimize the user experience and interaction in virtual space, offering a flexible and user-centered approach to content selection and insertion. In this case, context control includes selecting content and creating content layouts, as well as the specific layout according to individual needs. The content must not only adapt to the user's preferences but also to the temporality of the workflow, the missing document work environment, and the geometry of the 3D space.

The associative semantic search system we propose relies on the following concepts, which are based on the general principles of data structure and indexing techniques:

- Spatial Structure Nodes: These refer to individual units or elements of the spatial data structure representing the 3D geometric space. These nodes can store information about their location, their relationship with other nodes, or other spatial attributes.
- Spatial Index Nodes: These refer to nodes within a spatial index, a data structure that allows for accessing data based on its spatial coordinates. Typical spatial indexes include R-trees, k-d trees, and Quad-trees, where the tree nodes represent defined spatial areas and contain information (such as pointers) on current data entries or other nodes.
  Note: Spatial index nodes are essential for improving the efficiency of spatial queries, such as range queries or nearest neighbor searches, in large spatial databases.
- Temporal Structure Nodes: These refer to elements of the data structure that record temporal information or time-based relationships. They can represent specific instances in time or periods and can store information about their temporal position, their relationship with other temporal nodes, or other time-related attributes.
- Temporal Index Nodes: These are nodes within a time index, a data structure that allows for efficient time-based access to data. For example, a B-tree can be used as a time index, where each node corresponds to a specific time range and contains pointers to data entries or other nodes within that range. Temporal index nodes help speed up time-consuming queries, such as identifying all events that occurred within a specific timeframe.
- Semantic Structure Nodes: These are the nodes of a semantic data structure representing digital entities or concepts in a semantic model (e.g., knowledge graph). They can contain information about their semantic meaning, their relationship with other nodes, or other semantically relevant attributes. Note: This concept plays a central role in natural language processing, where relationships between various entities or concepts are often modeled as graphs.

- Semantic Index Nodes: These could be elements of a semantic index, a data structure that allows for efficient querying based on semantic meaning or relationships. Such index nodes correspond to specific semantic concepts or relationships and can point to data entries or other nodes in the semantic space. These indexes can be used in semantic databases or contexts where data is queried based on meaning rather than specific values or identifiers.

Considering our hypothesis that the configuration of the virtual reality (VR) environment at the panel level is predetermined, and the overarching goal of the associative semantic search service is to improve the handling of separate two-dimensional (2D) contents such as documents, the following sections present on the discourse about solutions related to the handling of said documents through two examples:

Document Insertion: In this scenario, the user aims to integrate the necessary document as a fresh content element into the virtual workspace. Practically, users often add multiple new digital content in succession, frequently related to each other. Consequently, it becomes imperative to consider users' previous activities, including actions and refinements performed during the insertion of previous digital content into the content system. The proposed associative semantic search system operates through three distinct threads: the Spatializer, the Temporalizer, and the Content Vector, each synergistically facilitating the execution of this task. The Spatializer and the Temporalizer share similar responsibilities, including the creation of new spatial index nodes. These nodes are then linked to the appropriate spatial index nodes for previous insertions, followed by the identification and linking of relevant structural nodes, whether spatial or temporal, with the new index node. The task of the Content Vector, however, is significantly more complex. It involves performing a similarity search on every node, followed by a thorough similarity assessment. Subsequently, a tensor index associated with a node is either inserted or a new node and tensor index are generated, depending on whether a match is found during the search process. It then examines the similarities of the existing node, and if present, establishes a connection with the new node. The execution workflow is illustrated in Fig. 16.5, which visually presents the aforementioned process.

Document Search: The process of searching for 2D documents within the system, as an associative semantic search process, involves a significantly simpler set of operations, and the specifics of this sequence.

16.5 Experimental Validation of Personalized Workspaces

The content of this section is based on a previous paper by the authors [12]. It outlines an experiment designed to assess the effectiveness of personalized workspace layouts in a 3D environment, specifically developed to align with the individual learning styles of users, as detailed in. Conducted with ethical approval, the study involved 52 electrical engineering students from the University of Pécs, with a mix of Generation Z and older participants, none of whom had prior experience with 3D VR workspaces.

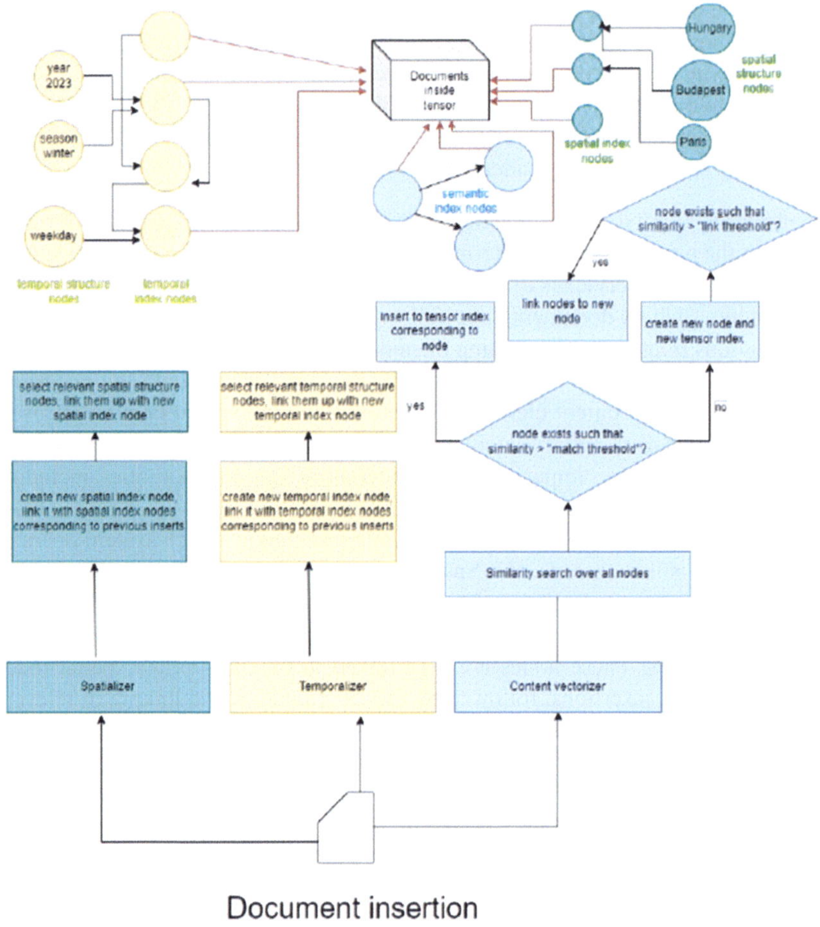

Fig. 16.5 Document insertion process [17]

The experiment began with participants completing Kolb's Learning Styles Questionnaire, followed by a task selection aligned with their identified learning style. Participants were then asked to choose from one of seven layout schemas—four designed based on learning styles and three common layouts from the MaxWhere VR platform—first with blank monitors and then with monitors displaying content relevant to their task. This design aimed to investigate the influence of 2D digital content on the selection of personalized spatial layouts without the bias of content-related details.

The study hypothesized that participants would prefer layouts matching their learning styles and that specific 2D digital content would significantly impact their choices. For our primary hypothesis, we posited that the selection among the seven offered layouts by users was not merely by chance. To test this, an analytical technique was applied to calculate the p-value, which represents the likelihood of obtaining results as significant as observed under the assumption that choices were made at random. With a one in seven chance (p = 1/7) for a correct selection and 52 participants, we expected the distribution of correct choices to follow a binomial pattern.

$$P(x) = \binom{n}{k} k^p (n-k)^{(1-p)} \tag{16.5}$$

The associated p-value can be computed as

$$p\_value(x) = \sum_{i=k}^{n} \binom{n}{i} x^p (n-i)^{(1-p)} \tag{16.6}$$

This analysis method demonstrated conclusively that participants chose layouts aligned with their learning styles far more frequently than would be expected by chance, as indicated by p-values below 5%. Moreover, the presence of specific content within these layouts significantly increased the likelihood of participants choosing those that matched their learning styles, as evidenced by the p-value comparison between selections made with and without content. This strongly supports the hypothesis that personalized layouts, especially those containing relevant content, are preferred over random selections or blank layouts.

Results indicated a strong preference for layouts that matched participants' learning styles, especially when the layouts included relevant digital content.

In conclusion, the experiment validated the effectiveness of personalized workspace layouts in enhancing user experience in 3D environments, demonstrating the importance of aligning digital content and layout design with individual learning styles.

16.6 Conclusion

This chapter successfully demonstrates the effectiveness of a systematic framework for personalizing workspace layouts in 3D Virtual Reality (VR) environments. By introducing a structured template language and hierarchical system for content layout specification, along with an associative semantic search, it paves the way for dynamic adaptation of content and layout according to individual user preferences and task requirements. The proposed innovative service for efficient digital content management, fitted to workflow demands and user preferences, gives a significant advancement in optimizing content selection, layout creation, and aligning these elements with the workflow's temporality and the 3D space's geometry.

The research underlines the importance of context control through algorithmic solutions and user-interaction designs that curate digital content in a way that resonates with users' working and learning styles within the constraints of 3D environments. The detailed methodology—spanning user profiling, workflow analysis, construction of workflow graphs, content significance analysis, to the development of spatial arrangements—reveals how personalized VR workspaces can significantly improve the efficiency and precision of digital task execution.

Empirical evidence from the analysis method and the observed strong preference for layouts that match participants' learning styles, especially when including relevant digital content, strongly supports the hypothesis that personalized layouts enhance user experience by improving task performance in speed and accuracy. This chapter not only validates the effectiveness of personalized workspace layouts in 3D VR environments but also highlights the critical role of aligning digital content and layout design with individual learning styles, contributing to the future of VR technology design.

References

1. Truong, P., Hölttä-Otto, K., Becerril, P., Turtiainen, R., Siltanen, S.: Multi-user virtual reality for remote collaboration in construction projects: a case study with high-rise elevator machine room planning. Electronics **10**(22), 2806 (2021)
2. Weiss, P.L., Jessel, A.S.: Virtual reality applications to work. Work **11**(3), 277–293 (1998)
3. Horváth, I., Csapó, Á.B., Berki, B., Sudár, A., Baranyi, P.: Definition, background and research perspectives behind 'cognitive aspects of virtual reality'(cvr). Infocommunications J. **15**(Special Issue), 9–14 (2023)
4. Berki, B.: Role of presence, memory and spatial ability in a desktop virtual reality. In: Accentuated Innovations in Cognitive Info-Communication, pp. 79–97. Springer International Publishing (2022)
5. Chiquet, S., Martarelli, C.S., Mast, F.W.: Eye movements to absent objects during mental imagery and visual memory in immersive virtual reality. Virtual Real. **25**, 655–667 (2021)
6. Huang, J., Klippel, A.: The effects of visual realism on spatial memory and exploration patterns in virtual reality. In: Proceedings of the 26th ACM Symposium on Virtual Reality Software and Technology, pp. 1–11 (2020)
7. Smith, S.A., Mulligan, N.W.: Immersion, presence, and episodic memory in virtual reality environments. Memory **29**(8), 983–1005 (2021)
8. Cho, B.-H., Ku, J., Jang, D.P., Kim, S., Lee, Y.H., Kim, I.Y., Lee, J.H., Kim, S.I.: The effect of virtual reality cognitive training for attention enhancement. CyberPsychology & Behav. **5**(2), 129–137 (2002)
9. Upenik, E., Ebrahimi, T.: A simple method to obtain visual attention data in head mounted virtual reality. In: 2017 IEEE International Conference on Multimedia & Expo Workshops (ICMEW), pp. 73–78. IEEE (2017)
10. Lanyi, C.S., Szücs, V.: Games applied for therapy in stroke tele-rehabilitation. In: Int. J. Stroke **9**, 300–300 (2014). Sage publications ltd 1 Olivers yard, 55 city road, London ec1y 1sp, England
11. Jerald, J.: The VR book: Human-centered design for virtual reality. Morgan & Claypool (2015)
12. Horváth, I., Csapó, Á.B.: Motivations and tools relevant to personalized workspaces in VR environments. Electronics **12**(9), 2059 (2023). [Online]. https://www.mdpi.com/2079-9292/12/9/2059

13. Horváth, I.: Personalized learning opportunity in 3D VR. In: 2020 11th IEEE International Conference on Cognitive Infocommunications (CogInfoCom), pp. 425–430. IEEE (2020)
14. Chen, Z.-H., Lu, H.-D., Lu, C.-H.: The effects of human factors on the use of avatars in game-based learning: customization vs. non-customization. Int. J. Hum. Comput. Interact. **35**(4–5), 384–394 (2019)
15. Dadi, G.B., Goodrum, P.M., Taylor, T.R., Carswell, C.M.: Cognitive workload demands using 2d and 3d spatial engineering information formats. J. Constr. Eng. Manag. **140**(5), 04014001 (2014)
16. Sweany, J., Goodrum, P., Miller, J.: Analysis of empirical data on the effects of the format of engineering deliverables on craft performance. Autom. Constr. **69**, 59–67 (2016)
17. Horváth, I., Csapó, Á.B.: Structured template language and generative AI driven content management for personalized workspace reconfiguration. In: 2023 IEEE 2nd International Conference on Cognitive Aspects of Virtual Reality (CVR). IEEE (2023)
18. Horváth, I.: An analysis of personalized learning opportunities in 3d VR. Front. Comput. Sci. **3**, 673826 (2021)
19. Gauch, S., Speretta, M., Chandramouli, A., Micarelli, A.: User profiles for personalized information access. In: The Adaptive Web: Methods and Strategies of Web Personalization, pp. 54–89 (2007)
20. El Allioui, Y., El Beqqali, O.: User profile ontology for the personalization approach. Int. J. Comput. Appl. **41**(4), (2012)
21. Fawcett, T., Provost, F.J.: Combining data mining and machine learning for effective user profiling. In: KDD, vol. 96, pp. 8–13 (1996)
22. Kanoje, S., Girase, S., Mukhopadhyay, D.: User profiling trends, techniques and applications (2015). arXiv:1503.07474
23. Coffield, F., Moseley, D., Hall, E., Ecclestone, K.: Learning Styles and Pedagogy in Post-16 Learning. Learning and Skills Research Centre (2004)
24. Kolb, D.A.: Learning styles and disciplinary differences. Mod. Am. Coll. **1**(January 1981), 232–235 (1981)
25. Kayes, D.C.: Internal validity and reliability of Kolb's learning style inventory version 3 (1999). J. Bus. Psychol. **20**, 249–257 (2005)
26. McLeod, S.: Kolb-learning styles. In: Simply Psychology (2017)
27. Vogel, C., Koutsombogera, M., Reverdy, J.: Aspects of dynamics in dialogue collaboration. Electronics **12**(10), 2210 (2023). May

SPRINGER NATURE

GPSR Compliance

The European Union's (EU) General Product Safety Regulation (GPSR) is a set of rules that requires consumer products to be safe and our obligations to ensure this.

If you have any concerns about our products, you can contact us on ProductSafety@springernature.com

In case Publisher is established outside the EU, the EU authorized representative is:

Springer Nature Customer Service Center GmbH
Europaplatz 3
69115 Heidelberg, Germany

The manufacturer's authorised representative in the EU is Springer Nature Customer Service Centre GmbH, Europaplatz 3, 69115 Heidelberg, Germany. If you have any concerns regarding our products, please contact ProductSafety@springernature.com

Printed and bound by CPI Group (UK) Ltd, Croydon, CR0 4YY
01/10/2026
02225858-0010